Mini-Micro Fuel Cells

NATO Science for Peace and Security Series

This Series presents the results of scientific meetings supported under the NATO Programme: Science for Peace and Security (SPS).

The NATO SPS Programme supports meetings in the following Key Priority areas: (1) Defence Against Terrorism; (2) Countering other Threats to Security and (3) NATO, Partner and Mediterranean Dialogue Country Priorities. The types of meeting supported are generally "Advanced Study Institutes" and "Advanced Research Workshops". The NATO SPS Series collects together the results of these meetings. The meetings are co-organized by scientists from NATO countries and scientists from NATO's "Partner" or "Mediterranean Dialogue" countries. The observations and recommendations made at the meetings, as well as the contents of the volumes in the Series, reflect those of participants and contributors only; they should not necessarily be regarded as reflecting NATO views or policy.

Advanced Study Institutes (ASI) are high-level tutorial courses intended to convey the latest developments in a subject to an advanced-level audience

Advanced Research Workshops (ARW) are expert meetings where an intense but informal exchange of views at the frontiers of a subject aims at identifying directions for future action

Following a transformation of the programme in 2006 the Series has been re-named and re-organised. Recent volumes on topics not related to security, which result from meetings supported under the programme earlier, may be found in the NATO Science Series.

The Series is published by IOS Press, Amsterdam, and Springer, Dordrecht, in conjunction with the NATO Public Diplomacy Division.

Sub-Series

A.	Chemistry and Biology	Springer
B.	Physics and Biophysics	Springer
C.	Environmental Security	Springer
D.	Information and Communication Security	IOS Press
E.	Human and Societal Dynamics	IOS Press

http://www.nato.int/science
http://www.springer.com
http://www.iospress.nl

Series C: Environmental Security

Mini-Micro Fuel Cells

Fundamentals and Applications

edited by

S. Kakaç
TOBB University of Economics and Technology,
Söğütözü, Ankara, Turkey

A. Pramuanjaroenkij
Kasetsart University,
Chalermphrakiat Sakonnakhon Province Campus,
Sakonnakhon, Thailand

and

L. Vasiliev
Luikov Heat & Mass Transfer Institute,
Minsk, Belarus

Published in cooperation with NATO Public Diplomacy Division

Proceedings of the NATO Advanced Study Institute on
Mini-Micro Fuel Cells – Fundamentals and Applications
Çesme – Izmir, Turkey
July 22 – August 3, 2007

A C.I.P. Catalogue record for this book is available from the Library of Congress.

ISBN 978-1-4020-8294-8 (PB)
ISBN 978-1-4020-8293-1 (HB)
ISBN 978-1-4020-8295-5 (e-book)

Published by Springer,
P.O. Box 17, 3300 AA Dordrecht, The Netherlands.

www.springer.com

Printed on acid-free paper

All Rights Reserved
© 2008 Springer Science + Business Media B.V.
No part of this work may be reproduced, stored in a retrieval system, or transmitted
in any form or by any means, electronic, mechanical, photocopying, microfilming,
recording or otherwise, without written permission from the Publisher, with the exception
of any material supplied specifically for the purpose of being entered
and executed on a computer system, for exclusive use by the purchaser of the work.

CONTENTS

Preface... ix

List of Participants.. xi

Some Introductory Technical Remarks: the NATO School
on Micro-Fuel Cells.. 1
    S. Gottesfeld

Fuel Cell Basic Chemistry, Electrochemistry and Thermodynamics........ 13
    F. Barbir

Fuel Cell Stack Design Principles with Some Design
Concepts of Micro-Mini Fuel Cells... 27
    F. Barbir

Performance Analysis of Microstructured Fuel Cells for Portable
Applications... 47
    S. Litster and N. Djilali

Engineering Durability of Micro-Miniature Fuel Cells................. 75
    K. Reifsnider and X. Huang

Portable Fuel Cells – Fundamentals, Technologies
and Applications... 87
    C.O. Colpan, I. Dincer, and F. Hamdullahpur

Autonomous Test Units for Mini Membrane Electrode Assemblies....... 103
    E. Budevski, I. Radev, and E. Slavcheva

Heat Pipes in Fuel Cell Technology.. 117
    L. Vasiliev and L. Vasiliev Jr.

Heat Transfer Enhancement in Confined Spaces of Mini-Micro
Fuel Cells... 125
    L. Vasiliev and L. Vasiliev Jr.

Performance Characteristics of Membrane Electrode Assemblies
Using the *EasyTest Cell*... 133
    E. Budevski, I. Radev, and E. Slavcheva

Water Transport Dynamics in Fuel Cell Micro-Channels 153
    G. Minor, X. Zhu, P. Oshkai, P.C. Sui, and N. Djilali

Polymer Electrolyte Fuel Cell Systems for Special Applications........... 171
    U. Pasaogullari

Sorption and Diffusion Selectivity of Methanol/Water Mixtures
in NAFION... 189
    D. T. Hallinan Jr. and Y. A. Elabd

Kinetics and Kinetically Limited Performance in PEMFCs
and DMFCs with State-of-the-art Catalysts..209
    H.A. Gasteiger, Y. Liu, D. Baker, and W. Gu

Catalyst Degradation Mechanisms in PEM and Direct Methanol
Fuel Cells... 225
    H.A. Gasteiger, W. Gu, B. Litteer, R. Makharia, B. Brady,
    M. Budinski, E. Thompson, F.T. Wagner, S.G. Yan, and P.T. Yu

Principles of Direct Methanol Fuel Cells for Portable
and Micro Power... 235
    C.-Y. Wang

Computational Modeling of Two-Phase Transport in Portable
and Micro Fuel Cells.. 243
    C.-Y. Wang

Optimization of Direct Methanol Fuel Cell Systems
and Their Mode of Operation... 257
    S. Gottesfeld and C. Minas

Development of A 5-W Direct Methanol Fuel Cell Stack
for DMB Phone..269
    Y.-C. Park, D.-H. Peck, S.-K. Kim, and D.-H. Jung

CONTENTS

Dynamic Characteristics of DMFC/Battery System
for Notebook PC.. 291
    Y.-R. Cho, M.-S. Hyun, and D.-H. Jung

A Review on Miniaturization of Solid Oxide Fuel Cell Power
Sources-I: State-of-the-art Systems.. 303
    X.Y. Zhou, A. Pramuanjaroenkij, and S. Kakaç

A Review on Miniaturization of Solid Oxide Fuel Cell
Power Sources-II: From System to Material............................ 319
    X.Y. Zhou, A. Pramuanjaroenkij, and S. Kakaç

Planar Solid Oxide Fuel Cells: From Materials to Systems.......... 349
    Y. Mizutani

Mathematical Analysis of Planar Solid Oxide Fuel Cells............. 359
    A. Pramuanjaroenkij, S. Kakaç, and X.Y Zhou

The Properties and Performance of Micro-Tubular (Less Than
1 mm OD) Anode Supported SOFC for APU-Applications......... 391
    N. Sammes, A. Smirnova, A. Mohammadi,
    F. Serincan, Z. Xiaoyu, M. Awano, T. Suzuki,
    T. Yamaguchi, Y. Fujishiro, and
    Y. Funahashi

Current State of R&D on Micro Tubular Solid Oxide
Fuel Cells in Japan... 407
    Y. Mizutani

Index... 419

PREFACE

This volume contains an archival record of the NATO Advanced Institute on Mini – Micro Fuel Cells – Fundamental and Applications held in Çesme – Izmir, Turkey, July 22–August 3, 2007. The ASIs are intended to be a high-level teaching activity in scientific and technical areas of current concern. In this volume, the reader may find interesting chapters on Mini-Micro Fuel Cells with fundamentals and applications. In recent years, **fuel-cell development**, modeling and performance analysis has received much attention due to their potential for distributed power which is a critical issue for energy security and the **environmental** protection. Small fuel cells for portable applications are important for the security. The portable devices (many electronic and wireless) operated by fuel cells for providing all-day power, are very valuable for the security, for defense and in the war against terrorism.

Many companies in NATO and non-NATO countries have concentrated to promote the fuel cell industry. Many universities with industrial partners committed to the idea of working together to develop fuel cells. As technology advanced in the 1980s and beyond, many government organizations joined in spending money on fuel-cell research. In recent years, interest in using fuel cells to **power portable electronic devices** and other small equipment (cell phones, mobile phones, lab-tops, they are used as micro power source in biological applications) has increased partly due to the promise of fuel cells having higher energy density. In developing these smaller fuel cell systems, one cannot simply use scaled-down system architectures and components used in their larger counterparts. Small fuel cell systems may be divided into portable fuel cells (>100 W), miniature fuel cells (10–100 W) and micro fuel cells (0–10 W). These types of **mini/micro** fuel cells importance are evident for security in the war against terror. This volume on Mini-Micro Fuel Cells as Electric Energy Generators provides a comprehensive state-of-the-art assessment of this technology by treating the subject in considerable depth through lectures from eminent professionals in the field, discussions, and panel sessions. Main lectures will include Fuel cell basic electrochemistry and thermodynamics, Design concepts of micro-mini fuel cells, Fundamental merit parameters of micro fuel cells and barriers to meeting target values for such parameters, Direct methanol fuel cell technology platforms for micro fuel cell applications, Electrocatalysis state of the art in methanol fuel cells and possible future directions, Fundamental comparison of high temperature vs. low temperature micro/miniature fuel cells, Durability of micro/miniature fuel cells, Accelerated characterization and life prediction of micro/miniature fuel cells, Dynamic characteristic of 30 W class DMFC system for notebook PC, Design and preparation of micro fuel cell using for DNB phone, Principal of direct methanol fuel cell for portable and micro power, Computational modelling of two-phase transport in portable and micro fuel cells, Current state of R&D on micro tubular solid oxide fuel cells in Japan, Toho's solid oxide fuel cells from material to systems, Micro-tubular SOFC systems, Mechanical properties and modeling of micro-tubes used in SOFC systems, Inter-mediate-temperature Solid oxide fuel cell systems used for APU units, Analysis and principals of methanol reformers for Fuel cells, The easy test concept, Autonomous test units for mini membrane assemblies, PEM systems for special applications, Heat pipe concept in fuel cell technology and others. During the ten working days of the Institute, the invited lecturers covered fundamentals and applications of Mini – Micro Fuel Cells. The sponsorship of the NATO Scientific Affairs Division is gratefully acknowledged; in person, we are very thankful to Dr. Fausto Pedrazzini director of ASI programs and his executive secretary Alison Trapp who continuously supported and encouraged us at every phase of our organization of this Institute. We are also very grateful to Annelies Kersbergen publishing editor of Springer Science; our special gratitude goes to Drs. Nilufer Egrican, Hafit Yuncu, Sepnem Tavman, Almila Yazicioglu, Tuba Okutucu, Ahmet Yozgatligil, Derek Baker and Gratiela Tarlea for coordinating sessions and we are very thankful to Basar Bulut and Ertan Agar for their very valuable help for smooth running the Institute with the assistant secretary Cahit Koksal.

S. Kakaç
A. Pramuanjaroenkij
L. Vasiliev

NATO ADVANCED STUDY INSTITUTE

MINI-MICRO FUEL CELLS AS ELECTRIC ENERGY GENERATORS
GOLDEN DOLPHIN HOTEL, CESME-IZMIR, TÜRKIYE (TURKEY)
JULY 22–AUGUST 3, 2007
ASI NO: 982648
Co-Directors: Professor S. Kakaç (USA)
Professor L. Vasiliev (Belarus)

SECTION 1. LIST OF PARTICIPANTS FROM NATO COUNTRIES

LECTURERS

BELARUS
Leonard Vasiliev (Co-Director)
Luikov Heat & Transfer Institute
P. Brovka 15, 220072, Minsk,
Belarus
Tel: (375)-17-284-21-33
E-mail: LVASIL@ns1.htmi.aac.by

BULGARIA
Evgeni Budevski
Institute of Electrochemistry and
Energy Systems
Bulgarian Academy of Sciences
BG-1113 Sofia,
G. Bonchevstr. 10
Tel: Office: 003592723454
Mobile: 00359(0)899845119
E-mail: budevski@bas.bg

CANADA
Ned Djilali
Department of Mechanical
Engineering
University of Victoria
P.O. BOX 3055, Victoria
BC V8W 3P6 Canada
Tel: (250)721-6034
E-mail: ndjilali@uvic.ca

CROATIA
Frano Barbir
Faculty of Electrical, Mechanical
Engineering & Naval
Architecture
University of Split
R. Boskovica b.b.
21000 Split
Tel: +388-21-305-889
E-mail: Frano.Barbir@fesb.hr

ITALY
Hubert Gasteiger
Acta S.p.A.
Via di Lavoria 56/G
56040 Crespina
Italy
E-mail: hubert.gasteiger@gmail.com

JAPAN
Yasunobu Mizutani
Fundamental Research
Department
Toho Gas Co., Ltd.
507-2 Shinpo-Machi, Tokai-City
Aichi Pref. 476-8501
Tel: 052-689-1616
E-mail: master@tohogas.co.jp

KOREA
Jung Doo-Hwan
Advanced Fuel-Cell Research
KIER, 71-2 Jangdong,
Yousung-Gu
Daejon City, S. Korea
Tel: +82 42-860-3577, 3180
E-mail: doohwan@kier.re.kr

TURKEY
Sadik Kakac (Co-Director)
Professor of Mechanical
Engineering
TOBB University of Economics
and Technology
06530 Söğütözü, Ankara, Turkey
Tel: +90(312)292-40-87
E-mail: skakac@etu.edu.tr

USA
Shimshon Gottesfeld
Fuel Cell Consulting, LLC
3404 Rosendale Road
Niskayuna, NY 12309
Tel: +1(518)533-2204
E-mail: shimson.gottesfeld@gmail.com

Kenneth Reifsnider
University of South Carolina
Mechanical Engineering
Department
500 Main St.
Columbia, SC 99208
Tel: (803)777-0084
E-mail: reifsnid@engr.sc.edu

Nigel Sammes
Herman F. Coors Distinguished
Professor of Ceramic
Engineering,
Department of Metallurgical and
Materials Engineering,
Colorado School of Mines,
1500 Illinois Street,
Golden, Colorado 80401
Tel: (303)273-3344
Fax: (303)273-3755
E-mail: nsammes@mines.edu

Chao-Yang Wang
Mechanical Eng. & Materials
Science
Electrochemical Engine Centre
Pennsylvania State University
338 Reber Building
University Park, PA 16802
Tel: (814)863-4762
Fax: (814)863-4848
E-mail: cxw31@psu.edu

Xiang Yang Zhou
Department of Mechanical
Aerospace Engineering
University of Miami
Coral Gables, FL 33124
Tel: (305)284-2571
E-mail: xzhou@miami.edu

ASI STUDENTS

BULGARIA
Ivan D. Radev
Institute of Energy Systems
Bulgarian Academy of Sciences
(IEES-BAS) 1113 Sofia, Bulgaria

Georgi Topalov
Institute of Energy Systems
Bulgarian Academy of Sciences
(IEES-BAS)1113 Sofia, Bulgaria

LIST OF PARTICIPANTS

CANADA
Gessie M. Brisard
Department of Chemistry
University of Sherbrooke
Sherbrooke, P. Quebec J1K 2R1,
Canada

Can Ozgur Colpan
Mechanical and Aerospace
Engineer
Carleton University
1125 Colonel by Drive
Ottawa, Ontario K15 5B6,
Canada

Abdulmajeed Mohamad
Dept. of Mechanical Engineering
The University of Calgary
Calgary, Alberta T2N 1N4,
Canada

Ibrahim Dincer
Faculty of Engineering and
Applied Science
University of Ontario
Institute of Technology
Oshawa, Ontario L1H 7K4,
Canada

FRANCE
Jacques Padet
University of Reims
UTAP-Laboratoire de
Thermomecanique
Faculte de Sciences
51687 Reims, France

GERMANY
Franz Mayinger
Munich Technical University
Institute of Thermodynamics A
85748 Garching, Germany

ITALY
Stefano Galli
ENEA, Energy Technologies
Dept.
301 Via Anguillarese
00123 Rome, Italy

LATVIA
Janis Rimshans
Institute of Mathematics
& Computer Science
University of Latvia
Riga LV 1459 Riga, Latvia

POLAND
Lukasz Sekiewicz
Academia Gorniczo – Hutnicza,
AL
Mickieswicza 30, 30-059 Krakow
Poland

Janusz Szmyd
AGH-University of Science
& Technology
30059 Krakow, Poland

PORTUGAL
Alexandra F. Rodrigues Pinto
FEUP – Faculty of Engineering
of Porto University
4200-465 Porto, Portugal

Jose Manuel de Sousa
LEPAE – Laboratory
Universidade de Tras-os-Montes
E – Alto Douro
Apartado 202, 5001-911
Vila-Real
Codex - Portugal

PUERTO RICO
Zahilia Cabán Huertas
Urb Santa Rita Calle Janer 106
Apt 4, San Juan, PR. 00925

ROMANIA

Eden Mamut
Center for Advanced Engineering Science
Ovidius University of Constanza
124 Mamaia Ave.
Tel: +40-21-457-49 49/160
900527 Constanza, Romania

Miu Mihaela
National Institute for R&D
in Microtechnologies
Erou Iancu Nicolae Str. 12A
Bucharest, Romania

Gheorghe Popescu
Department of Applied Thermodynamics
Polytechnic University of Bucharest
060042 Bucharest, Romania

Gratiela Tarlea
Department of Thermodynamics
Building Systems Faculty
Technical University of Civil Engineering
Bucharest, Romania

SPAIN

Antonio Echarri
E.P.S., E.U.P Linares
University of Jaen
C/ Bernarda Alba 12
Los Rebites, Huetor Vega
Granada, Spain

Marcos Vera
Dept. of Thermal & Fluid Engineering
Carlos III University of Madrid
28911 Leganes, Madrid, Spain

TURKEY

Faruk Arinc
Department of Mechanical Engineering
Middle East Technical University
06531 Ankara, Turkey

Kamil Arslan
Department of Mechanical Engineering
Gazi University
Maltepe 06570 Ankara, Turkey

Irfan Ar
Department of Chemical Engineering
Gazi University
Maltepe 06570 Ankara, Turkey

Havva Akdeniz
Dept. of Electronics
Namik Kemal University
59860 Corlu, Tekirdag, Turkey

Mehmet Bozoglu
Research Engineer
Demir Dokum A.S.
4 Eyul Mah.
Ismet Inonu Cad.
No. 2455 11300 Bozuyuk,
Bilecik
Turkey

Basar Bulut
Department of Mechanical Engineering
Middle East Technical University
06531 Ankara, Turkey

Caner Cicek
Department of Physics
Faculty of Arts & Sciences
Onsekiz Mart University
Canakkale, Turkey

LIST OF PARTICIPANTS

Levent Colak
Department of Mechanical Engineering
Baskent University
Ankara, Turkey

Derek Baker
Department of Mechanical Engineering
Middle East Technical University
06531 Ankara, Turkey

Nilufer Egrican
Department of Mechanical Engineering
Yedi Tepe University
Istanbul, Turkey

Serdar Erkan
Department of Chemical Engineering
Middle East Technical University
06531 Ankara, Turkey

Kemal Ersan
Faculty of Technical Education
Department of Mechanical Education
Gazi University
06510 Ankara, Turkey

Ilknur Kayacan
Chemical Engineering
Gazi University
Maletepe 06570, Ankara, Turkey

Suleyman Kaytakoglu
Chemical Engineering
Anadolu University
26555 Eskisehir, Turkey

Tuba Okutucu
Department of Mechanical Engineering
Middle East Technical University
06531 Ankara, Turkey

Leyla Ozgener
Department of Mechanical Engineering
Celal Bayar University
TR-45140, Muradiye, Manisu, Turkey

Mehmet Sankir
Dept. of Electrical Engineering
TOBB University Economics & Technology
06530 Söğütözü, Ankara, Turkey

Sebnem Tavman
Food Engineering Department
Ege University
35100 Bornova-Izmir, Turkey

Levent Yalcin
Chemical Engineering
Anadolu University
26555 Eskisehir, Turkey

Suha Yazici
UNIDO-ICHET
Sabri Ulker Sk. 38/4
Cevizlibag-Zeytinburnu
Istanbul, Turkey

Almila Yazicioglu
Department of Mechanical Engineering
Middle East Technical University
06531 Ankara, Turkey

LIST OF PARTICIPANTS

Ahmet Yozgatligil
Department of Mechanical Engineering
Middle East Technical University
06531 Ankara, Turkey

Hafit Yüncü
Department of Mechanical Engineering
Middle East Technical University
06531 Ankara

UK
Kevin Hughes
University of Leeds
SPEME, CFD Centre, Leeds
LS2 9JT, UK

Ma Lin
Energy Resources Research Institute/CFD Centre
University of Leeds
Wood House Lane
Leeds, L52 9JT, UK

Martin Thomas
CFD Center
The Houldsworth Building
Clarendon Rd, Leeds L53 1JS, UK

Yasemin Vural
Centre for CFD
The Houldsworth Building,
Clarendon Road
Leads LS3 1JS, West Yorkshire, UK

USA
Kathleen Allen
Dept. of Mechanical Engineering
Drexel University
3141 Chestnut Street
Philadelphia, PA 19104

Mehmet Arik
General Electric Global Research Center
One Research Circle, ES-102
Niskayuna, NY 12309

Ozer Arnas
US Military Academy at West Point
Department of C/ME
West Point, NY 10996

Yossef Elabd
Dept. of Chemical Engineering
Drexel University
3141 Chestnut Street
Philadelphia, PA 19104

Selcuk Guceri
Dean, College of Engineering
Drexel University
3141 Chestnut Street
Philadelphia, PA 19104

Ugur Pasaogullari
CT Global Fuel Cell Center
University of Connecticut
Storrs, CT 06269

Amy Peterson
Drexel University
3141 Chestnut Street
Philadelphia, PA 19104

Anchasa Pramuanjaroenkij
Department of Mechanical Engineering
University of Miami
Coral Gables, Florida 33124

LIST OF PARTICIPANTS xvii

Alan Tkaczyk
Department of Mechanical Engineering
University of Michigan
Ann Arbor, MI 48109

Kenneth Wilson
Rice University
6100 Main Street
Houston, TX 77005

Yaman Yener
School of Engineering
Northeastern University
Boston, MA 02115-5000

SECTION 2. LIST OF PARTICIPANTS FROM PARTNER AND MEDITERRANEAN DIALOGUE COUNTRIES

ASI STUDENTS

ALGERIA
Nouara Rassoul
Energy & Mechanical
BP72 Village Universitaire
University Houari Boumediene
Algeria

BELARUS
Leonid Vasiliev
Luikov Heat & Mass Transfer Institute
P. Brovka 15, 220072, Minsk, Belarus

CROATIA
Ivan Tolj
Faculty of Electrical, Mechanical Eng. & Naval Architecture,
University of Split
R.Boskovića b.b. 21000, Split

GEORGIA
Irakli Kordzakhia
L.E.P.L. Institute OPTICA
(Legal Entity, Public Low)
0193, Tbilisi, Georgia

ISRAEL
Leonid Pismen
Department of Chemical Engineering
Technion, 32000 Haifa, Israel

KYRGYZ REPUBLIC
Razia Gainutdinova
Institute of Physics
National Academy of Sciences
Chui Prosp., 265-A, Bishkek
720071
Kyrgyz Republic

Kazimir Karimov
Institute of Physics
National Academy of Sciences
Chui Prosp. 265-A, Bishkek
720071
Kyrgyz Republic

MOROCCO
Ismail M. Alaoui
Physics Department
Faculty of Sciences Semlalia
Cadi Ayyed University
BP 2390 Marrakesh 4000,
Morocco

RUSSIA
Robert I. Nigmatulin
P.P. Shirshov Institute of Oceanology
Russian Academy of Sciences
117997 Moscow, Russia

Karina Nigmatulin
P.P. Shirshov Institute of Oceanology
Russian Academy of Sciences
117997 Moscow, Russia

TUNISIA
Kamel Halouani
Micro Electro Thermal Systems Research Unit
METS-IPEIS-Route Menzel Chaker
BP. 805, Sfax

UKRAINE
Boris Kosoy
Computer Science Department
Odessa State Academy of Refrigeration
1/3 Dvoryanskaya Str.
65082 Odessa, Ukraine

Victor Mazur
Odessa State Academy of Refrigeration
1/3 Dvoryanskaya Str.
65082 Odessa, Ukraine

SOME INTRODUCTORY TECHNICAL REMARKS: THE NATO SCHOOL ON MICRO-FUEL CELLS

SHIMSHON GOTTESFELD[*]
*STA, MTI Microfuel Cells, Albany, NY, USA & President,
Fuel Cell Consulting, LLC, Niskayuna, NY, USA*

1. Introduction

To introduce the background for micro fuel cells in terms of target applications and key technology facets, the following specific aspects were chosen for discussion:

- The Military Perspective
- The Consumer Electronics Market Perspective
- The key merit parameter(s) of a micro-fuel cell system
- State of the art and remaining challenges
- Some perspective of the future of this technology.

2. The Military Perspective

Figure 1 shows a 2002 "notional concept" by the US Army, of the possible replacement of primary Li batteries (BA-5590) by a hybrid of a direct methanol fuel cell (DMFC) fueled by 100% methanol and a rechargeable battery. For a mission of 72 hours, the figure describes the possibility of getting the weight of such a FC/Battery hybrid down to 7.5 lbs, compared with the present solution based on BA-5590 batteries weighing 12 lbs. The weight reduction expected on re-supply is significantly higher: from 12 lbs down to 4.5 lbs, because at this point the FC/battery hybrid requires only fuel cartridge replacement whereas the primary batteries employed have to fully replenish.

---

[*]Shimshon Gottesfeld, STA, MTI Microfuel Cells, Albany, NY, USA and Fuel Cell Consulting, LLC, Niskayuna, NY, USA; e-mail: shimshon.gottesfeld@gmail.com

The choice of DMFC technology for the presentation of this concept system by the US Army was based on the understanding that this technology provides important advantages for portable power applications, of a fuel cell stack based on direct oxidation of an energy rich fuel (methanol) which is liquid under ambient conditions and is easily transportable in simple plastic containers. As of today, the DMFC still remains at the center of the US Army programs of fuel cell system testing and evaluation in the 20–30 W power range, with other technologies considered including the RMFC, a reformed methanol fuel cell system, and a hydrogen/air fuel cell system with "on-demand" supply of hydrogen achieved by passing an alkaline solution of sodium borohydride over a metal catalyst to release H_2 gas.

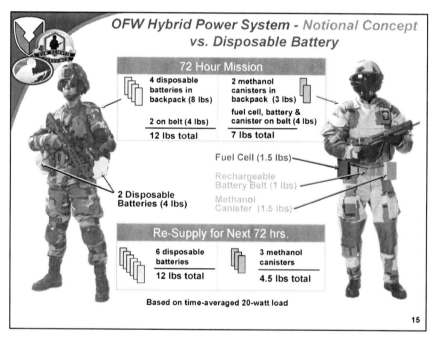

Figure 1. A 2002 presentation from the US Army; describing a "notional concept" of replacing soldier carried primary BA-5590 batteries by a DMFC/rechargeable battery hybrid, to achieve significant weigh reduction on a 72 hour long mission.

It can be clearly seen that significant gaps remained as of December 2006 on the road to implementation of this hybrid concept, mainly in system weight and in operation below freezing. The system conversion efficiency, reflected by the number of watt-hours to the load per liter of fuel in the tank, also still requires a boost by 30%. It is important to realize that, as of today, advanced DMFC stack technology demonstrated has in fact approached power density 100 W per kg of the stack and, consequently, a weight of around 0.5 lb should be achievable for a 25 W DMFC stack, potentially reducing the total weight of the system by around 1 lb vs. the 30 W/kg stack considered in Table 1. A cautionary counter comment is that performance parameters are ordinarily quoted based on measurements performed for not more than several days and still require evaluation of the degree of performance loss with longer operation time.

To have the discussion of the potential military applications more complete, two other types of fuel cell systems tested recently by the US military at the 20–40 W power level, have to be mentioned. One is based on a stack fueled by hydrogen derived from steam reforming of methanol within the system, the so called Reformed Methanol Fuel Cell (RMFC), developed by Ultracell (CA, USA). This platform attempts to combine the high energy density and liquid form characteristics of methanol fuel with the higher power density of a hydrogen fuel cell. The barriers of such, indirect fuel cell system, is a more complex system comprising a series combination of two catalytic reactors, reformer and fuel cell. The other fuel cell system of power output 20–40 W under evaluation and testing by the US military, achieves relatively high energy density for mission duration of, at least, 72 hours, using controlled catalytic decomposition of an alkaline, borohydride solution to release "hydrogen on demand" (Protonex and Millenium Cell, MA, USA).

TABLE 1. 20 W DMFC System: achieved in 2006 vs. Required for fielding by US Army.

	20 W system 2005	M-25 Target
Mission Time (hours)	72	72
Total energy (Wh)	420	440
Wh/lfuel; Wh/lsystem	1,100; 400	1,440; 700
W/kg stack	10	30 @ $\eta = 25\text{–}30\%$
System weight (lb)	4.2	2.5
Temp Envelope (°C)	0–50	–20–55

3. The Consumer Electronics Industry Perspective

A very significant feature of micro-fuel cell technology is that it is driven to large degree by a significant market pull, rather than by industry and/or government push. The reason for such market pull, is the limitations of advanced batteries, typically rechargeable Li-ion batteries, in providing from the volume and/or weight typically allocated the energy required for hours of operation of a modern day, multi-functional hand-set. A pictorial summary of the "barrage of demands" on the battery by the ever crowding

Cellular Phone battery capacity becomes insufficient

Presentation by H. Yomogita, Nikkei Business Electronics Apr. '05
Intl. Battery Seminar, Fort Lauderdale, Florida

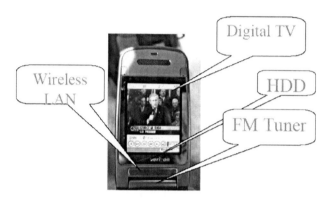

Figure 2. The energy challenge of the multifunctional handset.

Power Eaters Drive Battery Evolution

Presentation by H. Yomogita, Nikkei Business Electronics Apr. '05
Intl. Battery Seminar, Fort Lauderdale, Florida

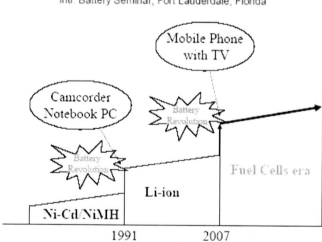

Figure 3. The projected solution: the dawn of the "fuel cell era" of power sources for handsets.

applications within a handset is given in Figure 2 which is taken from a presentation by Nikkei Business Electronics, a professional magazine catering for the EE community in Japan.

Whereas Figure 2 presents the challenge, Figure 3, taken from the same source, presents the projected solution: the dawn of the "fuel cell era" of power sources for handsets. Figure 3 suggests a parallel between the drive of the last "battery revolution" by emerging "Power Eater" applications in the early 1990 – entry of Li-ion batteries to respond to the power need of Notebook PCs and Camcorders – and the drive today for another "battery revolution" to satisfy the needs of multi-functional handsets, such new revolution projected to open the micro-fuel cell era.

The world-wide power-pack market is very large and is divided between various applications as shown in the pie-chart in Figure 4.

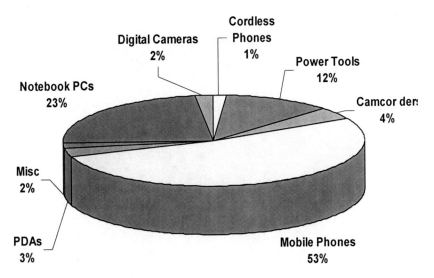

Figure 4. The world-wide power-pack market is very large and is divided between various applications.

Most efforts to-date to develop and demonstrate micro-fuel-cell systems for consumer applications have concentrated on a DMFC-based technology. Beneficial *Methanol Fuel* Properties mentioned in the context of military applications are relevant here as well and include:

- Liquid fuel under ambient pressure
- High energy density in neat liquid form
- Relatively negligible safety issues: approved for carrying in small cartridges on board passenger airplanes
- Amenable to effective distribution to consumers, e.g. methanol cartridge to be sold in convenience stores.

Figure 5. 50 W DMFC power system with plastic container of methanol fuel (Smart Fuel Cells, Germany).

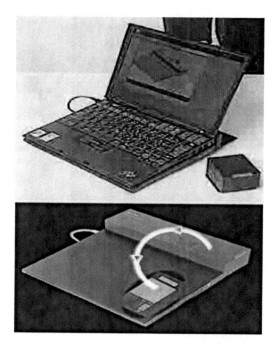

Figure 6. Early prototype of DMFC power pack added under a laptop computer (Sanyo and IBM).

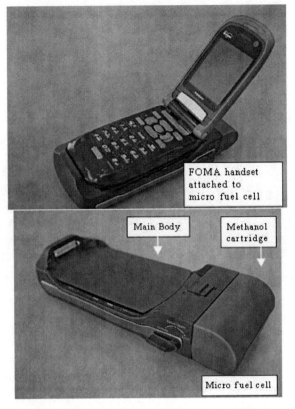

Figure 7. DMFC charger cradle for a FOMA handset (DoCoMo, Japan).

4. Analysis of Key Parameters and Key Requirements for Micro DMFCs and Other Micro Fuel Cell Systems

To further discuss DMFC technology, Figure 8 provides a schematic of such a cell, highlighting key components and processes.

The process on the anode side of the cell is:

$$CH_3OH + H_2O = CO_2 + 6H^+ + 6e^- \quad (1)$$

and the cathode process

$$1.5O_2 + 6H^+ + 6e^- = 3H_2O \quad (2)$$

Summing up (1) and (2) yields the overall DMFC cell process:

$$CH_3OH + 1.5O_2 = CO_2 + 2H_2O \\ + (\eta \times 5Wh \text{ per cc of neat methanol consumed}) \quad (3)$$

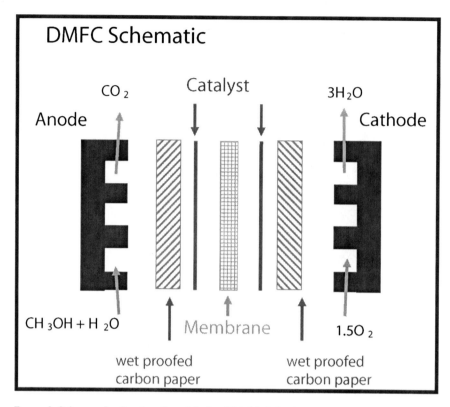

Figure 8. Scheme of a direct methanol fuel cell highlighting key components and processes.

In Eq. (3), the first line describes the chemical process in the DMFC and the second line describes the main intended product of the DMFC, i.e. the electric energy generated, expressed as the product of the theoretical energy content (heating value) of the volume of neat methanol consumed, multiplied by the overall conversion efficiency of the DMFC system, η.

Figure 9 below, describes the potential advantage in energy density over incumbent Li-ion batteries that can be reached using a DMFC power system of conversion efficiency 20–30%, provided the fuel occupies 75% of the volume of the power pack, i.e. the non-fuel components can be packaged into 25% of the overall volume allotted. Whereas the state-of-the-art of DMFC technology has reached conversion efficiencies assumed in Figure 9, the other assumption made in deriving the power system parameters in Figure 9, i.e. packaging of all non-fuel containing components into only 25% of the total volume of the power-pack, is not easily reachable in DMFC targeted applications where the limit on system volume is severe e.g. when the power pack needs to be <25 CC overall.

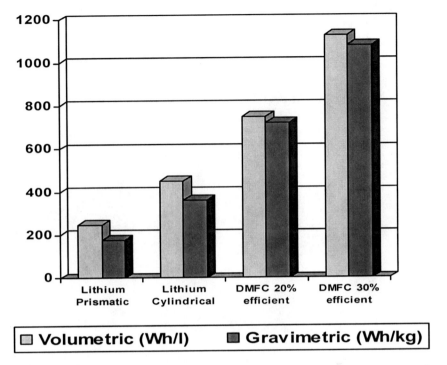

Figure 9. Calculated energy density of DMFC Micro Fuel Cell Systems at conversion efficiencies of 20% and 30% and assuming the volume fraction of the system occupied by (100%) methanol fuel, is 75%.

To elucidate some general micro-fuel cell system performance demands, that allow packaging into some volume limit while securing the advantage over a Li-ion battery, the following schematic (Figure 10) is used.

It can be understood from this figure, that the actual number of Wh delivered to the load per CC of the system, is given by:

$$Wh_{to\ the\ load} / CC_{system} = (Wh/CC)^*_{fuel} \times Fr._{fuel} \times \eta_{system} \qquad (4)$$

where $(Wh/CC)^*_{fuel}$ is the heating value of the fuel, $Fr._{fuel}$ is the fraction of the system volume occupied by fuel and η_{system} is the overall system conversion efficiency, chemical to electrical energy. Equation (4) teaches that, given a fuel of some heating value, the effective energy density of the fuel cell power system is defined to the same degree by the conversion efficiency of the fuel cell and by the fraction of the volume which contains chemical energy, i.e. the volume fraction filled by fuel. The obvious way to maximize $Fr._{fuel}$ is to minimize $1 - Fr._{fuel}$, the non-fuel fraction of the total volume of the system which implies, in turn, minimizing auxiliaries such as those required for water management and achieving high stack power density to have it take only a small fraction of the total volume in order to provide the average power demand for the micro fuel cell/battery hybrid system considered.

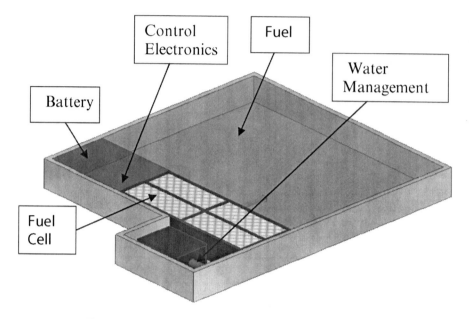

Figure 10. Building block diagram of a micro fuel cell system.

A simple way to present the relative advantage of a microfuel system with specific combined characteristics of: (a) Wh to the load per unit inner volume of the fuel cartridge, and (b) power density of the fuel cell stack, is to plot the volume of the system as function of use time at the relevant average power demand. The lower is the volume accommodating all fuel + non-fuel components of the system, the higher the effective energy density of the power system for the specific use time, i.e. the specific energy content of the system. As can be seen from Figure 11, contributions towards a smaller system volume, can be made by a higher stack power density, reflected in the plots as a lower system volume at zero energy content, and by a lower volume of fuel required to deliver a unit electric energy to the load, reflected by a lower slope of the linear plot. Such plots are shown in Figure 11 for a DMFC system assuming 1.5 Wh to the load per CC of neat methanol fuel, and for hydrogen fueled system where, assuming a metal hydride fuel cartridge, the energy to the load is 0.6 Wh per CC of the hydride fuel. Three possible power densities, 50, 75 and 100 mW/CC are considered in the figure for the DMFC micro fuel cell stack and 300 mW/CC is the power density assumed for the hydrogen fueled microfuel cell stack. The above numbers reflect state-of-the–art micro fuel cell systems demonstrated to date. The total volume of non-fuel containing components is considered in all cases in Figure 11 to be twice the volume of the stack itself.

Comparative evaluation of methanol vs. hydrogen based micro fuel cell systems, has been at the center of many discussions of micro fuel cell technology since it is inception. What Figure 11 shows, is that, in fact, the relative advantage of one or the other could change as function of the demand use time per fuel cartridge. As the energy density of methanol fuel is significantly higher than that of any benign form of hydrogen storage,

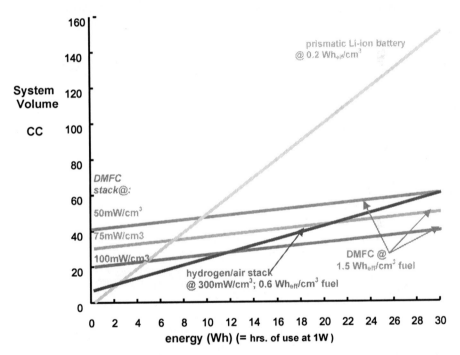

Figure 11. System volume for a 1 W micro fuel cell system and a Li-ion battery, as function of run time demanded per full cartridge. Based on key parameters reported, higher energy density will be obtained with the battery for use times of less than two hours (energy content of less than 2 Wh), with the hydrogen fueled micro fuel cell for use times of 2–12 hours (2–12 Wh of energy) and with a micro-DMFC for use times higher than 12 hours (>12 Wh of energy) per single fuel cartridge.

the longer the use time demanded per cartridge of fuel the larger is the advantage of the DMFC system, as seen from the lower system volumes calculated for it. This advantage of the DMFC system depends critically, however, on the power density of the DMFC stack: at 100 mW/CC of stack, the advantage over the hydrogen (hydride) fueled system, is reached for use times per cartridge larger than 11 hours, whereas at the lower, 50 mW/CC of the DMFC stack, the DMFC advantage is expected only for use times longer than 30 hours per cartridge. Clearly, from Figure 11, a prime challenge of enhancing the energy density advantage of the micro fuel cell system over the Li-ion battery is higher stack power density in the case of the DMFC system and higher energy density of the stored fuel in the case of the system based on hydrogen fuel.

The conclusion from the above discussion, is that the *combination* of average power demand and energy content per full cartridge, i.e. the demand combination {W; Wh} for the system, is the only way to define the system of highest energy density per some given application. Consideration of stack power density alone, or of the effective fuel energy density alone, would not suffice.

References

1. P. Piela and P. Zelenay, Researchers redefine the DMFC roadmap. *The Fuel Cell Review* 1, 17–23 (2004).
2. S. Gottesfeld, DMFCs power up for portable devices. *The Fuel Cell Review* 1, 25–29 (2004).
3. Proceedings of the Knowledge Foundation's Annual International Symposium on Small Fuel Cells, New Orleans, LA, USA.
4. M. Cappadonia, M. Eikerling, S. Gottesfeld, in: *Encyclopedia of Electrochemistry*, edited by A.J. Bard and M. Stratmann (Wiley-VCH Verlag GmbH & Co. KGaA, Weinheim, Germany, 2007), pp. 425–662.

FUEL CELL BASIC CHEMISTRY, ELECTROCHEMISTRY AND THERMODYNAMICS

FRANO BARBIR
Faculty of Electrical Engineering, Mechanical Engineering and Naval Architecture, University of Split, R. Boskovica bb, 21000 Split, Croatia

1. Introduction

A fuel cell is an electrochemical energy converter – it converts chemical energy of fuel, typically hydrogen, directly into electrical energy. Its theoretical efficiency is determined by thermodynamics of its (electro) chemical reactions. Because its efficiency is not limited by Carnot efficiency, the theoretical efficiency of low temperature fuel cells is higher than the theoretical efficiencies of internal combustion engines. Fuel cell actual efficiency is, of course, always lower than the theoretical one, and it depends on the rate of the electrochemical reactions, their kinetics, as well as the properties of the materials and design of the components used to direct the reactants to and take the products away from the reaction sites.

2. Theoretical Fuel Cell Potential, Power and Efficiency

2.1. BASIC REACTIONS AND THEIR THERMODYNAMICS

The electrochemical reactions in fuel cell happen simultaneously on both sides of the membrane – the anode and the cathode. The basic fuel cell reactions are:

At the anode: $\quad H_2 \rightarrow 2H^+ + 2e^-$ \hfill (1)

At the cathode: $\quad \tfrac{1}{2}O_2 + 2H^+ + 2e^- \rightarrow H_2O$ \hfill (2)

Overall: $\quad H_2 + \tfrac{1}{2}O_2 \rightarrow H_2O$ \hfill (3)

The above reactions are for fuel cells with acid electrolyte, such as the proton exchange membrane fuel cells (PEMFC) or phosphoric acid fuel cells (PAFC). Other types of fuel cells have different anodic and cathodic reactions but the overall reaction is always the same. The overall fuel cell reaction, Eq. (3), is exactly the same as the reaction of hydrogen combustion. Combustion is an exothermic process, which means that there is energy released in the process:

$$H_2 + \tfrac{1}{2}O_2 \rightarrow H_2O + heat \qquad (4)$$

The heat (or enthalpy) of a chemical reaction is the difference between the heats of formation of products and reactants. For hydrogen/oxygen fuel cells generating liquid water the enthalpy of the reaction is $\Delta H = -286$ kJ mol^{-1} (at 25°C). This is also by definition the hydrogen's higher heating value ($H_{HHV} = -\Delta H$), i.e. the amount of energy that can be generated by combustion of hydrogen.

In a fuel cell, not the entire enthalpy can be converted to useful work, i.e. electricity. The portion of the reaction enthalpy that can be converted to electricity corresponds to Gibbs free energy, and is given by the following equation:

$$\Delta G = \Delta H - T\Delta S \qquad (5)$$

In other words, there are some irreversible losses in energy conversion due to creation of entropy, ΔS. The values of ΔG, ΔH and ΔS at 25°C are given in Table 1 (data compiled from reference[1]).

TABLE 1. Enthalpies, entropies and Gibbs free energy for hydrogen oxidation process (at 25°C); (data compiled from reference[1]).

	ΔH (kJ mol^{-1})	ΔS (kJ mol^{-1} K^{-1})	ΔG (kJ mol^{-1})
$H_2 + \tfrac{1}{2}O_2 \rightarrow H_2O(l)$	−286.02	−0.1633	−237.34
$H_2 + \tfrac{1}{2}O_2 \rightarrow H_2O(g)$	−241.98	−0.0444	−228.74

The theoretical or reversible potential of fuel cell is:

$$E_r = \frac{-\Delta G}{nF} \qquad (6)$$

Number of electrons, n, involved in the fuel cell reaction, Eqs. (1–3), is 2 per molecule of hydrogen. With numerical values of ΔG, n and F, the theoretical fuel cell potential is 1.23 V (at 25°C).

2.2. EFFECT OF TEMPERATURE

Because both ΔH and ΔS depend on temperature[2], ΔG depends on temperature too, and consequently the theoretical cell voltage depends on temperature. Table 2 shows the values of enthalpy, Gibbs free energy and entropy of hydrogen/oxygen fuel cell reaction at various temperatures.[3] It should be noted that the theoretical fuel cell potential decreases with temperature, and at typical fuel cell operating temperatures (for PEM fuel cells) of 60–80°C, the theoretical cell potential is 1.20–1.18 V.

TABLE 2. Enthalpy, Gibbs free energy and entropy of hydrogen/oxygen fuel cell reaction with temperature and resulting theoretical cell potential.[3]

T K	ΔH (kJ mol^{-1})	ΔG (kJ mol^{-1})	ΔS (kJ mol^{-1} K^{-1})	Eth V
298.15	−286.02	−237.34	−0.16338	1.230
333.15	−284.85	−231.63	−0.1598	1.200
353.15	−284.18	−228.42	−0.1579	1.184
373.15	−283.52	−225.24	−0.1562	1.167

2.3. EFFECT OF PRESSURE

All of the above equations were valid at atmospheric pressure. However, a fuel cell may operate at any pressure, typically from atmospheric all the way up to 6–7 bar. In that case the theoretical cell potential will be higher at elevated pressure according to the Nernst equation[2]:

$$E_r = E_0 + \frac{RT}{nF} \ln\left(\frac{\Pi_{H_2} \Pi_{O_2}^{0.5}}{\Pi_{H_2O}}\right) \tag{7}$$

where Π is the partial pressure of the reactants and products relative to atmospheric pressure. Note that the above equations are only valid for gaseous products and reactants. When liquid water is produced in a fuel cell, $\Pi_{H2O} = 1$. From Eq. (9) it follows that at higher reactant pressures the cell potential is higher. Also, if the reactants are diluted, for example if air is used instead of pure oxygen, their partial pressure is proportional to their concentration, and consequently the cell potential is lower.

2.4. THEORETICAL FUEL CELL EFFICIENCY

The efficiency of any energy conversion device is defined as the ratio between useful energy output and energy input. In case of a fuel cell, the useful energy output is the electrical energy produced, and energy input is hydrogen's higher heating value. Assuming that all of the Gibbs free energy can be converted into electrical energy, the maximum possible (theoretical) efficiency of a fuel cell is:

$$\eta = \Delta G/\Delta H = 237.34/286.02 = 83\% \qquad (8)$$

Very often, hydrogen's lower heating value is used to express the fuel cell efficiency, not only because it results in a higher number, but also to compare it with fuel cell's competitor – internal combustion engine, whose efficiency has traditionally been expressed with lower heating value of fuel. In that case the maximum theoretical fuel cell efficiency would be:

$$\eta = \Delta G/\Delta H_{LHV} = 228.74/241.98 = 94.5\% \qquad (9)$$

The use of lower heating value, both in the fuel cell and especially in the internal combustion engine, is justified by water vapor being produced in the process, so the difference between higher and lower heating value (heat of evaporation) cannot be used anyway.

3. Actual Fuel Cell Potential, Power and Efficiency

The actual fuel cell potential, and the actual efficiency are lower than the theoretical ones due to various losses associated with kinetics and dynamics of the processes, reactants and the products.

$$V_{cell} = E_r - \Delta V_{loss} \qquad (10)$$

3.1. VOLTAGE LOSSES

If a fuel cell is supplied with reactant gases, but the electrical circuit is not closed, it will not generate any current, and one would expect the cell potential to be at the theoretical cell potential for given conditions (temperature, pressure and concentration of reactants). However, in practice this potential, called the open circuit potential, is significantly lower than the theoretical potential, usually less than 1 V. This suggests that there are some losses in the fuel cell even when no external current is generated. When the electrical circuit is closed with a load (such as a resistor) in it, the potential is expected to drop even further as a function of current being generated, due to unavoidable losses. There are different kinds of voltage losses in a fuel cell caused by several factors listed below:

- Kinetics of the electrochemical reactions
- Internal electrical and ionic resistance
- Reduced reactants concentration at the reaction sites
- Internal (stray) currents
- Crossover of reactants.

While mechanical and electrical engineers prefer to use voltage losses (electro)chemical engineers use terms such as polarization or overpotential. They all have the same physical meaning – difference between the electrode potential and the equilibrium potential. From the electrochemical engineer's point of view this difference is the driver for the reaction, and from a mechanical or electrical engineer's point of view this represents the loss of voltage and power.

3.1.1. Activation Polarization

Some voltage difference from equilibrium, called overpotential, is needed to get the electrochemical reaction going. The relationship between current density and this voltage difference is given by Butler-Volmer equation for any electrode[4, 5, 6]:

$$i = i_0 \left\{ \exp\left[\frac{-\alpha_{Rd} F(E - E_r)}{RT}\right] - \exp\left[\frac{\alpha_{Ox} F(E - E_r)}{RT}\right] \right\} \quad (11)$$

where E_r is the reversible or equilibrium potential. Note that the reversible or equilibrium potential at the fuel cell anode is 0 V by definition,[5] and the reversible potential at the fuel cell cathode is 1.230 V (at 25°C and atmospheric pressure) and it does vary with temperature and pressure as shown above.

Two factors in the above equation that determine the sluggishness of an electrode are the transfer coefficient, α, and exchange current density, i_0.

The transfer coefficients in above equation are the coefficients for forward and backward reactions on the electrode, i.e. reduction and oxidation. There is a fair amount of confusion in the literature concerning the transfer coefficient, α, and the symmetry factor, β, that is sometimes used. The symmetry factor, β, may be used strictly for a single step reaction involving a single electron (n = 1). Its value is theoretically between 0 and 1, but most typically for the reactions on a metallic surface it is around 0.5. The way in which β is defined requires that the sum of the symmetry factors in the anodic and cathodic direction be unity; if it is β for the reduction reaction it must be $(1 - \beta)$ for the reverse, oxidation reaction. However, both electrochemical reactions in a fuel cell, namely oxygen reduction and hydrogen oxidation involve more than one step and more than one electron. In that case, at steady state, the rate of all steps must be equal, and it is determined by the slowest step in the sequence, which is referred to as the rate-determining step. In order to describe a multi step process, instead of the symmetry factor, β, a rather experimental parameter is used, which is called the transfer coefficient, α. Note that in this case $\alpha_{Rd} + \alpha_{Ox}$ does not necessarily have to be equal to unity. Actually, in general $(\alpha_{Rd} + \alpha_{Ox}) = n/v$, where n is the number of electrons transferred in the overall reaction and v is the stoichiometric number defined as the number of times the rate-determining step must occur for the overall reaction to occur once.[6]

Exchange current density, i_0, in electrochemical reactions is analogous to the rate constant in chemical reactions. Unlike the rate constants, exchange current density is concentration dependent and it is also a function of temperature. The effective exchange current density (per unit of electrode geometrical area) is also a function of electrode catalyst loading and catalyst specific surface area. If the reference exchange current density (at reference temperature and pressure) is given per actual catalyst surface area, then the effective exchange current density at any temperature and pressure is given by the following equation:[7]

$$i_0 = i_0^{ref} a_c L_c \left(\frac{P_r}{P_r^{ref}}\right)^{\gamma} \exp\left[-\frac{E_C}{RT}\left(1 - \frac{T}{T_{ref}}\right)\right] \quad (12)$$

where:

i_0^{ref} = reference exchange current density (at reference temperature and pressure, typically 25°C and 101.25 kPa) per unit catalyst surface area, A cm^{-2} Pt

a_c = catalyst specific area (theoretical limit for Pt catalyst is 2,400 cm^2 mg^{-1}, but state-of-the-art catalyst has about 600–1,000 cm^2 mg^{-1}, which is further reduced by incorporation of catalyst in the electrode structures by up to 30%)

L_c = catalyst loading (state-of-the-art electrodes have 0.3–0.5 mg Pt cm^{-2}; lower loadings are possible but would result in lower cell voltages)

P_r = reactant partial pressure, kPa

P_r^{ref} = reference pressure, usually atmospheric pressure, kPa

γ = pressure coefficient (0.5–1.0)

E_C = activation energy (for example[7] E_C = 66 kJ mol^{-1} for oxygen reduction on Pt)

R = gas constant, 8.314 J mol^{-1} K^{-1}

T = temperature, K

T_{ref} = reference temperature, i.e. 298.15 K.

The product a_cL_c is also called electrode roughness, meaning the catalyst surface area, cm^2, per electrode geometric area, cm^2. Instead of the ratio of partial pressures, a ratio of concentrations at the catalyst surface may be used as well.

Exchange current density is a measure of an electrode's readiness to proceed with the electrochemical reaction. If the exchange current density is high, the surface of the electrode is more active. In a hydrogen/oxygen fuel cell, the exchange current density at the anode is much larger (several orders of magnitudes) than at the cathode. The higher the exchange current density the lower the energy barrier that the charge must overcome moving from electrolyte to the catalyst surface and vice versa. In other words the higher the exchange current density the more current is generated at any given overpotential. Or the other way around, the higher the exchange current density, the lower the activation polarization losses are. These losses happen at both anode and cathode, however oxygen reduction requires much higher overpotentials, i.e. it is much slower reaction than hydrogen oxidation.

At relatively high negative overpotentials (i.e. potentials lower than the equilibrium potential), such as those at the fuel cell cathode, the first term in Butler-Volmer Eq. (11) becomes predominant, which allows for expression of potential as a function of current density:

$$\Delta V_{act,c} = E_{r,c} - E_c = \frac{RT}{\alpha_c F} \ln\left(\frac{i}{i_{0,c}}\right) \tag{13}$$

Similarly, at the anode at positive overpotentials (i.e. higher than the equilibrium potential) the second term in Butler Volmer Eq. (11) becomes predominant:

$$\Delta V_{act,a} = E_a - E_{r,a} = \frac{RT}{\alpha_a F} \ln\left(\frac{i}{i_{0,a}}\right) \tag{14}$$

A simplified way to show the activation losses is to use the so called Tafel equation, which although empirical, has the same form and theoretical background:

$$\Delta V_{act} = a + b \, log(i) \tag{15}$$

Note that Eqz. (13) or (14) and (15) are identical when $a = -2.3 \frac{RT}{\alpha F} \log(i_o)$, and $b = 2.3 \frac{RT}{\alpha F}$. Term b is called the Tafel slope. Note that at any given temperature the Tafel slope depends solely on transfer coefficient, α. For $\alpha = 1$, the Tafel slope at 60°C is ~60 mV per decade, what is typically found for oxygen reduction on Pt.

If the activation polarizations were the only losses in a fuel cell, the cell potential would be:

$$V_{cell} = E_r - \Delta V_{act,c} - \Delta V_{act,a} \tag{16}$$

$$V_{cell} = E_r - \frac{RT}{\alpha_c F} \ln\left(\frac{i}{i_{0,c}}\right) - \frac{RT}{\alpha_a F} \ln\left(\frac{i}{i_{0,a}}\right) \tag{17}$$

3.1.2. Ohmic (Resistive) Losses

Ohmic losses occur because of resistance to the flow of ions in the electrolyte and resistance to the flow of electrons through the electrically conductive fuel cell components. These losses can be expressed by Ohm's law:

$$\Delta V_{ohm} = iR_i \tag{18}$$

Ri is the total cell internal resistance, which includes ionic, electronic and contact resistance, namely:

$$R_i = R_{i,i} + R_{i,e} + R_{i,c} \tag{19}$$

Electronic resistance is almost negligible, even when graphite or graphite/polymer composites are used as current collectors. Ionic and contact resistances are approximately of the same order of magnitude.[7,8] Ionic resistance mostly depends on the state of hydration of the polymer membrane, while the contact resistance greatly depends on the materials used for GDL and bipolar plates and on the clamping pressure applied. Typical values for R_i in well designed, assembled and operated fuel cell are between 0.1 and 0.2 Ω cm^2.

3.1.3. Concentration Polarization

Concentration polarization occurs when a reactant is rapidly consumed at the electrode by the electrochemical reaction so that concentration gradients are established. The electrochemical reaction potential changes with partial pressure of the reactants, and this relationship is given by Nernst equation:

$$\Delta V = \frac{RT}{nF} \ln\left(\frac{C_B}{C_S}\right) \tag{20}$$

where, C_B is bulk concentration of reactant and C_S is concentration of reactant at the surface of the catalyst.

According to Fick's Law, the flux of reactant is proportional to concentration gradient:[2]

$$N = \frac{D \cdot (C_B - C_S)}{\delta} A \qquad (21)$$

In steady state, the rate at which the reactant species is consumed in the electrochemical reaction is equal to the diffusion flux, which as per Faraday's Law is:

$$N = \frac{I}{nF} \qquad (22)$$

By combining Eqs. (21) and (22) the following relationship is obtained:[2]

$$i = \frac{nF \cdot D \cdot (C_B - C_S)}{\delta} \qquad (23)$$

The reactant concentration at the catalyst surface, thus, depends on current density – the higher the current density the lower the surface concentration. The surface concentration reaches zero when the rate of consumption exceeds the diffusion rate – the reactant is consumed faster than it can reach the surface. Current density at which this happens is called the limiting current density. A fuel cell cannot produce more than the limiting current because there are no reactants at the catalyst surface. Therefore, for $C_S = 0$, $i = i_L$, and the limiting current density is then:[2]

$$i_L = \frac{nFDC_B}{\delta} \qquad (24)$$

By combining Eqs. (20), (23) and (24) a relationship for voltage loss due to concentration polarization is obtained:

$$\Delta V_{conc} = \frac{RT}{nF} \ln\left(\frac{i_L}{i_L - i}\right) \qquad (25)$$

The above equation would result in a sharp drop of cell potential as the limiting current is approached. However, due to non-uniform conditions over the porous electrode area, the limiting current is almost never experienced in practical fuel cells. In order to experience a sharp drop of cell potential when the limiting current density is reached, the current density would have to be uniform over the entire electrode surface, which is almost never the case. There will be some areas that would reach the limiting current density sooner than the other areas. Limiting current density may be experienced at either cathode or anode.

An empirical equation better describes the polarization losses, as suggested by Kim, et al.:[9]

$$\Delta V_{conc} = c \cdot \exp\left(\frac{i}{d}\right) \qquad (26)$$

where c and d are empirical coefficients (values of $c = 3 \times 10 - 5$ V and $d = 0.125$ A cm^{-2} have been suggested,[10] but they clearly depend on the cell design and operating conditions).

3.1.4. Internal Currents and Crossover Losses

Although the electrolyte, a polymer membrane, is not electrically conductive and it is practically impermeable to reactant gases, some small amount of hydrogen will diffuse from anode to cathode, and some electrons may also find a "shortcut" through the membranes. Since each hydrogen molecule contains two electrons, this fuel crossover and so called internal currents are essentially equivalent. Each hydrogen molecule that diffuses through the polymer electrolyte membrane and reacts with oxygen on the cathode side of the fuel cell results in two fewer electrons in the generated current of electrons that travels through an external circuit. These losses may appear insignificant in fuel cell operation, since the rate of hydrogen permeation or electron crossover is several orders of magnitude lower than hydrogen consumption rate or total electrical current generated, however, when the fuel cell is at open circuit potential or when it operates at very low current densities, these losses may have a dramatic effect on cell potential. The total electrical current density is the sum of external (useful) current and current losses due to fuel crossover and internal currents. Therefore, even if the external current is equal to zero, such as at open circuit, the internal current and crossover losses combined with the activation losses, are sufficient to bring the cell voltage significantly lower than the reversible cell potential for given conditions. Indeed, open circuit potential of hydrogen/air fuel cells is typically below 1 V, most likely about 0.94 to 0.97 V (depending on operating pressure).

3.2. ACTUAL FUEL CELL POTENTIAL – POLARIZATION CURVE

The cell voltage is therefore:

$$V_{cell} = E_r - (\Delta V_{act} + \Delta V_{conc})_a - (\Delta V_{act} + \Delta V_{conc})_c - \Delta V_{ohm} \qquad (27)$$

By introducing Eqs. (13), (14), (18) and (25) into Eq. (27), a relationship between fuel cell potential and current density, so called fuel cell polarization curve, is obtained:

$$E_{cell} = E_{r,T,P} - \frac{RT}{\alpha_c F}\ln\left(\frac{i}{i_{0,c}}\right) - \frac{RT}{\alpha_a F}\ln\left(\frac{i}{i_{0,a}}\right)$$
$$-\frac{RT}{nF}\ln\left(\frac{i_{L,c}}{i_{L,c}-i}\right) - \frac{RT}{nF}\ln\left(\frac{i_{L,a}}{i_{L,a}-i}\right) - iR_i \quad (28)$$

Figure 1 shows how the cell polarization curve is formed, by subtracting the activation polarization losses, ohmic losses and concentration polarization losses from the equilibrium potential. Anode and cathode activation losses are lumped together, but, as mentioned above, a majority of the losses occur on the cathode due to sluggishness of the oxygen reduction reaction.

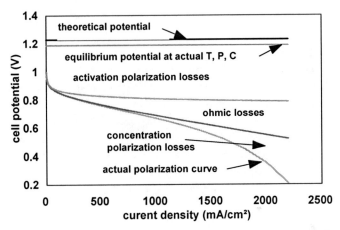

Figure 1. Voltage losses in fuel cell and resulting polarization curve.[3]

3.3. ACTUAL FUEL EFFICIENCY

As defined above, the fuel cell efficiency is a ratio between useful energy output, i.e. electricity produced, and energy input, i.e. hydrogen consumed. Of course, both must be in same units, such as Watts or kilowatts.

$$\eta = \frac{W_{el}}{W_{H2}} \quad (29)$$

Electricity produced is simply a product between voltage and current.

$$W_{el} = I \cdot V \quad (30)$$

Hydrogen consumed is (according to Faraday's Law) directly proportional to current:

$$N_{H2} = \frac{I}{nF} \tag{31}$$

The "energy value" of consumed hydrogen is:

$$W_{H2} = \Delta H \frac{I}{nF} \tag{32}$$

It should be noted that $\Delta H/nF$ has dimension of Volts, and for $\Delta H = 286$ kJ/mol it has a value of 1.482 V, which is the so called thermoneutral potential.

By combining Eqs. (29) through (32) the fuel cell efficiency is simply directly proportional to the cell potential:

$$\eta = \frac{V}{1.482} \tag{33}$$

If some hydrogen is lost either due to hydrogen diffusion through the membrane, or due to combining with oxygen that diffused through the membrane or due to internal currents, hydrogen consumption will be higher than that corresponding to generated current (Eq. (32)). Consequently, the fuel cell efficiency would be somewhat lower than given by Eq. (33). Typically this loss is very low, in the order of magnitude of a few mA/cm^2, and therefore it affects the fuel cell efficiency only at very low currents, i.e. current densities (as shown in Figure 2). The fuel cell efficiency is then a product of voltage efficiency and current efficiency:

$$\eta = \frac{V}{1.482} \frac{i}{(i + i_{loss})} \tag{34}$$

If hydrogen is supplied to the cell in excess of that required for the reaction stoichiometry, this excess will leave the fuel cell unused. In case of pure hydrogen this excess may be recirculated back into the stack so it does not change the fuel cell efficiency (not accounting for the power needed for hydrogen recirculation pump), but if hydrogen is not pure (such as in reformate gas feed) unused hydrogen leaves the fuel cell and does not participate in the electrochemical reaction. The fuel cell efficiency is then:

$$\eta = \frac{V}{1.482} \eta_{fu} \tag{35}$$

where η_{fu} is fuel utilization, which is equal to 1/SH2, where SH2 is the hydrogen stoichiometric ratio, i.e. the ratio between the amount of hydrogen actually supplied to the fuel cell and that consumed in the electrochemical reaction:

$$S_{H2} = \frac{N_{H2,act}}{N_{H2,theor}} = \frac{nF}{I} N_{H2,act} \qquad (36)$$

Well designed fuel cells may operate with 83–85% fuel utilization when operated with reformate and above 90% when operated with pure hydrogen. Note that the current efficiency term in Eq. (34) is included in fuel utilization, η_{fu} in Eq. (35).

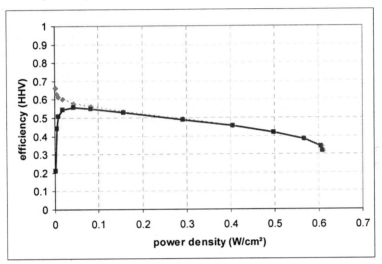

Figure 2. Fuel cell efficiency vs. power density curve; solid line with and dashed line without internal current and/or hydrogen crossover losses.[3]

NOMENCLATURE

a	parameter in Tafel equation, V	j	current density, A/m²
a_c	catalyst specific area, cm² mg⁻¹	R_i	areal resistance, Ω cm²
b	Tafel slope, V/decade	S	entropy, J mol⁻¹ K⁻¹
E	potential, V	T	temperature, K
F	Faraday constant, C mol⁻¹	V	potential, V
G	Gibbs free energy, J mol⁻¹	W	work, W
H	enthalpy, J mol⁻¹		
I	current, A	***Greek***	
i	current density, A cm⁻²	Δ	change
io	exchange current density, A cm⁻²	Π	pressure ratio
L_c	catalyst loading, mg cm⁻²	α	transfer coefficient
n	number of electrons per molecule	β	symmetry factor
P	pressure, Pa	γ	pressure coefficient
R	gas constant, J mol⁻¹ K⁻¹	η	efficiency

References

1. R.C. Weast, *CRC Handbook of Chemistry and Physics* (CRC Press, Boca Raton, 1988).
2. J.H. Hirschenhofer, D.B. Stauffer, and R.R. Engleman, *Fuel Cells A Handbook* (Revision 3) (U.S. Department of Energy, Morgantown Energy Technology Center, DOE/METC-94/1006, January 1994).
3. F. Barbir, *PEM Fuel Cells: Theory and Practice* (Elsevier/Academic, Burlington, 2005).
4. J.O'M. Bockris and S. Srinivasan, *Fuel Cells: Their Electrochemistry* (Mc Graw-Hill, New York, 1969).
5. A.J. Bard and L.R. Faulkner, *Electrochemical Methods* (Wiley, New York, 1980).
6. E. Gileadi, *Electrode Kinetics for Chemists, Chemical Engineers and Material Scientists* (VCH, New York, 1993).
7. H.A. Gasteiger, W. Gu, R. Makharia, and M.F. Mathias, Tutorial: catalyst utilization and mass transfer limitations in the polymer electrolyte fuel cell, *The 2003 Electrochemical Society Meeting* Orlando, FL (2003).
8. F. Barbir, J. Braun, and J. Neutzler, Properties of Molded Graphite Bi-Polar Plates for PEM Fuel Cells, *International Journal on New Materials for Electrochemical Systems*, 2, 197–200 (1999).
9. J. Kim, S.-M. Lee, S. Srinivasan, and C.E. Chamberlain, Modeling of Proton Exchange Fuel Cell Performance with and Empirical Equation, *J. of Electrochemical Society*, 142(8), 2670–2674 (1995).
10. J. Larminie and A. Dicks, *Fuel Cell Systems Explained*, 2nd edn. (Wiley, Chichester, 2003).

FUEL CELL STACK DESIGN PRINCIPLES WITH SOME DESIGN CONCEPTS OF MICRO-MINI FUEL CELLS

FRANO BARBIR
Faculty of Electrical Engineering, Mechanical Engineering and Naval Architecture, University of Split, R. Boskovica bb, 21000 Split, Croatia

1. Stack Design Principles

A single H_2/Air fuel cell has potential of about 1 V at open circuit, which decreases to 0.6–0.7 V in operation as a function of current density. In order to increase the potential to some practical levels the cells are connected in a stack. A fuel cell stack consists of a multitude of single cells stacked up so that the cathode of one cell is electrically connected to the anode of the adjacent cell. In that way exactly the same current passes through each of the cells. Note that the electrical circuit is closed with both electron current passes through solid parts of the stack (including the external circuit) and ionic current passes through electrolyte (ionomer), with the electrochemical reactions at their interfaces (catalyst layers).

The bi-polar configuration is the best for larger fuel cells because the current is conducted through relatively thin conductive plates, thus travels a very short distance through a large area (Figure 1). This causes minimum electro-resistive losses, even with a relatively bad electrical conductor such as graphite (or graphite polymer mixtures). For smaller cells it is possible to connect the edge of one electrode to the opposing electrode of the adjacent cell by some kind of connectors.[1-3] This is applicable only to very small active area cells because current is conducted in plane of very thin electrodes, thus traveling relatively long distance through very small cross-sectional area (Figure 2).

The main components of a fuel cell stack are the membrane electrode assemblies or MEAs (membranes with porous electrodes on each side with a catalyst layer between them), gaskets at the perimeter of the MEAs, bi-polar plates, bus plates (one at each end of the active part of the stack) with

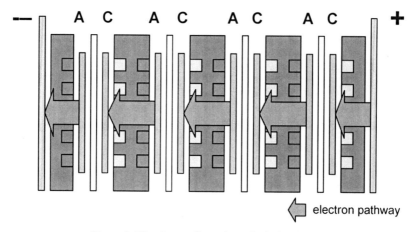

Figure 1. Bi-polar configuration of a fuel cell stack.

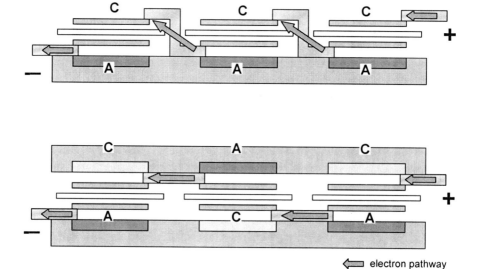

Figure 2. Examples of side-by-side configurations (a) zig-zag connections with open air cathode (b) flip-flop configuration.

electrical connections, and the endplates (one at each end of the stack), with fluid connections.[4] Cooling of the stack, i.e. its active cells, must be arranged in some fashion as discussed below. The whole stack must be kept together by tie-rods, bolts, shroud or some other arrangements (Figure 3).

STACK DESIGN PRINCIPLES

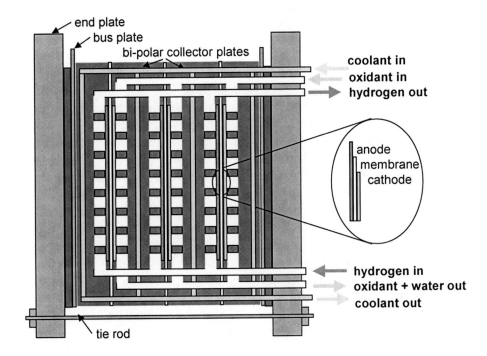

Figure 3. Stack (bi-polar configuration) schematic with main components.[4]

The following are the key aspects of a fuel cell stack design:

- Uniform distribution of reactants to each cell
- Uniform distribution of reactants inside each cell
- Maintenance of required temperature in each cell
- Minimum resistive losses (choice of materials, configuration, uniform contact pressure)
- No leak of reactant gases (internal between the cells or external)
- Mechanical sturdiness (internal pressure including thermal expansion, external forces during handling and operation, including shocks and vibrations).

1.1. UNIFORM DISTRIBUTION OF REACTANTS TO EACH CELL

Since fuel cell performance is sensitive to the flow rate of the reactants, it is absolutely necessary that each cell in a stack receives approximately the same amount of reactant gases. Uneven flow distribution would result in uneven performance between the cells. Uniformity is accomplished by feeding each cell in the stack in parallel, through a manifold which can be either external or internal. External manifolds can be made much bigger to insure uniformity, they result in a simpler stack design, but they can only be used in a cross-flow configuration, and are, in general, difficult to seal. Internal manifolds are more often used in PEM fuel cell design not only because of better sealing but also because they offer more versatility in gas flow configuration.

It is important that the manifolds that feed the gases to the cells and the manifolds that collect the unused gases are properly sized. The cross-sectional area of the manifolds determines the velocity of gas flow and the pressure drop. As a rule of thumb, the pressure drop through the manifolds should be an order of magnitude lower than the pressure drop through each cell in order to ensure uniform flow distribution.

The flow pattern through the stack can be either a "U" shape, where the inlet and outlet are at the same side of the stack and the flows in inlet and outlet manifolds are in opposite direction from each other, or a "Z" shape where the inlets and outlets are on opposite sides of the stack and the flows in inlet an outlet manifolds are parallel to each other (Figure 4). If properly sized both should result in uniform flow distribution to individual cells. The stacks with more than a hundred of cells have been successfully built.

In both U and Z configurations the flow of reactant(s) is parallel in each cell. However, it is possible to arrange the cells of a stack in segments, feed the cells in each segment in parallel ("Z" configuration), but connect the segments in series. In that case the depleted gas from the first segment is fed to the cells of the second segments (Figure 4c). In such parallel-serial arrangement all the cells of the stack operate at higher stoichiometry than if they have been fed in parallel.[5]

1.2. UNIFORM DISTRIBUTION OF REACTANTS INSIDE EACH CELL

Once the reactant gases enter the individual cell they must be distributed over the entire active area. This is typically accomplished through a flow field, which may be in a form of channels covering the entire area in some pattern or porous structures. The following are the key flow field design variables:

1.2.1. *Shape of the Flow Field*

The flow fields come in different shapes and sizes. The shape is the result of positioning the inlet and outlet manifolds, flow field design, heat management, and manufacturing constraints. The most common shapes of the

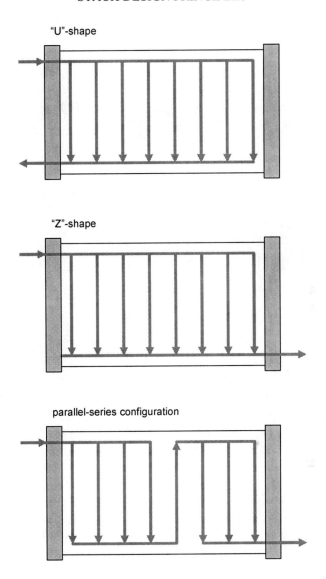

Figure 4. Stack flow configurations.

flow field are square and rectangular, but circular, hexagonal, octagonal or irregular shapes have been used or at least tried.[6]

1.2.2. *Flow Field Orientation*

The orientation of the flow field and positions of inlet and outlet manifolds are important only because of the gravity effect on water that may con-

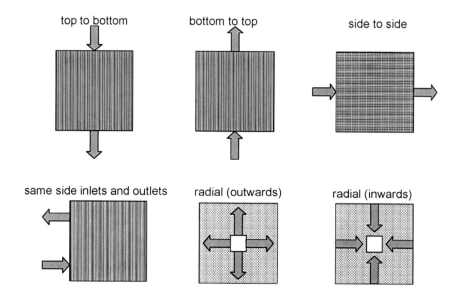

Figure 5. Possible flow field orientations.

dense inside the flow field (the effect of gravity on the reactant gases is negligible). Condensation may take place either during operation depending on the choice of operational conditions, or after shutdown. Numerous combinations are possible, some of which are shown in Figure 5.

Anode and cathode may be oriented in the same direction, in opposite direction or in cross-configuration. Position of anode vs. cathode may have some effect on fuel cell performance because of varied concentration of reactant gases and water. In some cases, the flow fields are oriented so that the cathode outlet is next to the anode inlet and vice versa, allowing water exchange through the membrane due to water concentration gradient (i.e. the exiting gas has much higher temperature and water content).

1.2.3. *Configuration of Channels*

There are many configurations of channels that have been tried in PEM fuel cells, all with the same goal – to ensure uniform reactant gases distribution and product water removal. Some most common designs are shown in Figure 6 and discussed below.

STACK DESIGN PRINCIPLES

Figure 6. Various flow field configurations.

- *Straight channels with large manifolds* – although this appears to ensure uniform distribution, it actually does not work in PEM fuel cells. Distribution is indeed uniform, but only under ideal conditions. Any water droplet that develops in the channel would effectively block the entire channel, and the velocity would not be sufficient to push the water out.

- *Straight channels with small manifolds* – this design has the same shortcomings, and on top of it, has inherent maldistribution of reactant gases, since the channels immediately below or above the manifold receive most of the flow. The early fuel cells built with such a flow field exhibited low and unstable cell voltages.

- *Criss-cross configuration* – this flow field attempts to eliminate the shortcomings of the straight channel flow field by introducing traversal channels allowing the gas to by-pass any "trouble" spot, i.e. coalescing water droplets. The problems of low velocities and uneven flow distribution due to positioning of the inlet and outlet manifolds are not reduced with this design.

- *Single channel serpentine* – as described by Watkins et al.[7] is the most common flow field for small active areas. It ensures that entire area is covered, although the concentration of reactants decreases along the channel. There is a pressure drop along the channel due to friction on the walls and due to turns. The velocity is typically high enough to push any water condensing in the channel. Attention must be paid to pressure differential between the adjacent channels, which may cause significant by-passing of channel portions.

- *Multi-channel serpentine* – A single channel serpentine configuration would not work for the large flow field areas, because of a large pressure drop. Although a pressure drop is useful in removing the water, excessive pressure drop may generate larger parasitic energy losses. Watkins et al.[8] proposed a flow field that has a multitude of parallel channels meandering through the entire area in a serpentine fashion. Except the lower pressure drop, this flow field has the same features, advantages and shortcomings of the single channel serpentine. The fact that there are parallel channels means that there is always a possibility that one of the channels may get blocked as discussed above with straight channels.

- *Multi-channel serpentine with mixing* – as suggested by Cavalca et al.[9] this flow field design allows gases to mix at every turn in order to minimize the effect of channel blocking. This does not reduce the chance of channel blockage, but it limits its effect only to a portion of the channel, since the flow field is divided in smaller segments each with its own connecting channel to both the inlet and the outlet.

- *Subsequent serially linked serpentine* – This flow field also divides the flow field into segments in an attempt to avoid the long straight channels and relatively large pressure differentials between the adjacent sections, thus minimizing the by-passing effect.[10]

- *Mirror serpentine* – This is another design to avoid large pressure differentials in adjacent channels, particularly suited for a larger flow fields with multiple inlets and outlets. These are arranged so that the resulting serpentine patterns in adjacent segments are mirror images of each other, which results in balanced pressures in adjacent channels, again minimizing the by-passing effect.[11]

- *Interdigitated* – first described by Ledjeff,[12] advocated by Nguyen,[13] and successfully employed by Energy Partners in their NG-2000 stack series,[14,15] this flow field differs from all of the above because the channels are discontinued, i.e. they do not connect inlet to the outlet manifolds. This way the gas is forced from the inlet channels to the outlet channels through the porous back-diffusion layer. Wilson et al.[16] suggested a variation of interdigitated flow field where the channels are made by cutting out the strips of the gas diffusion layer. Convection through the porous layer shortens the diffusion path and helps remove any liquid water that otherwise may accumulate in the gas diffusion layer, resulting in better performance, particularly at higher current densities. However, depending on the properties of the gas diffusion layer this flow field may result in higher pressure drops. Because of the fact that the most of the pressure drop occurs in the porous media, the uniformity of flow distribution between individual channels and between individual cells strongly depends on uniformity of gas diffusion layer thickness and effective porosity (after being squeezed). One of the problems with this flow field is inability to remove liquid water from the inlet channels. Issacci and Rehg[17] suggested the porous blocks at the end of the inlet channels allowing water to be removed.

- *BioMimetic* – suggested by Morgan Carbon[18] is a further refinement of the interdigitated concept. Larger channels branch to smaller side channels further branching to really tiny channels interweaving with outlet channels that are arranged in the same fashion – tiny channels leading to larger side channels leading to the large channels. This type of branching occurs in nature (leaves or lungs) hence the name biomimetic.

- *Fractal* – this flow field suggested by Fraunhoffer Institute[19] is essentially the interdigitated flow field concept, but the channels are not straight and they have branches.

- *Mesh* – Metallic meshes and screens of various sizes are successfully being used in the electrolyzers. The uniformity may greatly be affected by positioning of the inlet manifolds. The researchers at Los Alamos National Lab successfully incorporated metal meshes in fuel cell design.[20] The problems with this design are introduction of another component with tight tolerances, corrosion, and interfacial contact resistance.

- *Porous media flow field* – is similar to mesh flow field – the difference is in pore sizes and material.[21] The gas distribution layer must be sufficiently thick and have enough pores sufficiently large to permit a substantially free flow of reactant gas through both perpendicular to and parallel to the catalyst layer. While metallic porous materials (foams) are brittle, carbon based ones may be quite flexible. This type of flow field may only be applicable for smaller fuel cells because of the high pressure drop.

1.2.4. Channels Shape, Dimensions and Spacing

The flow field channels may have different shape, often resulting from the manufacturing process rather than from functionality. For example, slightly tapered channels would be very difficult to be obtained by machining, but are essential if the bi-polar plate is manufactured by molding. Channel geometry may have effect on water accumulation. In the round-bottomed channel condensed water forms a film of water at the bottom, while in the channel with tapered walls condensed water forms small droplets (Figure 7). The sharp corners at the bottom of the channel help to break the surface tension of the water film, resisting film formation.[22]

 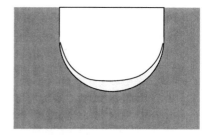

Figure 7. The shape of the channel cross-section affects the form of liquid water formation.

Typical channel dimensions are around 1 mm, but may vary from 0.4 to 4 mm. The spacing between the channels is similar. With today's advances in micro-manufacturing techniques (MEMS, photolithography) it is possible to produce channels of 0.1 mm and even smaller. The dimensions of the channels and their spacing affect the reactant gas access to the gas diffusion layer and pressure drop, but also electrical current and heat conduction. Wider channels allow more direct contact of the reactants gas to the gas diffusion layer and also providing wider area for water removal from the gas diffusion layer. Oxygen concentration, and therefore current density, is higher in the area directly above the channel and it is significantly lower in the area above the land between the channels.[23]

However, if the channels are too wide there will be no support for the MEA which will deflect into the channel. Wider spacing enhances conduction of electrical current and heat, however they reduce the area directly exposed to the reactants and promote the accumulation of water in the gas diffusion layer adjacent to these regions.

In, general, as landing width narrows, the fuel cell performance improves, until there is either MEA deflection into the channel or the gas diffusion layer crashes due to excessive force applied. The optimum channel size and spacing is therefore a balance between maximizing the open area for the reactant gas access to the gas diffusion layer and providing sufficient mechanical support to the MEA and sufficient conduction paths for electrical current and heat.

1.3. HEAT REMOVAL FROM A FUEL CELL STACK

In order to maintain the desired temperature inside the cells, the heat generated as a by-product of the electrochemical reactions must be taken away from the cells and from the stack. Some heat is removed by convection and radiation to the surrounding and some by the reactant gases and product water exiting the stack, but majority of the heat generated inside a stack must be removed by some active system. Different heat management schemes may be applied as shown in Figure 8.[6]

1.3.1. *Cooling with a Coolant Flowing Between the Cells*

Coolant may be de-ionized water, anti-freeze coolant or air. Cooling may be arranged between each cell, between the pair of cells (in such configuration one cell has the cathode and other cell has the anode next to the cooling arrangement), or between a group of cells (this is feasible only for low power densities since it results in a higher temperatures in the center cells). Equal distribution of coolant may be accomplished by the manifolding arrangement similar to that of reactant gases. If air is used as a coolant equal distribution may be accomplished by a plenum.

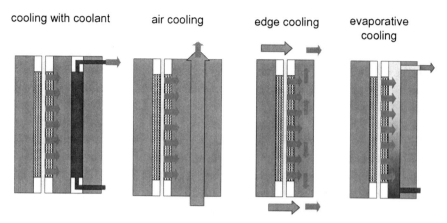

Figure 8. Different cell/stack cooling options.

1.3.2. *Cooling with Coolant at the Edge of the Active Area*

The heat is conducted through the bi-polar plate and then transferred to the cooling fluid, typically air. In order to achieve relatively uniform temperature distribution within the active area, the pi-polar plate must be very good thermal conductor. In addition, the edge surface may not be sufficient for heat transfer and fins may need to be employed. This method results in much simpler fuel cell stack, fewer parts, but is has heat transfer limitations, and is typically used for low power outputs.

1.3.3. *Cooling with Phase Change*

Coolant may be water or another phase change medium. Use of water simplifies the stack design since water is already used in both anode and cathode compartments.

1.3.4. *Cooling with Reactant Air*

Air is already passing through the cathode compartment in excess of oxygen exact stoichiometry. Theoretically, this air flow may be used as a coolant, but the flow rate would have to be much higher. From a simple heat balance, namely heat generated by fuel cell must be equal to heat taken away by the flow of air, it is possible to calculate the required oxygen stoichiometric ratio:[6]

$$S_{O2} = \frac{M_{O2} + \dfrac{4F(1.254 - V_{cell})}{c_p \Delta T}}{M_{O2} + \dfrac{1 - r_{O2in}}{r_{O2in}} M_{N2}} \qquad (1)$$

Where:
- M_{O2} = molecular mass of oxygen, g mol^{-1}
- F = Faraday's constant, C mol^{-1}
- 1.254 = potential corresponding to the lower heating value of hydrogen, $H_{LHV}/(2F)$, V
- V_{cell} = cell potential, V
- C_p = specific heat of air, J g^{-1} K^{-1}
- ΔT = temperature difference between air at the stack inlet and outlet, °C
- r_{O2in} = oxygen content in air by vol
- M_{N2} = molecular mass of nitrogen, g mol^{-1}

Equation (1) takes into account the heat removed from the stack by evaporation of product water. The only variables in the above equation are the cell potential, V_{cell}, and temperature difference between air at the stack inlet and outlet, ΔT. The cell potential determines the cell efficiency, and at a lower efficiency more heat is generated. The air temperature difference is determined by the ambient temperature and the stack operating temperature. From Eq. (1) it follows that a very large air flow rates would be required, with stoichiometric ratio higher than 20. Product water is not sufficient to saturate these large amounts of air and therefore this cooling scheme would cause severe drying of the anode, and would not be practical. However, if additional liquid water is injected at the stack cathode inlet it would be possible to prevent drying, and at the same time dramatically reduce the air flow rate requirement, since cooling would be achieved by evaporation of injected water. Wilson et al.[24] suggested, adiabatic cooling of an atmospheric stack with water introduced on the anode side and then relying on wicking through the gas diffusion layer (using hydrophilic thread) and electroosmotic drag to saturate the ambient air on the cathode side.

1.4. STACK CLAMPING

The individual components of a fuel cell stack, namely membrane electrode assemblies (MEAs), gas diffusion layers and bi-polar plates must be somehow held together with sufficient contact pressure to (i) prevent leaking of the reactants between the layers, and (ii) to minimize the contact resistance between those layers. This is typically accomplished by sandwiching the stacked components between the two end plates connected with several tie-rods around the perimeter (Figure 9) or in some cases through the middle. Other compression and fastening mechanisms may be employed too, such as snap-in shrouds or straps.

The clamping force is equal to the force required to compress the gasket plus the force required to compress the gas diffusion layer plus internal force (for example the internal operating pressure). The pressure required to prevent the leak between the layers depends on the gasket material and design. Various materials ranging from rubber to proprietary polymer configurations are used for fuel cell gaskets. The designs also vary from manufacturer to manufacturer, including flat or profile gaskets, gaskets as individual components or molded on bipolar plate, or on or around the gas diffusion layer. A so-called seven layer MEA includes the catalyzed membrane, two gas diffusion layers – one on each side, and the gasket that keeps the entire MEA together.

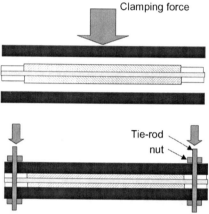

Figure 9. Compression of fuel cell components with tie-rods.[6]

Too much force on the perimeters may cause bending of the end-plates (as shown in Figure 10), which has an adverse effect on the compression over the active area. Compression distribution inside the cell may be monitored by pressure sensitive films (which register only the highest force applied) or by pressure sensitive electronic pads, which connected to a monitor allow inspection of compression force distribution in real time throughout the assembly process. Because of possibility of bending the end-plates must be designed with sufficient stiffness. Alternatively, the end-plates with hydraulic or pneumatic piston that applies a uniform force throughout the active area may be used. Another alternative is to put the tie-rods through the center of the plate, and then design the flow field around it.

A pressure of 1.5–2.0 MPa is required to minimize the contact resistance between a gas diffusion layer and bi-polar plate.[23] The gas diffusion layer is compressible and the required "squeeze" may be determined by the cell design, i.e. by carefully matching the thicknesses of the gas diffusion layer, gaskets, and hard-stops or recesses on the bi-polar plate.

If the gas diffusion layer is compressed too much, it will collapse and it will lose its main function – gas and water permeability, Optimum compression must be experimentally determined for each gas diffusion media.

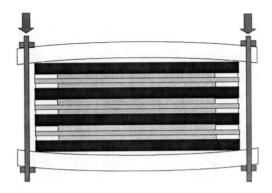

Figure 10. Bending of the end-plates if too much force is applied on tie-rods around the Perimeter.[6]

2. Some Design Concepts of Miniature Fuel Cells

The above design concepts should also apply to miniature fuel cells in order to maximize their performance. However, in many cases miniaturization is more important than maximizing the power output which results in greater freedom in design of miniature fuel cells which in turn may result in non-conventional, i.e. non stackable bi-polar configurations. Several miniature fuel cell configurations reported in literature are described below.

2.1. FREE CONVECTION (OPEN CATHODE) FUEL CELL CONCEPT

For very small power outputs it is possible to design and operate a fuel cell with passive air supply, relying only on natural convection due to temperature and concentration gradients. Such a fuel cell typically has either the front of the cathode directly exposed to the atmosphere (Figure 11),

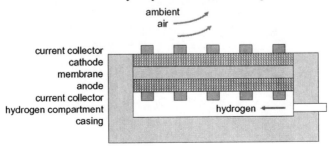

Figure 11. Free-convection fuel cell.

therefore without the bipolar plates, or in bi-polar configuration the cathode flow field is sideways opened to the atmosphere.[25] In either case an oxygen concentration gradient is formed between the open atmosphere and the catalyst layer where oxygen is being consumed in the electrochemical reaction. The performance of such fuel cells is typically not limited only by the oxygen diffusion rate, but also by water and heat removal, both dependent on the temperature gradient.[26] Maximum current density that can be achieved with free-convection fuel cells is typically limited to about 0.15–0.25 A cm^2.[27]

These ambient or free-convection fuel cells result in a very simple system, needing only hydrogen supply. Several cells may be connected sideways in series, anode of one cell electrically connected to the cathode of the adjacent cell (as shown in Figure 2a) to get the desired voltage output. The cells can be connected in combination of parallel and serial configurations in order to get the desired voltage.

2.2. NON-PLANAR CONFIGURATION

Litster et al.[28] reported non-planar, air-breathing microstructured cell architecture (as shown in Figure 12) developed by Angstrom Power Inc. In this configuration the electrodes (and the membrane) are perpendicular to the open surface. In such an electrode, gases must traverse the catalyst layer as they diffuse through the nano-porous gas diffusion layer in a path parallel to the active area. The micro-structured electrodes employ a nano-porous gas diffusion layer featuring a mean pore diameter on the order of 100 nm.

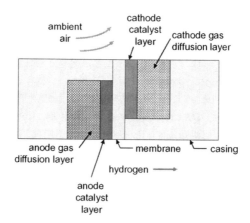

Figure 12. Non-planar configuration fuel cell concept.

2.3. PLANARIZED CONFIGURATION – FUEL CELL ON A CHIP

Researchers at University of Connecticut[29] reported a micro fuel cell configuration in which hydrogen and oxygen channels are in the same plane (Figure 13). The cannels may be as small as 5–10 μm. In this configuration there is no need for the gas diffusion layers as the catalyst layers are directly exposed to the gas in the corresponding channel. Electrical contacts are deposited on the opposite side of the catalyst layer. The two catalyst layers are separated by a layer of polymer membrane. Such planar configuration allows a multi-cell design to be realized as a single component instead of a stack of individual components, thus allowing the use of relatively simple manufacturing techniques such as photolithography, coating, and electrodeposition, among others. The entire cell may be made on a silicon substrate. As many as 25–50 micro-cells may be placed on a 1 cm^2 substrate. Applying a combination of serial-parallel connections desired voltage of such a "micro-stack on a chip" may be obtained, as high as 24–48 V.

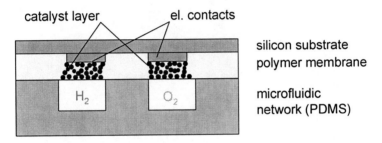

Figure 13. Planarized microfluidic fuel cell concept.

2.4. CYLINDRICAL CONFIGURATION

Yazici[30] utilized a novel gas diffusion media based on micro-perforated natural graphite based expanded graphite (GRAFCELL by GrafTech International Ltd.) in a cylindrical cell. In this configuration, the fuel (hydrogen or methanol) are inside a tube and the electrode at the outside of the tube is exposed to the ambient air. One of the key features of the expanded graphite is its excellent thermal conductivity (~200 W m^{-1} K^{-1}) which simplifies thermal management. GRAFCELL mass transfer layer technology allowed diameters smaller than 2 mm to be formed to construct cylindrical fuel cells with no additional gasket material. Assembly of individual cells into a stack is possible without bi-polar plates using wires for electrical connections and piping for fuel connections.

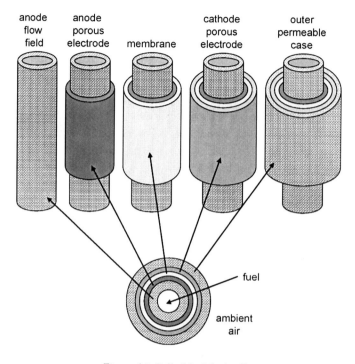

Figure 14. Cylindrical fuel cell.

3. Conclusions

Fuel cell stack is a simple, yet a complex device. It is simple because it is made of a repetitive unit cells. Each unit cell consists of several layers (bi-polar plate, gas diffusion layers, catalyst layers, membrane). The design of a fuel cell stack must ensure that each of the unit cells operates at exactly same conditions. It is therefore important to understand how operating conditions affect the unit cell performance. Complication arises from the fact that operating conditions (flow rates, pressure, temperature and humidity of reactant gases) are inter-related. Also, it is important that selection of the key parameters and conditions must be made from the system perspective. A stack is only a component of a complex system.

The same design concepts should also apply to miniature fuel cells in order to maximize their performance. However, in many cases miniaturization is more important than maximizing the power output which results in greater freedom in design of miniature fuel cells which in turn may result in non-conventional, i.e. non stackable bi-polar configurations. Often, the design of miniature fuel cells is driven by the choice of material and simplicity of the applicable manufacturing process.

References

1. F. Barbir, Development of an Air-Open PEM Fuel Cell, SBIR Phase I Final Technical Report, A report by Energy Partners, Inc. to US Army Research Laboratory, contract DAAL01-95-C-3511, 1995.
2. K. Ledjeff and R. Nolte, New SPFC-Technology with Plastics, in O. Savadogo, P.R. Roberge and T.N. Veziroglu (eds.) New Materials for Fuel Cell Systems, pp. 128–134, Editions de l'Ecole Politechnique de Montreal, Montreal, 1995.
3. Cisar, D. Weng and O.J. Murphy, Monopolar fuel cells for nearly passive operation, Proc. 1998 Fuel Cell Seminar, Palm Springs, CA, pp. 376–378, November 1998.
4. F. Barbir, Progress in PEM Fuel Cell Systems Development, in Hydrogen Energy System, in Y. Yurum (ed.) *Utilization of Hydrogen and Future Aspects*, NATO ASI Series E-295, pp. 203–214, Kluwer Academic, Dordrecht, The Netherlands, 1995.
5. P.A. Rapaport, J.A. Rock, A.D. Bosco, J.P. Salvador and H.A. Gasteiger, Fuel Cell Stack Design and Method of Operation, US Patent No. 6,794,068, 2004.
6. F. Barbir, *PEM Fuel Cell Theory and Practice*, Elsevier/Academic, Burlington, MA, 2005.
7. D.S. Watkins, K.W. Dircks and D.G. Epp, Novel fuel cell fluid flow field plate, US Patent No. 4,988,583, 1991.
8. D.S. Watkins, K.W. Dircks and D.G. Epp, Fuel cell fluid flow field plate, US Patent No. 5,108,849, 1992.
9. C. Cavalca, S.T. Homeyer and E. Walsworth, Flow field plate for use in a proton exchange membrane fuel cell, US Patent No. 5,686,199, 1997.
10. J.A. Rock, Serially-linked serpentine flow channels for PEM fuel cell, US Patent No. 6,309,773, 2001.
11. J.A. Rock, Mirrored serpentine flow channels for fuel cell, US Patent # 6,099,984, 2000.
12. K. Ledjeff, A. Heinzel, F. Mahlendorf and V. Peinecke, "Die Reversible Membran – Brennstoffzelle," in Elektrochemische Energiegewinnung, Dechema Monographien, Vol. 128, p. 103, 1993.
13. T.V. Nguyen, Gas distributor design for proton-exchange-membrane fuel cells, *Journal of Electrochemical Society*, Vol. 143, No. 5, pp. L103–105, 1996.
14. F. Barbir, J. Neutzler, W. Pierce and B. Wynne, Development and Testing of High Performing PEM Fuel Cell Stack, Proc. 1998 Fuel Cell Seminar, Palm Springs, CA, pp. 718–721, 1998.
15. F. Barbir, M. Fuchs, A. Husar and J. Neutzler, Design and Operational Characteristics of Automotive PEM Fuel Cell Stacks, Fuel Cell Power for Transportation, SAE SP-1505, pp. 63–69, SAE, Warrendale, PA, 2000.
16. M.S. Wilson, T.E. Springer, J.R. Davey and S. Gottesfeld, Alternative Flow Field and Backing Concepts for Polymer Electrolyte Fuel Cells, in S. Gottesfeld, G. Halpert, and A. Langrebe (eds.) *Proton Conducting Membrane Fuel Cells I*, The Electrochemical Society Proceedings Series, PV 95–23, p. 115, Pennington, NJ, 1995.
17. F. Issacci and T.J. Rehg, Gas block mechanism for water removal in fuel cells, US Patent No. 6,686,084, 2004.

18. A.R. Chapman and I.M. Mellor, Development of BioMimetic™ Flow-Field Plates for PEM Fuel Cells, Proc. 8th Grove Fuel Cell Symposium, London, September 2003.
19. K. Tüber, A. Oedegaard, M. Hermann and C. Hebling, Investigation of fractal flow fields in portable PEMFC and DMFC, Proc. 8th Grove Fuel Cell Symposium, London, September 2003.
20. C. Zawodzinski, M.S. Wilson and S. Gottesfeld, Metal screen and foil hardware for polymer electrolyte fuel cells, in S. Gottesfeld and T.F. Fuller (eds.) *Proton Conducting Membrane Fuel Cells II*, Electrochemical Society Proceedings, Vol. 98-27, pp. 446–456, 1999.
21. P.J. Damiano, Fuel cell structure, US Patent No. 4,129,685, 1978.
22. D.P. Wilkinson and O. Vanderleeden, Serpentine flow field design, in W. Vielstich, A.Lamm, and H. Gasteiger (eds.), *Handbook of Fuel Cell Technology – Fundamentals, Technology and Applications*, Vol. 3, Part 1, pp. 315–324, Wiley, New York, 2003.
23. M.F. Mathias, J. Roth, J. Fleming and W. Lehnert, Difusion media materials and characterization, in W. Vielstich, A.Lamm, and H. Gasteiger (eds.), *Handbook of Fuel Cell Technology – Fundamentals, Technology and Applications*, Vol. 3, Part 1, pp. 517–537, Wiley, New York, 2003.
24. M.S. Wilson, S. Moeller-Holst, D.M. Webb, and C. Zawodzinski, Efficient fuel cell systems, 1999 Annual Progress Report Fuel Cells for Transportation, pp. 80–83, US Department of Energy, Washington, DC, 1999.
25. P.-W. Li, T. Zhang, Q.-M. Wang, L. Schaefer and M.K. Chyu, The performance of PEM fuel cells fed with oxygen through the free-convection mode, *Journal of Power Sources*, Vol. 114, No. 1, pp. 63–69, 2003.
26. M.S. Wilson, D. DeCaro, J.K. Neutzler, C. Zawodzinski and S. Gottesfeld, Air-breathing fuel cell stacks for portable power applications, Proc. 1996 Fuel Cell Seminar, Orlando, FL, November 17–20, pp. 314–317, 1996.
27. T. Hottinen, M. Mikkola and P. Lund, Evaluation of planar free-breathing polymer electrolyte membrane fuel cell design, *Journal of Power Sources*, Vol. 129, pp. 68–72, 2004.
28. S. Litster, J.G. Pharoah, G. McLean and N. Djilali, Computational analysis of heat and mass transfer in a micro-structured PEMFC cathode, *Journal of Power Sources*, Vol. 156, pp. 334–344, 2006.
29. C.J. Lefaux, P.T. Mather and F. Barbir, Design and Processing of a Planarized Microfluidic Fuel Cell, Abstracts 2004 Fuel Cell Seminar, San Antonio, TX, November 2004.
30. M.S. Yazici, Passive air management for cylindrical cartridge fuel cells, *Journal of Power Sources*, Vol. 166, pp. 137–142, 2007.

PERFORMANCE ANALYSIS OF MICROSTRUCTURED FUEL CELLS FOR PORTABLE APPLICATIONS

S. LITSTER[1] AND N. DJILALI[2]*
[1]*Department of Mechanical Engineering, Stanford University, Stanford, CA, 95304-3030, USA*
[2]*Department of Mechanical Engineering & Institute for Integrated Energy Systems, University of Victoria, Victoria, BC, V8W 3P6, Canada*

1. Introduction

In this chapter we review the development of the theoretical models for analyzing the performance of microstructured, air-breathing fuel cells for portable devices. The ever increasing power density demands of portable consumer electronics are driving the integration of fuel cells into portable devices. The widespread integration of fuel cells into portable devices is expected to occur before automotive fuel cell systems are widely adopted.[1] The market for portable electronics is much more accessible to fuel cells than automotive applications because of the higher cost tolerance at this scale, as much as two orders of magnitude higher.[1]

The need for increased power density results from devices with new power hungry functionality and performance. For example, the development of mobile phones with digital broadcast reception may spur the integration of fuel cells into mobile phones. Kariatsumari and Yomagita[2] report that even if the capacity of Li-ion batteries grows at 10% per year, they will not be capable of powering future devices at their present volume. Thus, they have predicted that the integration of fuel cells into next-generation mobile phones will be the next battery revolution. This forthcoming battery revolution will be similar in process to the previous one. Previously, the development of power hungry camcorders and notebook PCs spawned the wide-spread adoption of Li-ion batteries when Ni-Cd2 and NiMH batteries could no longer meet the energy density demands. In a similar fashion, fuel cells could supplant Li-ion batteries. Figure 1 illustrates the series of battery revolutions discussed by Kariatsumari and Yomagita.

---

*Ned Djilali, Dept. of Mechanical Engineering & Institute for Integrated Energy Systems, University of Victoria, Victoria, BC, V8W 3P6, Canada, e-mail: ndjilali@uvic.ca

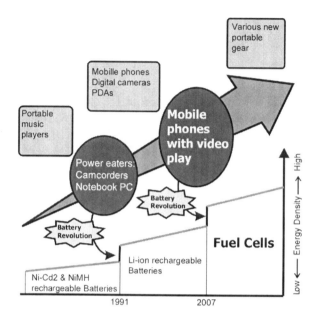

Figure 1. Evolution of energy density demands for portable consumer electronics. (Adapted from Kariatsumari and Yomogita[2]).

Ideally, fuel cells will be integrated seamlessly into or onto the surfaces of portable devices. Figure 2a illustrates the integration of a fuel cell into the back plate of a mobile phone. When designing such fuel cells, the system volume is at a premium and requires special considerations. To minimize the space occupied by the power supply, many of the ancillary devices used by large scale systems, such as air compressors and humidifiers, are forgone in favor of passive air-breathing operation. Air-breathing fuel cells use a combination of ambient air-currents and thermally-induced natural convection to deliver oxygen and cool the fuel cell.

1.1. OXYGEN DELIVERY

In large-scale systems, air is typically delivered by compressors or blowers at over-stoichiometric rates to ensure adequate oxygen supply to the cathode at all current densities. In the case of air-breathing fuel cells, oxygen delivery is limited by the strength of ambient air currents and the flows generated by natural convection (induced by the temperature rise in the fuel cell). The stack architecture of an air-breathing fuel cell requires special consideration. There are two prevalent architectures of air-breathing fuel cells. The first is the planar fuel cell arrangement depicted in Figure 2. The second is similar to conventional fuel cell stacks, but the ends of the cathode channels are open to the surrounding air rather connected to a pressurized manifold.

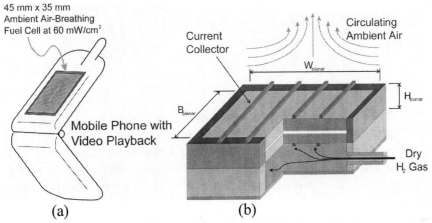

Figure 2. Ambient air-breathing fuel cell powering a mobile phone (a) and a schematic of a planar air-breathing fuel cell. (Reprinted from Litster and Djilali,[33] Copyright (2007). With permission from Elsevier.)

1.2. FUEL STORAGE AND DELIVERY

Fuelling for fuel cells in portable devices is an area of controversy and continuing development as it is central to achieving high *system* energy density. Two dominant schools of thought for fuel storage in portable devices are metal hydride hydrogen storage and direct methanol fuel cells. Another emerging storage methods are on-board reforming of methanol and sodium borohydride. Storing hydrogen within the lattice of metals is attractive because of the high fuel cell efficiency and volumetric storage capacity, whereas high gravimetric energy density and ease of liquid refueling make direct oxidation of liquid methanol a suitable option. With each approach there are distinct advantages and challenges in both implementation and analysis. The recent review by Kundu et al.[3] provides a thorough review and comparison of fuel storage and conversion technologies for micro fuel cells. In this chapter we specifically address fuel cells fed dry hydrogen that has been stored in metal hydrides or as compressed gas.

To further reduce the volume for hydrogen delivery, the devices and manifolding for recycling excess hydrogen can be removed and a "dead-ended" hydrogen supply is used.[4] The "dead-ended" arrangement for hydrogen supply is achieved by maintaining relatively constant hydrogen pressure at the anode GDL interface. Constant gas pressure can be achieved with a high-pressure hydrogen supply and a pressure-regulating valve, or through desorption of hydrogen from a metal hydride. Volume can be further reduced by eliminating gas conditioning components (such as heaters and humidifiers). Normally, conditioning improves fuel cell performance by maintaining the humidification of the membrane on the anode side of the MEA.

1.3. HEAT MANAGEMENT

In most full-size PEM fuel cells, active pumping of liquid coolant through the cooling channels of the bipolar plate maintains the optimum temperate of the stack. For device integrated fuel cells, a method of passively managing the heat is required. For small systems operating at low-current densities with low heat production, the excess heat can be removed by the surrounding air through convective heat transfer created by ambient air currents and natural convection. Natural convection is generated by the variation in the air density with temperature. Density stratification causes buoyancy forces that induce air flows. Fabian et al.[4] visualized the natural convection flows above their air-breathing fuel cell using shadow-graph imaging. Figure 3 presents their visualizations of the transient evolution of the thermal plume. This passive heat transfer process can be enhanced by the addition of high surface area cooling fins.

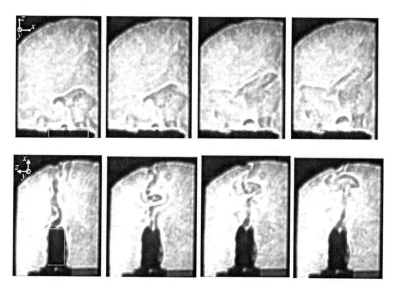

Figure 3. Shadowgraph visualizations, made by Fabian et al.[4] that show natural convection above an air-breathing fuel cell. In top row of images the fuel cell surface is horizontal and in the bottom row it is oriented vertically. (Reprinted from Fabian et al.[4] Copyright (2008). With permission from Elsevier.)

1.4. WATER MANAGEMENT

Water management is a critical consideration of most low-temperature fuel cell systems, especially for those employing polyflurosulfonic acid (PFSA) type membranes (i.e. Nafion). These membranes require a high level of hydration to achieve suitable levels of ionic conductivity and performance. Large scale systems humidify the membrane by heating and humidifying the gases prior to entering the fuel cell. However, membrane humidication in fuel cells for portable devices is achieved with the water content of the ambient air and the water produced by the oxygen reduction reaction at the cathode. Like larger systems, fuel cells for portable devices are also susceptible to the common problem of liquid water flooding, which introduces significant mass transfer limitations and performance loss. When air compressors are used to deliver air, liquid water accumulation is mitigated by the advective removal of water droplets. Air-breathing fuel cells, however, have no such mechanism for removing water and evaporation is the only way that liquid water is removed. One exception is the development of electro-osmotic pumps for active removal of liquid water from an open air-breathing cathode.[5]

The level of membrane hydration and liquid water accumulation strongly depends on the heat transfer characteristics of the fuel cell system. The maximum rate of water removal is proportional to the difference between the ambient water vapor concentration and the water vapor concentration at the external surface of the cathode, which depends on the temperature at the surface through the vapor pressure. If the fuel cell is operating at low-temperatures, either because of high heat transfer or during initial start-up, the fuel cell will accumulate liquid water. In contrast, if the fuel cell heats up significantly, the membrane will dry out and the fuel cell will suffer from severe Ohmic losses.

1.5. MICROFABRICATION

In order for fuel cells to be viable in portable devices, current PEM fuel cells must undergo significant miniaturization. The adaptation of conventional fuel cell designs for smaller applications is restricted by the macro-scale materials and manufacturing processes they utilize. Exploitation of microscale transport processes in conjunction with micro-manufacturing processes, such as those applied in the production of integrated-circuits, make it possible to conceive extremely high power density fuel cells.[6] Such fuel cells have the potential to be significantly cheaper, smaller, and lighter than planar plate and frame fuel cells. The implementation of thin layer manufacturing processes can also mass transport and conductive path length;

enhancing the volumetric power density. Micro-fabrication of flow fields, current collectors, and electrical interconnects has been reported in the literature.[7-9] In general, however, these fuel cell designs have relied on traditional planar MEA architecture. Four examples of planar air-breathing fuel cells are those developed by Wainright et al.,[9] Schmitz et al.,[10] Hahn et al.[7] and Fabian et al.[4]

1.6. FUEL CELL MODELING

Fuel cell researchers are increasingly turning to fuel cell models to improve their fundamental understanding of the transport phenomena present in PEM fuel cells and optimize their designs. The majority of the phenomena take place in regions of the fuel cell that are in general inaccessible to experimental measurement. Therefore, a mathematical model is vital in improving current understanding. The need for a physics-based model is compounded when considering device integrated fuel cells. As these forms of PEM fuel cells are emerging, the design process can be streamlined by the insight provided by a well-developed model.

Fuel cell models in the literature can be divided into two main categories: those that use computational fluid dynamics for multi-dimensional studies with high-spatial resolution[11-17] and models that use analytical, semi-analytical, and other novel solutions.[18-27] Analytical models can require significant assumptions to arrive at tractable solutions. However, analytical expressions can provide greater insight into the governing physics than large-scale computational models. Another major distinction between various PEM fuel cell models is whether liquid water accumulation and transport is simulated.

Currently, there are only a handful of theoretical studies that have considered the special case of an air-breathing PEM fuel cell.[28-40] Of those, only a few have considered a PEM fuel cell with an open-cathode architecture (versus open-ended channels) for oxygen delivery and cooling by natural convection. Litster et al.[33] and Rajani et al.[36] developed computational models that resolve the natural convection flow outside the cathode, whereas the studies Li et al.,[30] Litster and Djilali,[31, 32] O'Hayre et al.[35] and Ziegler et al.[40] use heat and mass transfer coefficients to model the interface between the cathode and the ambient air.

1.7. MICROSTRUCTURED FUEL CELL ARCHITECTURES

Fuel cells with the conventional planar plate and frame architectures have a power density that is ultimately constrained by their two-dimensional active area. Non-planar and microstructured designs enable fuel cells with high volumetric power density because of greater active area to volume ratios. This concept was demonstrated by the waved cell topology developed by Mérida et al.[41] The fuel cell developed by Mérida et al. featured a corrugated expanded metallic mesh structure that supported the flexible membrane-electrode assembly. However, conventional MEA materials restrict the pitch of the MEA corrugations. To circumvent this limitation, a microstructured fuel cell has been proposed.[42] Figure 1a shows a schematic of the microstructured fuel cell "on-end" arrangement of the electrodes. Similar to conventional stacks, electrically conductive separators collect current and separate the hydrogen from the air. Caps on the bottom and top surfaces also separate the gases. As shown in Figure 4a, gases diffuse parallel to the MEA through the GDL to reach the catalyst layer. Figure 4b depicts the architecture of the microstructured fuel cell stack. The air-breathing stack is formed by attaching unit cells at adjacent separators.

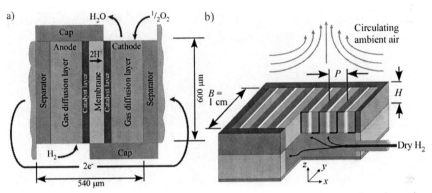

Figure 4. A schematic of a unit-cell of the microstructured fuel cell (a), and a schematic of the stack architecture (b). (Reprinted from The Electrochemical Society, *Electrochem. Solid-State Lett.*, **11**, B1 (2008). Copyright 2007. With permission.)

Fuel cells for portable devices are designed for high volumetric power density at a high voltage operating point. A high operating voltage increases efficiency and reduces fuel storage requirements. This objective is more critical than the common concern of reducing platinum loading for automotive systems because of the higher cost tolerance for portable devices. The microstructured architecture offers intrinsically greater volumetric power densities compared to planar fuel cell because the active area (catalyst layer area in y–z plane) can be greater than the planar area (surface area in the x–y plane). For planar fuel cells, the active area to planar area ratio is unity. This ratio can be greater than one for the microstructured fuel cells and increases as pitch (P in Figure 4b) decreases.

2. Heat Transfer

Experiments have shown that heat transfer has dominant effect on the performance of device integrated fuel cells.[4] They passively dissipate the by-product heat and do not necessitate the complex heat management systems of full-scale PEMFC stacks. During operation, device integrated fuel cells can experience large changes in temperature due to environmental changes and transient start-up. The thermal response time of a fuel cell decreases as the length scale is reduced. The theoretical heat transfer time constant for a solid body with a spatially lumped temperature is:

$$\tau = \frac{V}{A}\frac{\rho c}{h} \quad (1)$$

where V and A are the volume and surface area of the body, ρ is the density of the body, c is the specific heat, and h is the heat transfer coefficient. As the scale of the fuel cell is reduced, the temperature is expected to become more uniform. To model a solid body with a spatially lumped temperature, the Biot number (Bi) should meet the criterion:

$$Bi < 0.1 \quad (2)$$

The Biot number is the ratio of convective heat transfer to the conductive heat transfer within the solid body and is expressed as:

$$Bi = \frac{h\delta}{k_{sb}} \quad (3)$$

where δ is the conductive length scale and k_{sb} is the solid body conductivity. We can make a conservative estimate of the Biot number for device integrated fuel cell by assuming a low thermal conductivity of 1 W m^{-1} K^{-1}, a length scale of 3 mm, and a heat transfer coefficient of 10 W m^{-2} K^{-1} that is consistent natural convection. The Biot number of the fuel cell is then 0.03, which permits modeling heat transfer using a lumped parameter approach.

The heat generated by the fuel cell is rejected through the exterior surfaces of the fuel cell by convection and radiation. The heat flux from the fuel cell (q_t) can be expressed as:

$$q_t = h(T - T_\infty) + \varepsilon_r \sigma_r (T^4 - T_\infty^4) \quad (4)$$

where T_∞ is the temperature of the ambient environment, ε_r is the radiative emissivity of the fuel cell top surface, and σ_r is the Stefan-Boltzmann constant. The heat transfer coefficient can be related to the Nusselt number (Nu) which is a non-dimensional characterization of the heat transfer rate, i.e.

$$h = \frac{Nu \cdot k}{l} \quad (5)$$

where l is the length scale of the geometry and k is the thermal conductivity of air. A common characteristic length is surface area divided by the perimeter. There are many empirical and theoretical correlations for natural convection Nusselt numbers.[33] For example, Martorell et al.[43] developed a correlation for heated rectangular plates; a condition similar to an air-breathing fuel cell. This correlation agrees well with experiments and numerical calculations having Rayleigh numbers (Ra) from 200 to 20,000, and takes the form:

$$Nu = 1.20 Ra^{0.175} \quad (6)$$

The Rayleigh number characterizes the magnitude of buoyancy flows, and is expressed as the product of the Grashof number (ratio of buoyancy forces to viscous forces) and the Prandtl number (ratio of momentum and thermal diffusivities). The Rayleigh number is defined as:

$$Ra = \frac{g \beta C_p}{\mu k} \Delta T l^3 \quad (7)$$

where g is the gravity, β is the thermal expansion coefficient, C_p is the specific heat, and μ is the viscosity. Combining Eqs. (5)–(7), we can see that the effectiveness of natural convection to cool the fuel cell depends on the length scale according to:

$$\Delta T \propto l^{0.53} \quad (8)$$

Thus, smaller fuel cells will tend to feature a higher heat transfer coefficient and are more easily cooled by natural convection. In the following sections we investigate whether natural convection is sufficient for cooling.

3. CFD Modeling of Natural Convection

The following section presents our use of computational fluid mechanics (CFD) to explicitly model fethe natural convection above a microstructured fuel cell.[33] The model includes heat and mass transport in the ambient air as

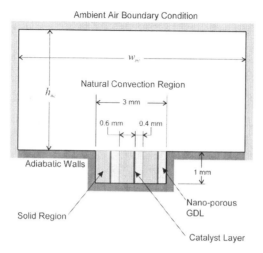

Figure 5. The two-dimensional domain and dimensions for the model of a microstructured air-breathing fuel cell. (Reprinted from Litster et al.[33] Copyright (2006). With permission from Elsevier.)

well as the heat transfer within the fuel cell and diffusion of oxygen and water in the cathodes. Figure 5 shows the domain of the model, which includes three unit cells of a microstructured stack. The anode, membrane and separators of the fuel cell are treated as solid region and are included for heat transfer modeling.

3.1. THEORY

The governing equations for this model include the continuity, momentum, energy and mass transport equations. The continuity equation is expressed as:

$$\nabla \cdot (\rho \mathbf{u}) = S_H^\& \quad (9)$$

where ρ is the gas mixture density, $\mathbf{u}$ is the gas velocity, and $S_H^\&$ We model momentum transport and buoyancy driven flows with the compressible momentum equation, including the gravitational body force term ($\rho \mathbf{g}$):

$$\nabla \cdot (\rho \mathbf{u} \otimes \mathbf{u} - \mu \nabla \mathbf{u}) = \rho \mathbf{g} - \nabla \cdot \left(P\delta + \frac{2}{3}\mu \nabla \cdot \mathbf{u} \right) + \nabla \cdot \left(\mu (\nabla \cdot \mathbf{u})^T \right) \quad (10)$$

where μ is the viscosity and P is the gas pressure.

The ideal gas equation of state relates gas composition and temperature to the density:

$$\rho_i = \frac{PM_i}{RT} \tag{11}$$

$$\frac{1}{\rho} = \sum \frac{y_i}{\rho_i} \tag{12}$$

Conjugate heat transfer is governed by the energy equation:

$$\nabla \cdot \left(\rho \mathbf{u}\left(h + \frac{1}{2}\mathbf{u}^2\right) - \lambda \nabla T\right) = \dot{S}_H \tag{13}$$

where h is the enthalpy, and λ is the thermal conductivity. The source term in term in the energy equation, $\dot{S}_H$, contains the heat released by reactions and Joule heating. The source term, applied at the catalyst layer, is:

$$\dot{S}_H = \left[\frac{T(-\Delta s)}{F(n_{e-})} + \eta\right] i \tag{14}$$

where η is the difference between reversible voltage and the assumed operating voltage.

Species transport is modeled with an advection-diffusion equation and the diffusion coefficient is assumed constant,

$$\nabla \cdot (\rho \mathbf{u} y_i) - \nabla \cdot (D_i \nabla y_i) = \dot{S}_i \tag{15}$$

where the species source term, $\dot{S}_i$, is zero everywhere except for the catalyst layer interface where the oxygen sink and water vapor sources are:

$$\dot{S}_{O_2} = -\frac{M_{O_2}}{4F} i \tag{16}$$

$$\dot{S}_{H_2O} = \frac{M_{H_2O}}{2F} i \tag{17}$$

In this model, a simplified electrochemistry approximation is used by assuming uniform membrane hydration and cathode activation overpotential. An average current density is then specified in conjunction with the following relation to determine the effect of oxygen concentration on the current density distribution:

$$i = I \frac{y_{O2}}{y_{O2}^{ave}} \tag{18}$$

where I is the average current density.

The electrode considered consists of a nano-porous substrate (65 nm pore diameter with 45% porosity) in which species transport is dominanted by Knudsen diffusion. Knudsen diffusion is driven by gradients in partial pressure and occurs when pore diameters approach the mean free path of the gas. We implement Knudsen diffusion transport by separating the gradient in partial pressure into a component driven by pressure gradients and another driven by mass fraction gradients: The pressure driven flow is cast in the form of Darcy's law:

$$\mathbf{u} = \frac{k_{Kn}}{\mu} \nabla P \qquad (19)$$

where k_{Kn} is an effective Knudsen permeability:[33]

$$k_{Kn} = \frac{4}{3}\frac{\varepsilon}{\tau}\frac{\mu d_{por}}{P}\sqrt{\frac{RT}{2\pi}}\sum\frac{y_i}{\sqrt{M_i}} \qquad (20)$$

Transport according to mass fraction gradient transport is characterized by the Knudsen diffusion coefficient:

$$D_{i,Kn} = \frac{4}{3}\frac{\varepsilon}{\tau}d_{por}\sqrt{\frac{RT}{2\pi M_i}} \qquad (21)$$

The governing equations were solved using CFX v4.3, which is computational fluid dynamics (CFD) code that uses the finite volume method. A higher-order upwind differencing scheme and SIMPLEC pressure correction are employed. The numerical solution is obtained with the algebraic multi-grid method. The ambient air region is meshed with 30,000 cells and each electrode has a 20 × 29 grid.

3.2. CFD RESULTS

We simulated fuel cell operation at 200 mA/cm^2 with an ambient temperature and relative humidity of 20°C and 80%, respectively. The current density of 200 mA/cm^2 was selected as it typically offers a high efficiency (high voltage) operating point. Litster et al.[33] provides more details on the parameters we used for this simulation. The objective of this simulation was to identify whether natural convection could sufficiently deliver oxygen to the cathode at this current density and to examine efficacy of passive cooling.

Figure 6a presents the steady-state velocity vectors of the flow generated by the elevated temperature of the fuel cell. This flow is the classic

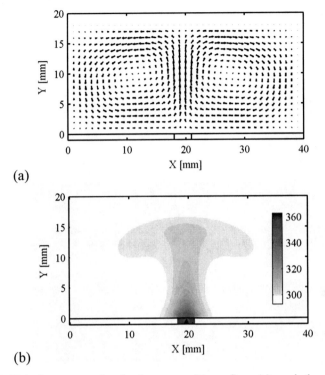

Figure 6. Velocity vectors for the buoyancy driven flow (a), and the corresponding temperature distribution [K] (b).[33] The ambient air temperature and relative humidity is 293 K and 60%, respectively.

Rayleigh-Bernard convection cells. This flow generates a thermal plume as shown in Figure 6b. The structure of the thermal plume is very similar to the experimental visualizations of Fabian et al.[4] shown in Figure 3. As we discuss in Litster et al.[33] natural convection readily supplies oxygen and there is only 7% decrease in the oxygen mass fraction at the exterior surface of the cathodes. However, as Figure 6b shows, the temperature within the fuel cell region is quite high: ranging between 85–91°C.

The high fuel cell temperature combined with low water production at 200 mA/cm^2 and the relatively low water content of the ambient air results in a very low relative humidity with the cathodes. Figure 7 displays contours of percent relative humidity in the three cathodes modeled in this simulation. Between the bottom and top of the cathode cavity the relative humidity decreases from 30% to 5%. The low effective diffusivity of the nano-porous GDL generates a large water vapor concentration gradient that creates this relative humidity distribution. Such low relative humidity results in severe membrane dry-out. Sone et al.[44] have shown that Nafion conductivity is almost completely lost below 20% relative humidity. These

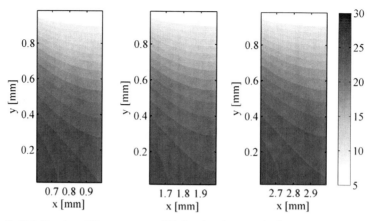

Figure 7. Relative humidity contours (%) in the three cathode gas diffusion layers.[33] The ambient air interface is at the top of the GDLs and the catalyst layer is on left edge of each GDL.

results motivated subsequent studies that focused on developing models capable of predicting electrolyte dry-out in air-breathing fuel cells. The following sections present those models and major findings.

4. Quasi-Analytical Modeling of Air Breathing Fuel Cells

When possible, the use of analytic techniques that do not require numerical discretization is advantageous as they allows fast turnaround simulation for design and analysis with minimal computing resources. Such a model is presented here to investigate electrolyte dry-out in air-breathing fuel cells. The following section outlines the equation set for modeling the electrochemical reactions, ion conduction, and heat and mass transport. We introduce two important features in this model for more precise predictions of dry-out behavior. First, conduction of H^+ ion is accounted for through the finite thickness of the cathode catalyst layer. Second, a higher-order empirical correlation for membrane conductivity is employed. In contrast to the commonly used Springer correlation,[18] this correlation offers substantially better predictions of membrane conductivity at low relative humidity.

In the following section, we discuss the derivation of the membrane electrode assembly (MEA) model and then apply it to a traditional, albeit air-breathing, planar geometry. We use the planar fuel cell model to validate the model against published experimental data and demonstrate the susceptibility of these fuel cells to dry-out. We also highlight significant effect of heat transfer characteristics of air-breathing fuel cells on performance.

Finally, we implement the MEA model in a quasi-2D model of the microstructured fuel cell.

4.1. THEORY

We solve for current as a function of cell voltage by summing the voltage losses, all a function current, and setting the voltages to the difference between the open-cell potential and simulated voltage. The potential summation through the MEA is:

$$(E_{OC} - E_{cell}) - \eta^C - \eta^A - W_{Mem} i_C(\eta^C)/\sigma_{Mem} = 0 \tag{22}$$

where E_{OC} is the reversible open-circuit voltage, E_{cell} is the operating voltage, η_C and η_A are the activation over-potential in the cathode and anode, respectively, is the membrane thickness, i is the current density, and σ_{Mem} is the membrane conductivity.

The solution is constrained by current conservation at the anode and cathode:

$$i_C(\eta^C) - i_A(\eta^A) = 0 \tag{23}$$

The Ohmic loss through the membrane, $W_{Mem} i_C(\eta^C)/\sigma_{Mem}$, depends on the current and the conductivity of the membrane. In this study we employ an empirical conductivity correlation obtained by Sone et al.[44] for a temperature of 303 K:

$$\sigma_{303K} = 3.46a^3 + 0.0161a^2 + 1.45a - 0.175 \tag{24}$$

where a is the activity of water vapor. The conductivity is corrected for temperature, T, with the activation energy used by Springer et al.:[18]

$$\sigma = \exp\left(1268\left(1/303 - 1/T\right)\right)\sigma_{303K} \tag{25}$$

The reversible potential is calculated with the change in Gibb's free energy for the fuel cell's reaction:

$$E = -\Delta G/nF \tag{26}$$

where ΔG at varying temperatures is interpolated from thermochemical tables.[45] The Nernst equation then determines the open-circuit potential based on gas activities:

$$E_{OC} = E + \frac{RT}{nF}\left(\ln(a_{H2}) + \frac{1}{2}\ln(a_{O2})\right) \tag{27}$$

As we are particularly concerned with electrolyte dry-out, we model ion conduction through the finite thickness of the cathode with hydration and temperature dependent ionic conductivity. However, we neglect variation in oxygen concentration within the cathode catalyst layer. Due to the relatively higher electrical conductivity, we assume uniform electric potential. We obtain an implicit solution for the over-potential distribution in the cathode by applying the current conservation and the Tafel equations:

$$\eta(x) = \frac{1}{b} \ln \left\{ \tan^2 \left[\sqrt{\frac{bA \exp(b\eta_o)}{2}} (x - W_{CL}) + \arctan\left(\frac{\sqrt{\exp(b\eta_1) - \exp(b\eta_o)}}{\sqrt{\exp(b\eta_o)}}\right) \right] + 1 \right\} + \eta_o \qquad (28)$$

where:

$$A = \frac{i_o^C}{\sigma} \left(\frac{\overline{c}_{O2}}{c_{O2}^{ref}}\right)^\gamma, \quad i_o^C = A_{Pt} \upsilon_{Pt} i_0^{Pt/N} \quad \text{and} \quad b = \frac{\alpha F}{RT}$$

and where x is the position in the catalyst layer, W_{CL} is the thickness, and η_o and η_1 are the overpotentials at the GDL and membrane interfaces, respectively. This relationship is similar to that obtained by Eikerling and Kornyshev.[46] Litster and Djilali[31] provides details on the macrohomogeneous model we use to determine the parameters in Eq. (28).

We find the ionic current at the membrane interface with the product of x-derivative of overpotential and the conductivity:

$$i = \sigma \sqrt{\frac{2A}{b}} \sqrt{\exp(b\eta_1) - \exp(b\eta_o)} \qquad (29)$$

In contrast, we treat the anode as an interface, due to the high exchange current density for the hydrogen oxidation reaction, using the Butler-Volmer equation:

$$i = i_o^A \left[\exp\left(\frac{\alpha_a F}{RT} \eta^A\right) - \exp\left(-\frac{\alpha_c F}{RT} \eta^A\right) \right] \qquad (30)$$

Mass fractions at the interface (y_A^o) between the GDL and ambient air are found using mass transfer coefficients (h_{O2}, h_{H2O}) that are consistent with ambient air currents and the current fuel cell geometry. A Nusselt number for mass transfer determines the coefficients ($h_A = Nu \cdot D_A / l$). The interface oxygen mass fractions are:

PERFORMANCE OF MICROSTRUCTURED FUEL CELLS   63

$$y^o_{O2} = y^{Amb}_{O2} - M_{O2}i/4F\rho h_{O2} \quad (31)$$

Considering a homogeneous and isotropic porous GDL, the species mass fractions at the interface between the GDL and the catalyst layer ($y^{GDL/CL}_A$) can be derived from Fick's law for binary gas diffusion. Multicomponent diffusion effects are neglected. The mass fraction at the interface of the GDL and the catalyst is then:

$$y^{GDL/CL}_{O2} = y^o_{O2} - \frac{W_{GDL}}{\rho D^{GDL}_{O2}} \frac{M_{O2}i}{4F} \quad (32)$$

The diffusivity has been corrected for the GDL's porosity and tortuousity ($D^{GDL}_i = \varepsilon D_i / \tau$). The water vapor mass fractions are found with similar expressions. The average oxygen mass fraction in the catalyst layer is corrected for the catalyst layer's lower effective diffusivity. A representative oxygen mass fraction in the catalyst layer is found with:

$$y^{CL}_{O2} = y^{GDL/CL}_{O2} - M_{O2}iw^2/32F\rho D^{CL} \quad (33)$$

We assume negligible water transport across the membrane because we are considering thin membranes, for which back-diffusion is almost completely balanced electro-osmotic drag.[47] This assumption is viable when considering fuel cells having a dead-ended supply of dry hydrogen.

The overall heat production (Q) is a product of the heat of reaction and reaction rate ($i/2F$) and the irreversible voltage loss multiplied by the current density. At steady-state, the heat produced is equal to the heat convected away from the surface of the fuel cell:

$$Q = A_A \left\{ \frac{T(-\Delta s)}{2F} + (E_{OC} - E_{cell}) \right\} i = A_s h(T - T_\infty) \quad (34)$$

We solve the coupled set of equations iteratively using Matlab software. The solution starts by solving for the current with fixed temperature and gas concentrations. The code then updates the gas concentrations and temperature for the new current density. This cycle repeats until the solution converges.

4.2. VALIDATION

We validated this model by comparing the predictions with the experimental polarization data of Fabian et al.[4] In particular, we compare the model's steady-state results with their long-time polarization data (each polarization point held for 2 hours). These long dwell times are necessary to achieve steady operation when dry-out occurs. Moxely et al.[48] have shown that the time-scale for membrane hydration equilibration during dry-out can be as long as 6 hours. To simulate the Fabian et al. fuel cell, we use a macro-homogeneous model (see Litster and Djilali[31]) to predict the electrochemical parameters from the MEA specifications. To model the experimental fuel cell's overall heat transfer properties, we calculated a bulk heat transfer coefficient from their temperature, voltage, and current data.

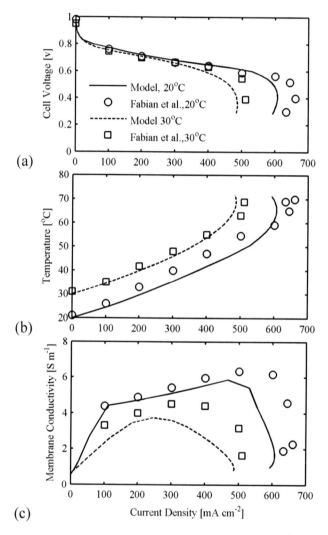

Figure 8. Comparison between model predictions and Fabian et al.'s[4] experimental results for polarization curves (a), temperature (b), and membrane conductivity (c) at ambient air temperatures of 20°C and 30 C.[31]

Besides matching the fuel cell heat transfer coefficient, we performed no further tuning or fitting of the model to match experimental data.

Figure 8a presents the experimental long-time polarization curves of Fabian et al.[4] for ambient temperatures of 20°C and 30°C and the model predictions. There is excellent agreement between the experimental and model curves. Unlike most PEM fuel cells, the limiting current density of this air-breathing fuel cell is caused by membrane dry-out. Figure 8b shows the increases in temperature that induce the membrane dry-out. Furthermore, Figure 8c shows the sharp decrease in membrane conductivity at the limiting current. The agreement between the experimental and predicted conductivity is compelling proof of the model's predictive capabilities.

4.3. EFFECT OF HEAT TRANSFER ON MEMBRANE DRY-OUT

We now present results for a 1 cm^2 planar air-breathing fuel cell to demonstrate the significance of heat transfer on performance. Further details on the model parameters are found in Litster and Djilali[31]. In Figure 9, we present four polarization curves generated by the model for Nusselt numbers ranging from 10 to 40. The results show that an increase in Nusselt number from 10 to 40 offers a four-times increase in the limiting current density. These limiting current densities are solely due to over-heating and membrane dehydration.

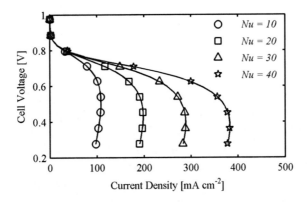

Figure 9. Polarization curves for four Nusselt numbers that range from 10 to 40.[31] Higher Nusselt numbers entail better cooling that reduces membrane dry-out and increases steady-state power density.

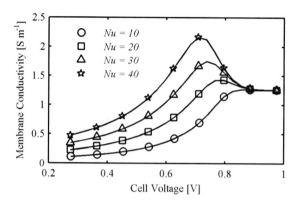

Figure 10. Membrane conductivity versus cell voltage for four Nusselt numbers ranging from 10 to 40.[31]

Figure 10 plots the membrane conductivity for the polarization curves in Figure 9 versus the cell voltage. For three of the polarization curves (20, 30, 40), there is an increase in conductivity with lowering voltage. This increase is caused by increased current density and greater self-hydration (i.e. the effect of water production outweighs the effect temperature increase on relative humidity). For the low Nusselt number of 10, there is no self-hydration and conductivity only decreases as voltage goes lower and current increases. These plots shows there exists a critical voltage, at which further voltage decreases will reduce the conductivity and cause membrane dry-out. Ideally, for long steady state operation the fuel cell should operated at voltages near or above this critical value.

4.4. MICROSTRUCTURED FUEL CELLS

We now present the use of the MEA model and heat transfer correlations developed for the planar fuel cell to evaluate the performance of the microstructured fuel cell. The details of this study can be found in Ref. 32. Figure 11 illustrates the structure of the model. The height of the MEA is discretized into vertical segments that are resolved with the previously described MEA model. We use an analytical 2D model of mass transport in the GDL to resolve the oxygen and relative humidity distribution over the height of the MEA interface. As before, heat and mass transfer coefficients allow predictions of the temperature in the fuel cell and the species concentration at the cathode surface.

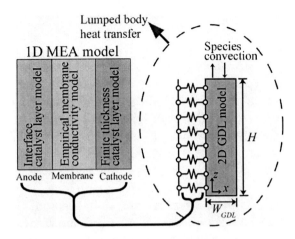

Figure 11. Schematic of the fuel cell model. Two-dimensional gas diffusion is modeled in the GDL, a one-dimensional model resolves the potential distribution and current through the MEA, and a lumped-parameter formulation resolves the mass and heat transfer between the fuel cell and the ambient air. (Reprinted from The Electrochemical Society, *Electrochem. Solid-State Lett.*, **11**, B1 (2008). Copyright 2007, With permission.)

Expressions providing a good approximation of the two dimensional oxygen and water vapor two-dimensional distributions in the cathode GDL were derived in Litster:[49]

$$y_{O2}(x,z) = \frac{M_{O2}}{4FW\rho D_{O2}^{eff}} \left[\int_0^z\int_0^{z'} i(z')dz'dz'' - \int_0^H\int_0^z i(z')dz'dz + K \atop i(z)\left(W_{GDL}x - \frac{x^2}{2} - \frac{3W_{GDL}^2}{8}\right) \right] + y_{O2}^o \qquad (34)$$

The mass fractions at the interface between the catalyst layer and GDL model is found by fitting an intermediate solution for the current distribution to a 3rd order polynomial. The order of the polynomial is constant and thus only the coefficients vary with each iteration. Then the integrals in Eq. (34) can be readily computed.

The solution procedure for this geometry is similar to the planar case, however the MEA equations are now discretized along the MEA/GDL boundary. For each point along the interface, the code solves the local current density and then updates the interface mass fractions and the temperature. The algorithm cycles through the discretized points until it obtains a converged solution.

4.5. MICROSTRUCTURED FUEL CELL RESULTS

Figure 12 illustrate the effect of the microstructured architecture on the relative humidity distribution in the GDL, the membrane conductivity profile, and the current density distribution. The results are for a cell voltage of 0.6 V. At this voltage, the model predicts an average current density of 145 mA cm^{-2}. As seen in the CFD results, the relative humidity at the top of the GDL is very low (<20%) and so are the membrane conductivity and current density. However, at the bottom of the GDL the relative humidity is higher because of the porous media lower effective diffusivity and the GDL's aspect ratio.

Figure 13 presents the polarization curves for two microstructured fuel cell with 150 and 250 μm wide GDLs. The figure also includes a polarization curve for a planar fuel cell having the same surface area. All three cases have surface area of 1 cm^2. The planar case also has an active area of 1 cm^2. However, the microstructured fuel cells with 150 and 250 μm wide GDLs have active areas of 1.36 and 0.93 cm^2, respectively, due to their non-planar architecture. The polarization curves of the microstructured fuel cells are shown with the current density normalized by the planar area of the fuel cell's top surface and the active area (MEA area). With the planar

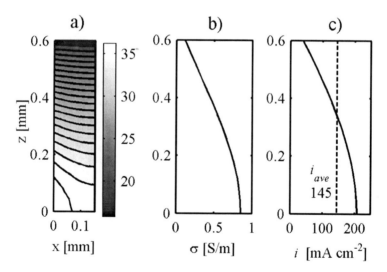

Figure 12. Cathode GDL relative humidity distribution (a), membrane conductivity profile (b), and current density profile (c) for the microstructured fuel cell at an operating voltage 0.6 V.[32]

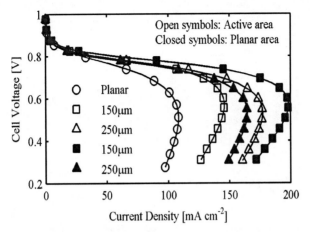

Figure 13. Polarization curves for microstructured fuel cell based on active area and planar surface area.[32]

surface area metric, the model predicts higher current densities for the fuel cell with 150 μm wide GDLs because of the high active area to planar surface area ratio.

The polarization curves in Figure 13 indicate that there are additional advantages to the microstructured design beyond the increased active area. Though, the 250 μm GDL case has slightly less active area than the planar, it still exhibits greater current densities because of improved membrane hydration. Figure 14 shows the maximum relative humidity in the GDL versus current density. The higher relative humidity and improved membrane hydration are attributed to the microstructured GDL's aspect ratio and increased diffusion length scales. Similarly, the even greater resistance of the

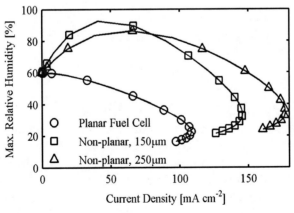

Figure 14. Maximum relative humidity in the cathode's GDL.[32]

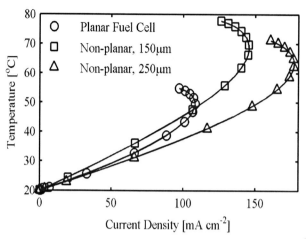

Figure 15. Temperature versus the active area current density.[32]

150 μm wide GDL results in relative humidities above 90% at low current densities. However, the relative humidity of the 150 μm case drops dramatically at higher currents because of increased temperatures. Figure 15 shows the increased temperatures of the 150 μm case caused by greater volumetric heating. The higher heating is a product of the reduced pitch between adjacent unit cells.

These results demonstrate that there are a number of competing effects that need to be considered when optimizing microstructured fuel cells and air-breathing fuel cell in general. For example the GDL width of the microstructured fuel cell alters the pitch between unit cells. The width would generally be reduced as much as fabrication methods allow for greater active area. However, if too narrow, the greater volumetric heating will increase membrane dry-out. Furthermore, thin GDLs could result in greater susceptibility to flooding during start-up and ramp-up in current density.

5. Conclusion

Fuel cells have the potential to supplant batteries in portable consumer electronics because of increased cost tolerances at this scale, rapid recharges, and higher energy densities. This chapter presented CFD and one- and two-dimensional quasi-analytical models of air-breathing fuel cells for portable devices. We examined two fuel cell architectures; a microstructured non-planar architecture and the more conventional planar arrangement. The studies we have performed with these models have elucidated the significant effect heat transfer has on membrane dry-out in ambient air-breathing fuel cells. The dry-out process, which manifests itself quite uniquely in air-breathing fuel cells, was evidenced by analyzing the underlying cause of limiting current. The study of the non-planar microstructured fuel cell gave evidence that significantly greater power densities and higher efficiencies are achievable with this architecture. The non-planar architecture exhibits better performance because of increased active area per unit volume. In addition, the "on-end" arrangement of the microstructured electrodes improves membrane hydration because of the diffusion barrier created by the GDLs high aspect ratio. Passive air-breathing fuel cells have a number of inherent advantages that will continue to attract development efforts. As new air-breathing architectures evolve, it is expected that physical features discussed here, and in particular the close and unique coupling between temperature and membrane hydration, will remain relevant and become even more pronounced with the higher heat fluxes expected with increasing miniaturization and power densities.

Acknowledgements

This work was made possible with the financial support of the Natural Sciences and Engineering Research Council of Canada (NSERC), the Canada Research Chairs Program, The MITACS Network of Centres of Excellence, and Angstrom Power Systems.

References

1. K. Kendall, N. Q. Minh, and S. C. Singhal, in: *High-Temperature Solid Oxide Fuel Cells-Fundamentals, Design and Applications*, edited by S. Singhal and K. Kendall (Elsevier Science, Oxford, England, 2003), pp. 197–229.
2. C. K. Dyer, Fuel cells for portable applications, *Journal of Power Sources* 106, 31–34 (2002).
3. H. Kariatsumari and H. Yomogita, Mobile Phones Drive Fuel Cell Development, *Nikkei Electronics Asia* (February 2005).
4. A. Kundu, J. H. Jang, J. H. Gil, C. R. Jung, H. R. Lee, S. H. Kim, B. Ku, and Y. S. Oh, Micro-fuel cells – Current development and applications, *Journal of Power Sources* 170, 67–78 (2007).
5. T. Fabian, J. D. Posner, R. O'Hayre, S. W. Cha, J. K. Eaton, F. B. Prinz, and J. G. Santiago, The role of ambient conditions on the performance of a planar, air-breathing hydrogen PEM fuel cell, *Journal of Power Sources* 161, 168–182 (2006).
6. T. Fabian, R. O'Hayre, S. Litster, F. B. Prinz, and J. G. G. Santiago, Water management at the cathode of a planar air-breathing fuel cell with an electroosmotic pump, *ECS Transactions* 3, 949 (2006).

7. G. McLean, N. Djilali, M. Whale, and T. Niet, Application of Micro-scale Techniques to Fuel Cell Systems Design, *Proceedings of the 10th Canadian Hydrogen Conference*, Quebec City (2000).
8. R. Hahn, S. Wagner, A. Schmitz, and H. Reichl, Development of a planar micro fuel cell with thin film and micro patterning technologies, *Journal of Power Sources* 131, 73–78 (2004).
9. S. J. Lee, A. Chang-Chien, S. W. Cha, R. O'Hayre, Y. I. Park, Y. Saito, and F. B. Prinz, Design and fabrication of a micro fuel cell array with "flip–flop" interconnection, *Journal of Power Sources* 112, 410–418 (2002).
10. J. S. Wainright, R. F. Savinell, C. C. Liu, and M. Litt, Microfabricated fuel cells, *Electrochimica Acta* 48, 2869–2877 (2003).
11. A. Schmitz, S. Wagner, R. Hahn, H. Uzun, and C. Hebling, Stability of planar PEMFC in Printed Circuit Board technology, *Journal of Power Sources* 127, 197–205 (2004).
12. T. Berning and N. Djilali, Three-dimensional computational analysis of transport phenomena in a PEM fuel cell – a parametric study, *Journal of Power Sources* 124, 440 (2003).
13. D. Natarajan and T. V. Nguyen, Three-dimensional effects of liquid water flooding in the cathode of a PEM fuel cell, *Journal of Power Sources* 115, 66–80 (2003).
14. P. T. Nguyen, T. Berning, and N. Djilali, Computational model of a PEM fuel cell with serpentine gas flow channels, *Journal of Power Sources* 130, 149 (2004).
15. S. Shimpalee, S. Greenway, D. Spruckler, and J. W. Van Zee, Predicting water and current distribution in a commercial-size PEMFC, *Journal of Power Sources* 135, 79–87 (2004).
16. B. Sivertsen and N. Djilali, CFD Based modelling of proton exchange membrane fuel cells, *Journal of Power Sources* 141, 65–78 (2005).
17. S. Um and C. Y. Wang, Three-dimensional analysis of transport and electrochemical reactions in polymer electrolyte fuel cells, *Journal of Power Sources* 125, 40–51 (2004).
18. L. Wang and H. Liu, Performance studies of PEM fuel cells with interdigitated flow fields, *Journal of Power Sources* 134, 185–196 (2004).
19. T. E. Springer, T. A. Zawodzinski, and S. Gottesfeld, Polymer electrolyte fuel cell model, *Journal of Electrochemical Society* 138, 2334–2342 (1991).
20. D. M. Bernardi and M. W. Verbrugge, A mathematical model of the solid-polymer-electrolyte fuel cell, *Journal of Electrochemical Society* 139, 2477–2491 (1992).
21. T. F. Fuller and J. Newman, Water and thermal management in solid-polymer-electrolyte fuel cells, *Journal of Electrochemical Society* 140, 1218–1225 (1993).
22. T. V. Nguyen and R. E. White, A water and heat management model for proton-exchange-membrane cells, *Journal of Electrochemical Society* 140, 2178–2186 (1993).
23. J. C. Amphlett, R. M. Baumert, B. A. Peppley, and P. R. Roberge, Performance modeling of the ballard mark IV solid polymer electrolyte fuel cell I. Mechanistic model development, *Journal of Electrochemical Society* 142, 1–8 (1995).
24. J. S. Yi and T. V. Nguyen, An along-the-channel model for proton exchange membrane fuel cells, *Journal of Electrochemical Society* 145, 1149–1159 (1998).
25. V. Gurau, F. Barbir, and H. Liu, An analytical solution of a half-cell model of PEM fuel cells, *Journal of Electrochemical Society* 147, 2468–2477 (2000).

26. R. Bradean, K. Promislow, and B. Wetton, Transport phenomena in the porous cathode of a proton exchange membrane fuel cell, *Numerical Heat Transfer, Part A* 42, 121–138 (2002).
27. P. Berg, K. Promislow, J. St. Pierre, and J. Stumper, Water management in PEM fuel cells, *Journal of Electrochemical Society* 151, A341–A353 (2004).
28. A. A. Kulikovsky, Semi-analytical 1D+1D model of a polymer electrolyte fuel cell, *Electrochemical Community* 6, 969–977 (2004).
29. J. J. Hwang, Species-electrochemical Modeling of an air-breathing cathode of a planar fuel cell, *Journal of the Electrochemical Society* 153, A1584–A1590 (2006).
30. J. J. Hwang, S. D. Wu, R. G. Pen, P. Y. Chen, and C. H. Chao, Mass/electron co-transports in an air-breathing cathode of a PEM fuel cell, *Journal of Power Sources* 160, 18–26 (2006).
31. P. W. Li, T. Zhang, Q. M. Wang, L. Schaefer, and M. K. Chyu, The performance of PEM fuel cells fed with oxygen through the free-convection mode, *Journal of Power Sources* 114, 63–69 (2003).
32. S. Litster and N. Djilali, Mathematical modelling of ambient air-breathing fuel cells for portable devices, *Electrochimica Acta* 52, 3849–3862 (2007).
33. S. Litster and N. Djilali, Theoretical Performance Analysis of microstructured air-breathing fuel cells, *Electrochemical and Solid-State Letters* 11, B1–B5 (2008).
34. S. Litster, J. G. Pharoah, G. McLean and N. Djilali, Computational analysis of heat and mass transfer in a micro-structured PEMFC cathode, *Journal of Power Sources* 156, 334–344 (2006).
35. T. Mennola, M. Noponen, M. Aronniemi, T. Hottinen, M. Mikkola, O. Himanen, and P. Lund, Mass transport in the cathode of a free-breathing polymer electrolyte membrane fuel cell, *Journal of Applied Electrochemistry* 33, 979–987 (2003).
36. R. O'Hayre, T. Fabian, S. Litster, F. B. Prinz, and J. G. Santiago, Engineering model of a passive planar air breathing fuel cell cathode, *Journal of Power Sources* 167, 118–129 (2007).
37. B. P. M. Rajani and A. K. Kolar, A model for a vertical planar air breathing PEM fuel cell, *Journal of Power Sources* 164, 210–221 (2007).
38. Y. Wang and M. G. Ouyang, Three-dimensional heat and mass transfer analysis in an air-breathing proton exchange membrane fuel cell, *Journal of Power Sources* 164, 721–729 (2007).
39. W. Ying, Y. J. Sohn, W. Y. Lee, J. Ke, and C. S. Kim, Three-dimensional modeling and experimental investigation for an air-breathing polymer electrolyte membrane fuel cell (PEMFC), *Journal of Power Sources* 145, 563–571 (2005).
40. W. Ying, T. H. Yang, W. Y. Lee, J. Ke, and C. S. Kim, Three-dimensional analysis for effect of channel configuration on the performance of a small air-breathing proton exchange membrane fuel cell (PEMFC), *Journal of Power Sources* 145, 572–581 (2005).
41. C. Ziegler, A. Schmitz, M. Tranitz, E. Fontes, and J. O. Schumacher, Modeling planar and self-breathing fuel cells for use in electronic devices, *Journal of the Electrochemical Society* 151, A2028–A2041 (2004).
42. W. R. Mérida, G. McLean, and N. Djilali, Non-planar architecture for proton exchange membrane fuel cells, *Journal of Power Sources* 102, 178–185 (2001).
43. G. F. McLean, Electrochemical cell, United States Patent 6872287 (2005).
44. I. Martorell, J. Herrero, and F. X. Grau, Natural convection from narrow horizontal plates at moderate Rayleigh numbers, *International Journal of Heat and Mass Transfer* 46, 2389–2402 (2003).

45. Y. Sone, P. Ekdunge and D. Simonsson, Proton conductivity of Nafion 117 as measured by a four-electrode AC impedance method *Journal of Electrochemical Society* 143, 1254–1259 (1996).
46. M. W. Chase, *JANAF Thermochemical Tables* (American Chemical Society and NIST, 1986).
47. M. Eikerling and A. A. Kornyshev, Modelling the performance of the cathode catalyst layer of polymer electrolyte fuel cells, *Journal of Electroanalytical Chemistry* 453, 89–106 (1998).
48. G. J. M. Janssen and M. L. J. Overvelde, Water transport in the proton-exchange-membrane fuel cell: measurements of the effective drag coefficient, *Journal of Power Sources* 101, 117–125 (2001).
49. J.F. Moxley, S. Tulyani, and J. B. Benziger, Steady-state multiplicity in the autohumidification polymer electrolyte membrane fuel cell, *Chemical Engineering Science* 58, 4705–4708 (2003).
50. S. Litster, *Mathematical Modelling of Fuel Cells for Portable Devices*, M.A.Sc. Thesis, University of Victoria (2005).

ENGINEERING DURABILITY OF MICRO-MINIATURE FUEL CELLS

K. REIFSNIDER[1,]* AND X. HUANG[2]
[1]Department of Mechanical Engineering
University of South Carolina, Columbia, SC 29209
[2]Global Fuel Cell Center, University of Connecticut
Storrs, CT 06269, USA

1. Introduction

Durability is a primary concern for fuel cells, and a major concern for fuel cells and fuel cell systems. Durability is not only a requirement, it is an engineering subject with science foundations. Fuel cells involve material systems, some of which operate at elevated temperatures, typically in the 600–1,000°C range for solid oxide fuel cells (SOFCs), for long periods of time, up to 40,000 hours of expected life for stationary power systems, for example. As a result, durability of fuel cell systems has been the subject of much research attention.[1-3, 9] Generally speaking, durability of fuel cells has typically been approached by focusing on some major failure modes, such as cracking or pin hole formation (mechanical failure), chemical degradation, and microstructural changes such as changes in porosity or grain boundaries. The focus of such work is most often to select materials and avoid failure. But degradation in these (and other) systems is an imperative; one cannot afford to design a system that lasts forever (even if it were possible).

We will develop the subject of fuel cell durability on the basis of two fundamental concepts. The first is the definition of engineering durability as *engineered performance histories*. This recognizes the requirement that the design and development of fuel cell systems requires that we be able to estimate the performance of a fuel cell (and the system it is part of) under any relevant prospective operating *history*. The second fundamental concept is the definition of a fuel cell as a functional composite material system, wherein the combined action and interaction of the constituent

---

*Kenneth Reifsnider, Department of Mechanical Engineering, University of South Carolina, Columbia, SC 29209, e-mail: REIFSNID@engr.sc.edu

material layers (and in some cases, the heterogeneous materials themselves in each region) control the performance of the system, and especially the long-term performance history of that system. The instantaneous interacttion of the constituents and the input/output materials is described by multiphysics equations.[1-5] The long term performance is controlled by the "aging" of the material properties in those equations. And the performance changes caused by that aging process can be represented by engineering metrics of performance calculated from the output of those equations as a function of operating history. In the present paper, we will outline a protocol for this approach, and include a few physical results to illustrate some of the steps.

2. Engineering Durability

Power systems (involving fuel cells, turbines, etc.) are too complex and expensive to permit "trial and failure" design. Modern engineering design and development of such systems makes extensive use of knowledge based design, and enterprise engineering for knowledge and engineering process management. An example of such a process is sketched in Figure 1, as represented by Pratt and Whitney.[8]

Figure 1. Knowledge based design and development using a standard work concept developed by Pratt and Whitney Corporation.[8]

If fuel cells are to be successfully introduced into our society, it is necessary to provide the understanding and engineering models needed for such knowledge based engineering. Durability is an essential element of that process. It is essential to develop and select materials and to create material systems in a fuel cell that last long enough for a given application, but it is even more essential to design and develop a fuel cell system with physics based models and related materials knowledge so that the engineer knows how long it will last, what it's performance is as a function of operation history, and when to service or retire the system based on it's monitored condition. Only under those conditions will the risk of building and using such a system be reduced to a level that enables it to become a commercial product. The current approach addresses that need. The paper begins with the engineering equations needed for the design and development enterprise, discusses the material inputs to those equations in the context of durability, and introduces the concept of metrics to interpret the results of the calculations and simulations.

3. Multiphysics Modeling

We will discuss engineering durability methodology for one low temperature type (PEM) and one high temperature type (solid oxide or SOFC) fuel cell, for which the authors have the most data and experience. A diagram of the essential features of those two types is shown in Figure 2.

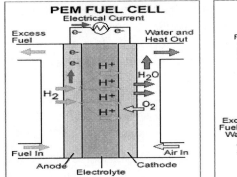 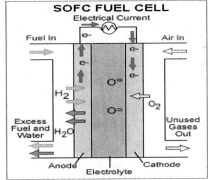

Figure 2. Polymer-based PEM fuel cell (left) and ceramic-based SOFC fuel cell (right) diagrams.

Figure 3 shows an example set of multiphysics equations for the case of SOFC, which represent conservation of mass, momentum, energy and charge (including stress and strain) as field equations, i.e. they produce point-wise information in a fuel cell or stack. There are many variations of how these equations are set, applied and solved, but even simple representations of PEM and SOFC fuel cells yield reasonable representations of the voltage-current relationship (polarization curves) for cells and simple stacks.[1, 2, 4, 9, 10, 11, 15] Predictions of instantaneous global performance can successfully be made with such equations. So a "static design" of a fuel cell can be constructed to satisfy specified performance requirements. But that does not answer the question of how to design for performance histories.

Engineering durability requires the prediction of performance histories, i.e. the prediction of long-term performance in the presence of certain prescribed histories of operation (applied conditions). Under those (changing, local) conditions, material 'constants,' material microstructure (or nanostructurre), and functional material properties that appear in the balance equations (and local boundary conditions such as surface transfer coefficients for heat, etc.) are generally not constant. If grain boundaries migrate in a crystalline SOFC electrolyte during aging, for example, then the resulting changes in the ionic conductivity of the electrolyte will enter the equations that predict performance, to cause predicted changes in SOFC performance at the global level. Each specific history of temperature, time, and any of the other driving forces that are known to influence the value of that conductivity will produce history-specific changes in conductivity, at the local level, as a point function. Changes in local properties in the constituents of PEM cells is equally complex.[12, 13]

We will discuss how to represent the local material changes in properties (which are observables) as a function of local applied conditions, represented as independent variables like time, temperature, hydration, etc. Then we will introduce the concept of metrics to help us to interpret the local changes at the global level.

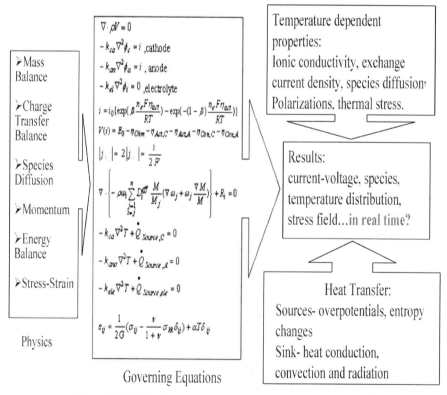

Figure 3. Multiphysics equations for SOFC fuel cells. (From Ju[2].)

4. Physical State Changes

Durability requires the prediction of performance histories, i.e. the prediction of long-term performance in the presence of certain prescribed histories of operation (applied conditions). If grain boundaries migrate in a crystalline electrolyte during aging, for example, then the resulting changes in the ionic conductivity of the electrolyte will enter the equations that predict performance, to cause predicted changes in SOFC performance at the global level. Each specific history of temperature, time, and any other driving forces that are known to influence the value of that conductivity will produce history-specific changes in conductivity, at the local level, as a point function. To represent the changes in properties as a function of local applied conditions, represented as independent variables like time, temperature, etc. (which are observables), we invoke the idea of metrics, as applied to the space of variables that define the changes in our material state.[16] For property changes, we require only a simple interpretation of this idea, namely that we be able to construct (based on experimental data) non-negative real number functions (or functionals) which serve as a measure of the changes in our material (or engineering) property as a function of the changes in material state induced by the variables that cause the changes. We can find simple examples of such metrics in the literature. Zhu and Miller provide examples of metrics that measure change in thermal conductivity and elastic stiffness of YSZ with time and temperature, as quoted below.[16]

$$\frac{k_c - k_c^0}{k_c^{inf} - k_c} = 102.2 \times exp\left(-\frac{68,228}{RT}\right)\left[1 - exp\left(-\frac{t}{\tau}\right)\right] \quad (1)$$

$$\frac{E_c - E_c^0}{E_c^{inf} - E_c} = C_E\left[1 - exp\left(-\frac{t}{\tau}\right)\right] \quad (2)$$

$$\tau = 572.5 \times exp\left(\frac{41,710}{RT}\right) \quad (3)$$

where k is the thermal conductivity, T is temperature in absolute, E is Young's modulus, R is the international gas constant, and t is sidereal time.

We will postulate such a metric for ionic conductivity in the electrolyte of a SOFC, in a subsequent paragraph. Equations (1–3) represent material property changes, which enter the coupled performance equations, e.g. Figure 3, noted earlier. When those metrics are used, the performance equations become representations of future performance, defined by the future values of the independent variables (determined by the expected history of applied conditions and the metric equations). So a method of predicting performance *histories* of a fuel cell with a mathematical simulation using known physical material long-term behavior has been constructed.

5. Functional State Changes

Functional performance is usually measured in terms of "engineering properties" such as strength, power, or system current at a given voltage, etc., that do not explicitly appear in the field equations discussed in Figure 3. Again, metrics can be used to combine the action and interaction of ageing or degradation processes to assess changes in performance. For composite materials, metrics for change in strength with operation history (known as damage tolerance) have been extensively examined and validated with experimental data. The book by Reifsnider, et al. on this subject provides a discussion of this established methodology[16] and a recent book chapter addresses the application of these concepts to SOFCs, directly.[1] An example of a metric used for strength is given in Eq. (4) were $\sigma(t)$ and $X(t)$ are the local point-wise stress and strength, n is the number of cycles, N is the life of the local material element if local conditions were to remain constant, and j is a material parameter of order one.

$$Fr(t) = 1 - \int_0^t \left(1 - Fa\left(\frac{\sigma(t)_{ij}}{X(t)_{ij}}\right)\right) j \left(\frac{n}{N}\right)^{j-1} d\left(\frac{n}{N}\right) \quad (4)$$

Fr(t) is the remaining strength as a function of time (history) as measured by the failure function, Fa, for a given failure mode; the equation is generally applied at a given material point (in a finite element code, for example). Fa is specific to an observed failure process, and several processes can be tracked at a given material point. Of course, Eq. (4) can be used to represent single crack growth, but it is most useful when there is distributed damage (which is the most common failure mode in composite materials, and fuel cells). Strength reduction due to distributed microcracking, growth of micro-porosity, or the accumulation of chemical strains associated with impurity migration (such as Cr) are examples of distributed damage. The value obtained from Eq. (4) for each history and material point is typically carried along as a global variable in a computational code. An interpretation of the output of Eq. (4) is shown in Figure 4. At the beginning of life, remaining strength, Fr(t = 0) = 1, since (for this normalized quantity) the initial remaining strength is equal to the initial strength, identically. For all other times, the remaining strength is reduced, according to what level of conditions we are applying, as represented by the history of Fa(t). For this strength discussion, the local failure function, Fa(t) is a function of the local stress level, and the local material strength, $X(t)_{ij}$, at every value of time, t, and the life at current conditions, N, is also a function of the applied conditions, i.e. $N = N(Fa(t))$. "Life" is defined as the time (or cycles) at which the remaining strength is reduced to the current level of applied conditions, i.e. $Fr(t_1) = Fa(t_1)$, which is the simple physical case of having the material strength reduced to the level of local stress. If all conditions were held constant, it is easily shown that the integral equation simply results in predicting that the life is equal to N. But the integral can handle time variations of all variables, and is sensitive to the sequence of events, i.e. a low-to-high load sequence does not generally give the same strength reduction as a high-to-low sequence, etc.

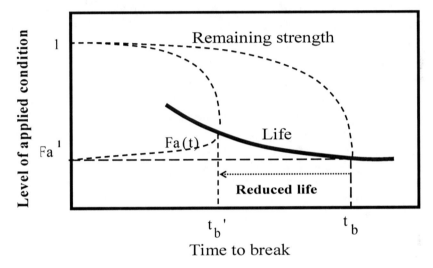

Figure 4. Interpretation of remaining strength predicted by Eq. (4).

The references cited earlier provide a more complete discussion of the efficacy of the metric approach (especially the integral metric approach) to construction of so-called evolution equations for material characteristics and engineering "properties." We only mention here a few advantages over purely empirical curve fits to data or "equivalency" principles such as time-temperature equivalence. Metrics are a familiar mathematical formalism, well established in fields as diverse as network analysis, mathematical statistics, geometry (including non-Euclidian and non-Newtonian geometry), and even astro-physics. They bring the attendant advantages of self consistency and established rigor to the question of how to construct a form that represents the change in a dependent variable in a space of independent variables (applied conditions in our case). They avoid artificial methods such as partitioning (e.g. energy or strain partitioning) when it is necessary to ask how the calculated quantity changes when more than one variable is changed during the same history, e.g. when temperature *and* time *and* current are changed. Integral forms like (4) have even greater advantages, since they are sensitive not only to histories that may be non-constant (stochastic, etc.), but also they are sequence dependent, i.e. they give differrent results if a low temperature is followed by a high one, etc. They simply follow the physics better in a simulation of history of operation.

In a previous paper,[15] we have discussed how metrics for power in an SOFC cell can be constructed, and how point-wise changes in conductivity of the electrolyte can be introduced into such a metric to assess changes in power of the fuel cell as the electrolyte ages. For the present discussion, we will focus on an example from PEM durability studies, relating to the combined effects of cyclic hydration and chemical degradation.

6. Performance Metrics for Combined Effects

For fuel cells, the engineering metric of interest may or may not be strength. Here we consider an example for PEM cells for which the strain to break is known to be proportional to the useful life of the fuel cell.[13]

The example illustrates how the combined effect of two degradation mechanisms, cyclic hydration and chemical degradation due to peroxides in this case, act and interact to cause eventual failure.

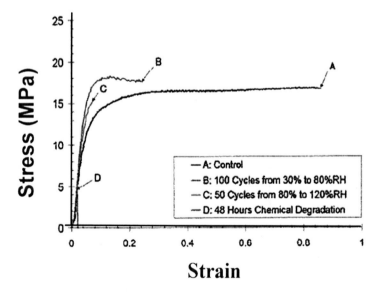

Figure 5. Interpretation of remaining strength predicted by Eq. (4).

Figure 5 shows some experimental data that indicates the changes in stress-strain response of a Nafion 111 membrane when those degrading conditions are separately. One can see in the figure that the strain to break is reduced from a value of the order of 100 percent to values of the order of 30 percent or less for cyclic hydration, and to values of the order of 10 percent for the chemical degradation conditions used in these experiments.[13] In order to interpret these changes in terms of the integral degradation metric in Eq. (4), a "failure function" that correctly represents this physics is needed. The authors have postulated several such metrics, one of which is quoted in Eq. (5) below. That metric, evaluated at every local point of interest in the fuel cell, is simply the ratio of the total range of strain caused by changes in hydration of the membrane, $\Delta\varepsilon$, to the current value of the material strain to break, ε_b. Both of these quantities are functions of time and of the history of operation of the fuel cell. The strain range can be determined from solutions of the field equations, discussed (albeit for SOFC) in Figure 3.

$$Fa = \frac{(\text{operating strain range})}{(\text{strain to break})} = \frac{\Delta\varepsilon}{\varepsilon_b} \quad (5)$$

where

$$\varepsilon_b(t,T,C) = (1-0.02t)\frac{C}{C_0}\frac{T-T_g}{T_0} \quad (6)$$

The material strain to break can be obtained for a given temperature, concentration, and time from a constitutive equation such as Eq. (6), wherein the material constants, C_0, T_0, and the coefficient of t are determined from experimental data. Then the failure function in Eq. (5) is used in the performance degradation metric, Eq. (4) to estimate the remaining strength and life of a given material point, or for a group of points in a simulation analysis of the entire membrane, fuel cell, or system.

Equation (6) suggests a relationship that can be used to represent the dependence of the strain to break on the concentration of a degrading environment (like hydrogen peroxide in the case of PEM cells) and time of exposure. But we also saw from Figure 5 that the strain to break may be degraded by cyclic hydration. Figure 6 shows an example of such data.[13]

We can see that strain to break is not only a function of time, temperature, chemical concentration, and hydration, but it is also a function of the number of times the hydration of the membrane cycles between specific limits (between relative humidities of 30 and 80 percent in this figure). For the case shown in Figure 6, it would appear that the strain to break is nearly a linear function of the number of cycles of hydration, at least for this range

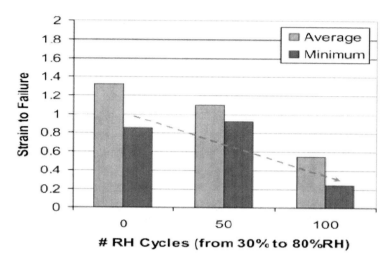

Figure 6. Degradation of strain to break caused by cyclic hydration. (From Huang et al.[13])

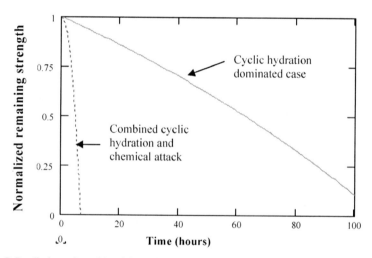

Figure 7. Prediction of combined degradation due to cyclic hydration and chemical attack on PEM membrane life.[13]

of hydration change. For the materials tested in reference 13, that is the case for several different ranges of change of hydration.

Among the most important advantages of this approach is the ability to estimate the combined effect, of cyclic hydration and chemical degradation in this case. Figure 7 shows an illustration of how that is done in the present case, although it should be noted that the details of the results for the actual materials tested and modeled are proprietary. As Figure 7 shows, the degradation effect of combined cyclic hydration and chemical environments is much greater than cyclic hydration alone. While one might have guessed that general result, a specific estimate, which is easily validated in the laboratory by stopping the degradation process and measuring membrane strength in this case, is essential as a foundation for rational design and development. Such validation of predictive capability for combined effects has been done for many situations as discussed above; final validation for PEM membranes (to define limitations of the method) are underway in the continuing work of the present investigators.

7. Conclusions

We have presented a protocol and outlined a methodology for estimating *engineered performance histories* in fuel cells and fuel cell systems, something we have called *engineering durability*. This method enables us to design and develop fuel cell systems for any relevant prospective operating *history* as a functional composite material system, wherein the combined action and interaction (and senescence) of the constituent material layers is represented in multiphysics analysis which estimates the performance of the fuel cell. The long term performance is controlled by the "aging" of the material properties in those equations. And the performance changes caused by that aging process can be represented by engineering metrics of performance calculated from the output of those equations as a function of operating history. This methodology enables accelerated characterization, life prediction, condition-based maintenance, virtual design in a knowledge-based environment, and simulation of long-term behavior. It is sensitive to sequence and history of applied conditions and compatible with modern engineering methodology.

References

1. K. Reifsnider, X. Huang, and G. Ju, Durability of Solid Oxide Fuel Cells, *Fuel Cell Technology*, edited by N. Sammes, 53–67 (Springer-Verlag, London, 2005).
2. G. Ju, Durability Study and Multi-Physics Based Life Prediction Method Investigation of a Solid Oxide Fuel Cell, Ph.D. Dissertation, College of Engineering, Department of Mechanical Engineering, University of Connecticut (July, 2006).
3. G. Ju, K. Reifsnider, and X. Huang, Time Dependent Properties and Performance of a Tubular Solid Oxide Fuel Cell, *ASME Journal of Fuel Cell Science and Technology*, 1, 35–41 (2004).

4. K. An and K. Reifsnider, A Multiphysics Modeling Study of (Pr0.7Sr0.3) Mn03/8 mol% Yttria-Stabilized Zirconia Composite Cathodes for Solid Oxide Fuel Cells, *Journal of Fuel Cell Science and Technology*, 2, 45–51 (2005).
5. K. An, K. Reifsnider, and D. Gao, Durability of $(Pr_{0.7}Sr_{0.3})MnO_{3\pm\delta}$/8YSZ Composite Cathodes for Solid Oxide fuel cells, *Journal of Power Sources*, 158 #1, 254–262 (2006).
6. K. Reifsnider, G. Ju, M. Feshler, and K. An, Mechanics of Composite Materials in Fuel Cell Systems, *Mechanics of Composites Materials*, 41, 1, 3–16 (2005).
7. K. Reifsnider, G. Ju, X. Huang, and Y. Du, Time Dependent Properties and Performance of a Tubular Solid Oxide Fuel Cell, *ASME Journal of Fuel Cell Science and Technology*, 1, 35–42 (2004).
8. K. Kent Bowen and C. Purrington, *Pratt & Whitney: Standard Work*, Harvard Business Online, 9-604-084 (2006).
9. A. Smirnova, N. Sammes, J. Pusz, J. Tang, X. Xue, and A. Mohammadi, Micro-Modeling and Electrochemical Testing of Small-Diameter Tubular SOFCs, *Proceedings of the Extended Abstracts of the First Symposium on Advanced Ceramic Reactor*, Tokyo, Japan, pp. 43–46 (eb. 24, 2006).
10. Solid Oxide Fuel Cells 10, K. E., S. Singhal, H. Yokokawa and J. Mizusaki, Eds., *ECS Transactions*, Vol. 7, No. 1 (July 2007).
11. N. Fekrazad, Effect of Compressive Load on PEM Fuel Cell and Stack Performance and Behavior, Ph.D. dissertation, Department of Mechanical Engineering, University of Connecticut (2007).
12. Y.H. Lai, et al. Viscoelastic Stress Model and Mechanical Characterization of Perfluorosulfonic Acid Polymer Electrolyte Membranes, in *Proceedings of the 3rd International Conference on Fuel Cell Science, Engineering and Technology*, ASME, Ypsilanti, MI (2005).
13. X. Huang, R. Solasi, Y. Zou, K. Reifsnider, D. Condit, S. Burlatsky, and T. Madden, Mechanical Endurance of Polymer electrolyte Membranes and PEM Fuel Cell Durability, *Journal of Polymer Science, Part B: Polymer Physics*, 44, 16, 2346–2357 (2006).
14. K. Reifsnider and S.W. Case, *Damage Tolerance and Durability of Material Systems*, Wiley (2003).
15. K. Reifsnider, F. Chen and C. Xue, Engineering Durability of Solid Oxide Fuel Cells: A Methodology, *Journal of Ceramic Society* (in press).
16. D. Zhu and R. Miller, Thermal-Barrier Coatings for Advanced Gas-Turbine Engines, *MRS Bulletin*, 25, 7, 43–47 (2000).

PORTABLE FUEL CELLS – FUNDAMENTALS, TECHNOLOGIES AND APPLICATIONS

C. O. COLPAN[1], I. DINCER[2*], AND F. HAMDULLAHPUR[1]
[1]*Mechanical and Aerospace Engineering Department, Carleton University, 1125 Colonel By Drive, Ottawa, Ontario, Canada K1S 5B6*
[2]*Faculty of Engineering and Applied Science, University of Ontario Institute of Technology, 2000 Simcoe Street North, Oshawa, Ontario, Canada L1H 7L7*

1. Introduction

Unlike transportation, small and large scale stationary applications, portable applications are generally considered to have power requirement less than 1 kW. PFCs are used in this kind of devices which are small and lightweight. Portable devices requiring low power have progressed from primary (disposable) and secondary (rechargeable) batteries to portable fuel cells. It is foreseen that PFCs will soon start replacing with Li-based or other rechargeable batteries since these battery systems are not suitable for high-power and long-lifespan portable devices due to their limited specific energy and operational time. In the case of high-power applications, PFCs will be preferred to internal combustion engines since they are more efficient, quiet and environmentally friendly. Some possible application areas that PFCS may be selected instead of internal combustion engines include, but not limited to, buildings and film sets.

Recently, many fuel cell companies have started paying the highest attention to portable fuel cells (PFCs) for micro applications and faster commercialization. Such fuel cells are especially crucial for the devices where high power density and long operation time are needed. Their application areas include, but not limited to, laptops, battery chargers, external power units and military applications. Proton exchange membrane fuel cells (PEMFCs) and direct methanol fuel cells (DMFCs) are considered the most feasible fuel cell types for niche applications.

---

*Ibrahim Dincer, Faculty of Engineering and Applied Science, UOIT, 2000 Simcoe Street North, Oshawa, Ontario, Canada L1H 7L7, e-mail: Ibrahim.Dincer@uoit.ca

On the other hand, Solid Oxide Fuel Cell (SOFC) is a very promising type that may penetrate the market in the future, especially for military applications. This study is intended to discuss the PFCs, current technologies and challenges, potential applications. Some illustrative examples are presented to highlight the importance of these PFCs.

2. Fundamentals

Many companies and research groups work on developing different types of PFC technologies to make them more advantageous than their competing technologies, i.e. batteries. However, battery manufacturers always seek solutions to increase the battery energy densities. Current Li-ion batteries are achieving 400 Wh/l.[1] But, it is very risky to increase battery capacity too much because a high capacity battery means a large amount of energy is packed into a small space which may dangerous like a bomb. On the other hand, in addition to the high power density and lifetime benefits of portable fuel cells, life cycle costs of these fuel cells are expected to be lower.

2.1. COMPARISON OF PORTABLE FUEL CELLS AND BATTERIES

Many properties and their associated variables should be taken into account in comparing different power sources for a specific application.[2,3] Among them, energy, size, weight, operation time, transient behavior and cost are the key properties; and volumetric energy density, gravimetric energy density, change of power density with time, specific cost and life cycle cost are their associated key variables. The linkages between these key variables and properties are illustrated in Figure 1. Most important ones of them are discussed in detail in the following subsections.

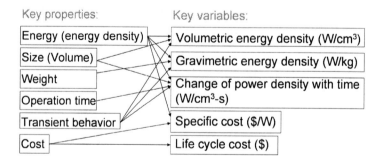

Figure 1. Key variables and their associated properties for portable fuel cells.

2.1.1. *System Size*

An active portable fuel cell system consists of three parts: fuel storage, stack, and balance of plant (peripherals); whereas the battery has just one part as shown in Figure 2.

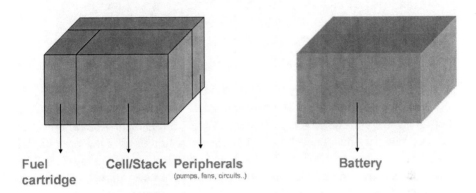

Fuel cartridge **Cell/Stack** **Peripherals** (pumps, fans, circuits..) **Battery**

Figure 2. Schematic of components of a portable fuel cell and a battery.

The power requirement of the device determines the size of the stack and balance of plant; whereas the operation time is related to the size of the storage. So, the volume of the fuel cell system, V_{FCS}, is written as

$$V_{FCS}(t) = V_{S+P} + V_f(t) \quad (1)$$

where V_{S+P} is the total volume occupied by the stack and balance of plant and V_f is the volume of the stored fuel.

The power required by the device, P, is defined as

$$P = I \cdot E - P_{BOP} = I \cdot E \cdot \eta_{BOP} = z \cdot F \cdot U_f \cdot \dot{n}_f \cdot E \cdot \eta_{BOP} \quad (2)$$

where z is the number of electrons produced for each mole of fuel, F is the Faraday constant, U_F is the fuel utilization ratio, $\dot{n}_f$ is the molar flow rate of fuel entering the fuel cell, E is the operating cell voltage, η_{BOP} is the efficiency of balance of plant accounting for the parasitic loads. Here, E depends on the fuel cell parameters such as geometry of the fuel cell and materials, type of the fuel and other operation variables such as temperature and pressure.

The volumetric discharge flow rate of the stored fuel is given as

$$\dot{V}_f = \frac{M_f \cdot P}{\rho_f \cdot z \cdot F \cdot U_f \cdot E \cdot \eta_{BOP}} \quad (3)$$

where M_f is the molar weight of the fuel and ρ_f is the density of the fuel.

Hence, the volume of the fuel cell system results in

$$V_{FCS} = V_{S+P} + \frac{M_f \cdot P}{\rho_f \cdot z \cdot F \cdot U_f \cdot E \cdot \eta_{BOP}} \cdot t_{op} \qquad (4)$$

where t_{op} is the operation time of the device.

In the case of batteries, the size is directly proportional to the energy density. The volume of the battery, V_B, is written as

$$V_B = \frac{P}{ED} \cdot t_{op} \qquad (5)$$

where ED is the energy density of the battery.

Here we conduct a comparison study on the system size required for both batteries and fuel cells as shown in Figure 3. As clearly seen, the sizes of both battery and fuel cell vary linearly as a function of the operational time. This shows that the fuel cells appear to be more advantageous than batteries for the operational time of the device greater than t_{min} since the system size becomes smaller for fuel cell after this time.

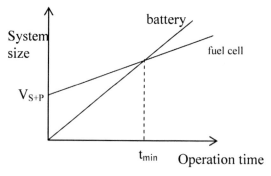

Figure 3. Comparison of battery and fuel cell in terms of system size.

Here, t_{min} may be calculated by equating V_{FCS} and V_B.

$$t_{min} = \frac{V_{S+P}}{P \cdot \left(\dfrac{1}{ED} - \dfrac{M_f}{\rho_f \cdot z \cdot F \cdot U_f \cdot E \cdot \eta_{BOP}} \right)} \qquad (6)$$

Illustrative Example-I:
Let's consider that we want to choose between Li-ion battery and DMFC to be used in a laptop requiring an average capacity of 13 W. For comparison, size of the system is considered as the design criterion. The input parameters that are assumed for the battery and the fuel cell are shown in Table 1.

TABLE 1. Input parameters for the Illustrative Example-I.

DMFC		Li-ion battery	
Volume of stack and BOP[a]	520 cm^3	Energy density	340 Wh/l
Fuel utilization	0.85		
Operating cell voltage	0.35 V		
Efficiency of Balance of Plant	0.90		

[a] The size of the DMFC is approximated as P/0.05 ^{4} and BOP is assumed to have same volume with the DMFC

Using Eq. (6), we find that the minimum operation time required to consider DMFC instead of Li-ion battery is 20 hours. For this time, the volume of the system is equal to 765 cm^3. For this example, we also see that the volume of the fuel cartridge of the DMFC system is 12.25 cm^3/hour.

Using efficient components for balance of plant plays a significant role in the preference of portable fuel cells instead of batteries. Its effect on the minimum time, t_{min}, is shown in Figure 4. As one can see from this figure, t_{min} decreases from 48 to 20 hours as the balance of plant efficiency increases from 40% to 90%, respectively.

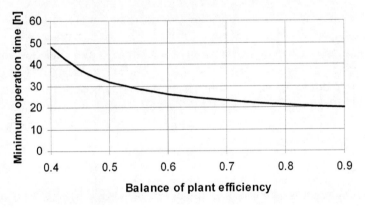

Figure 4. Effect of balance of plant efficiency on the minimum time to consider portable fuel cells instead of batteries.

2.1.2. Recharging

It can take several hours to recharge a battery depending on type of the battery. Unless you are carrying extra batteries, you have to find electricity outlet to recharge them. In the case of PFCs, there is no need for recharging. Instead, you have to change the fuel cartridge. Moreover, most fuel cells can run one or two minutes during cartridge change which is a sustainable operation due to fuel recirculation.

2.1.3. *Weight*

Weight is another important criterion in the selection of power source, especially for laptops, wearable devices and power devices carried by soldiers. PFCs should be preferred over batteries when the device is intended to be used for long operation time.

Illustrative Example-II:
As an example, let's consider that for a laptop we replace the power source from Li-ion battery to a DMFC cartridge and compare their weights for the same operation time. The input data for both battery and DMFC as assumed are given in Table 2.

TABLE 2. Input parameters for the example.

DMFC		Li-ion battery	
Weight of stack and BOP	1.1 kg	Weight	0.143 kg/hour
Weight of fuel cartridge	0.02 kg/hour		

Currently, PFCs are not manufactured to fit inside the laptop like Li-ion batteries since they will not be advantageous in terms of operation time. In this example, we consider that both DMFC and Li-ion battery are put outside the laptop, so we do not have size restrictions. We find that the minimum operation time needed to consider DMFC instead of Li-ion battery for a lighter device is 8.94 hours If we take a 20 hours of operation time, the weight becomes 1.5 kg for DMFC and 2.86 kg for Li-ion battery; which shows us that PFCs are more advantageous and should be preferred for longer operation time.

2.1.4. *Life Cycle Cost*

The economical benefits of PFCs become clear when the life cycle cost of them are considered. Compared to batteries, PFCs are expected to have lower life cycle costs, which is illustrated with the following example.

Illustrative Example-III:
Let's consider that a professional video camera manufacturer thinks of making a system operating with a PFC instead of a battery. The existing system works with a set of three batteries connected in series. As the running time capacity will decrease as the camera is used, the manufacturer suggests to change the batteries one by one at certain years, such as first battery after 1.6 years, second battery after 2.8 years and third battery after

3.7 years. The lifetime without losing its performance significantly is considered as 4 years. There is also an additional electricity cost to charge the battery in this system. If we use a PFC instead, there will be only initial cost and fuel cost. A comparison depending on a manufacturer's data[5] is illustrated in Table 3. As it can be seen here, the life cycle cost for a PFC system is lower, and hence becomes more cost effective.

TABLE 3. Life cycle cost comparison of a battery system and PFC for a video camera.

	Battery		PFC	
	Component	Cost	Component	Cost
Year: 0 (initial cost)	Battery pack	3 × $525 = $1,575	Fuel cell system	$4,000
	Charger	$1,595		
Year: 1.6	New battery	$525	Fuel cost	$269
Year: 2.8	New battery	$525	Fuel cost	$202
Year: 3.7	New battery	$525	Fuel cost	$151
Year: 4	–	–	Fuel cost	$50
Total cost of electricity		$175		–
Total life cycle cost		~$4,920		~$4,672

2.2. FUEL CHOICE

Depending on the technology and type employed, hydrogen, methanol, ethanol and hydrocarbons generally appear to be potential fuels for PFCs. Among them, hydrogen and methanol are considered the most promising ones. Safety, storage and distribution issues are compared below for these fuels.

2.2.1. *Safety*

In the case of an accident such as leak or fire, the severity of the situation depends upon the physical properties of the fuel such as flammability.

Also, the risk depends on the physical conditions, i.e. well-ventilated space or totally enclosed space. The flammability range is higher for hydrogen and it involves a higher risk than methanol in enclosed spaces, whereas its risk is lower than methanol in well ventilated space. Another problem with methanol is its toxicity. If it mixes with any drink, it may cause severe health problems. It may be concluded that as long as all the safety precautions are taken and safety standards are applied, fuel cells may be designed safely.

2.2.2. Storage and Distribution

Hydrogen storage methods may be classified into two general groups: direct storage of hydrogen for use and generation of hydrogen through chemical methods (e.g. reforming) on demand. The first one may include storage in compressed cylinders, metal hydrides and carbon nanotubes. The latter one includes reacting chemicals to liberate hydrogen and reforming liquid fuels. The substances used in this method include methanol, alkali metal hydrides, sodium borohydride and ammonia. For example, among these methods, compressed cylinders seem the cheapest and straightforward method; however the container of hydrogen is thick and heavy. Metal hydride is suitable for small systems, however the performance is low. In general, chemical reaction based subsystems provide higher energy storage compared to physical storage. Reforming offers even higher energy density and it is more economic compared to other options.

Methanol is more advantageous than hydrogen in terms of storage since it is a liquid. The main problem with methanol storage is its ability to mix with water, which may cause a corrosive environment eventually. Due to this, rather than ordinary steel, stainless steel or glass should be used as the material of the container. On the other hand, methanol can only be shipped by plane in the checked baggage compartment subject to dangerous goods regulations; but its existence in passenger compartment is expected to be allowed very soon.

3. Technologies

There are various types of fuel cells that may be used for portable applications. Except SOFCs, all of them may be regarded as low temperature fuel cells. These fuel cell types are discussed below.

3.1. PEMFC

This type of fuel cell also known as the polymer electrolyte membrane fuel cell, consists of a proton conducting membrane, such as Nafion, which is chemically highly resistant, mechanically strong, acidic, good proton conductor and water absorbent. The reactions occurring at anode, cathode and overall reaction are given in Eqs. (7)–(9), respectively.

$$H_2 \rightarrow 2H^+ + 2e^- \tag{7}$$

$$0.5O_2 + 2H^+ + 2e^- \rightarrow H_2O \tag{8}$$

$$H_2 + 0.5O_2 \rightarrow H_2O \tag{9}$$

Some main advantages of this type of fuel cell are:

- Fast startup capability since it works at low temperatures
- Compactness since thin MEAs can be made
- Elimination of corrosion problem since the only liquid present in the cell is water.

The main disadvantage of this type of fuel cell is the need for expensive catalysts as promoters for the electrochemical reaction. Additionally, carbon monoxide can not be used as a fuel since it poisons the cell. On the other hand, the main challenge for PEMFC is the water management which may be summarized as follows: The proton conductivity of the electrolyte is directly proportional to the water content and high enough water content is necessary to avoid membrane dehydration. Contrarily, low enough water should be present in the electrolyte to avoid flooding the electrodes. Hence, a balance between the production of water by oxidation of the hydrogen and its evaporation has to be controlled.

3.2. DMFC

This type of fuel cell also uses a proton conducting membrane like PEMFCs. Its main difference is the direct feeding of methanol to the fuel cell instead of reforming it before feeding. The electrochemical reactions occurring at this fuel cell are as follows: The reactions occurring at anode, cathode and overall reaction are given in Eqs. (10)–(12), respectively.

$$CH_3OH + H_2O \rightarrow CO_2 + 6H^+ + 6e^- \tag{10}$$

$$1.5O_2 + 6H^+ + 6e^- \rightarrow 3H_2O \tag{11}$$

$$CH_3OH + 1.5O_2 \rightarrow CO_2 + 2H_2O \tag{12}$$

The main advantages of this kind of fuel cell are as follows:

- Methanol as fuel is a readily available and less expensive
- High energy density of methanol
- Simple to use and very quick to refill.

The main disadvantage of this type of fuel cell is the slow reaction kinetics of the methanol oxidation, which results in a lower power for a given size. The second major problem is the fuel crossover which is summarized as follows: Polymer membrane of DMFC is permeable to methanol which means it may diffuse from the anode through the electrolyte to the cathode. Hence, migrated fuel is wasted which will decrease the amount of electron produced. It also reduces the cell voltage, hence the cell performance. The current approach to minimizing the methanol permeation rate is to limit the methanol concentration to approximately 5 wt%[6] despite the loss in performance.

There are two types of DMFC, active type and passive type and their schematics are shown in Figure 5. In the active type, fuel and air flows are controlled to get high performance. In the passive type, the air is introduced into the cell by natural flow, i.e. self breathing, and the fuel is infiltrated into the cells. There is less control over the variables of fuel and air stoichiometry in this type. The passive one is much simpler compared to the active type, but the performance is not that high. Active type is good for high power products, such as laptops, LCD-TVs, and digital cameras. Passive type is good for small and low power products such as fuel cell powered mp3 player.

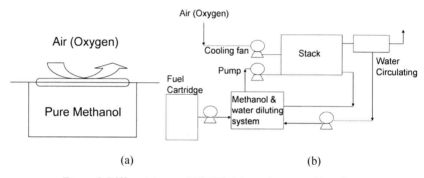

Figure 5. Different types of DMFC: (a) passive type, (b) active type.

3.3. AFC

These have become popular, particularly for space vehicles, but the success in other low-temperature fuel cells has declined the interest in this type of fuel cell. The main reasons for this were the issues with cost, reliability and ease of use. However, there is one type of AFC which still takes attention and is a candidate to be used in portable applications. It is the Direct Borohydride Fuel Cell (DBFC) which uses a solution of sodium borohydride as fuel. The reactions occurring at anode, cathode and overall reaction are given in Eqs. (13)–(15), respectively.

$$NaBH_4 + 8OH^- \rightarrow NaBO_2 + 6H_2O + 8e^- \quad (13)$$

$$2O_2 + 8e^- + 4H_2O \rightarrow 8OH^- \quad (14)$$

$$NaBH_4 + 2O_2 \rightarrow NaBO_2 + 2H_2O \quad (15)$$

The main advantages of DBFC are as follows:

- Eight electrons are formed from one mole of fuel
- Highly alkaline fuel and waste borax prevent the fuel cell from CO_2 poisoning
- It is very simple to make it as the electrolyte and the fuel are mixed

The main disadvantage of DBFC is the side reaction called hydrolysis reaction at which hydrogen is produced as $NaBH_4$ reacts with water. However, with modern techniques, hydrogen can be oxidized immediately giving eight electrons provided that hydrolysis reaction is not proceeding too quickly.

3.4. SOFC

SOFCs are mostly used for stationary power generation applications; however they may also be applied to portable applications. Unlike the other fuel cell types mentioned, they are high temperature fuel cells which may operate between 500°C and 1,000°C. The most common material used for electrolyte is yttria stabilized zirconia. The reactions occurring at anode, cathode and overall reaction are given in Eqs. (16)–(18), respectively when H_2 is used as fuel.

$$H_2 + O^{-2} \rightarrow H_2O + 2e^- \quad (16)$$

$$0.5O_2 + 2e^- \rightarrow O^{-2} \quad (17)$$

$$H_2 + 0.5O_2 \rightarrow H_2O \quad (18)$$

If CO is used as fuel, the reactions occurring at anode, cathode and overall reaction become as shown in Eqs. (19)–(21):

$$CO + O^{-2} \rightarrow CO_2 + 2e^- \quad (19)$$

$$0.5O_2 + 2e^- \rightarrow O^{-2} \quad (20)$$

$$CO + 0.5O_2 \rightarrow CO_2 \quad (21)$$

The main advantages of this fuel cell are as follows:

- Fuel flexibility (methane, propane, butane, JP-8 may be used as fuel)
- Direct reforming at the anode catalyst
- Elimination of precious metal electrocatalysts.

The main disadvantage of this fuel cell may be given as the challenges for construction and durability due to its high temperature. However, this issue also depends on the design type. For example, this problem is especially severe for planar design due to the sealing problem. In tubular design, the cells may expand and contract without any constraints. The second major problem is the deposition of carbon particles at the anode. This may be solved by sending sufficient amount of external water or recirculating the depleted fuel at the fuel channel exit to the inlet.

Another important consideration is the balance of plant which may include air pump, valves, sensors, piping, tank, recuperator, fan, etc. The main criteria for selecting these components are lightweight, efficient, low power consumption and low cost. Heat interaction of components between each other should be well designed to obtain a highly efficient system.

The energy and exergy efficiencies of a SOFC system may be defined as shown in Eqs. (22) and (23), respectively.

$$\eta_{sys} = \eta_{fp} \times \eta_{fc} \times \eta_{util} \times \eta_{BOP} \qquad (22)$$

$$\varepsilon_{sys} = \varepsilon_{fp} \times \varepsilon_{fc} \times \varepsilon_{util} \times \varepsilon_{BOP} \qquad (23)$$

Here, the subscripts fp, fc, util and BOP stands for fuel processor, fuel cell, fuel utilization and balance of plant, respectively.

Illustrative Example-IV:
As an example, let's look at the exergy efficiency of the system for a practical representation for two cases. Considering the best case, we take each term in Eq. (23) as 90%, the system exergy efficiency will become ~66%. On the other hand, for the worst case (by taking each term as 50%), the system exergy efficiency will result in 6%. So, one may expect that the system exergy efficiency vary between 6% and 66% for a portable SOFC system.

3.5. OTHER FUEL CELL TYPES

Direct Formic Acid Fuel Cell (DFAFC), Direct Ethanol Fuel Cell (DEFC) and biofuel cell (BFC) may be used in some of the portable applications. The first two uses a PEM where formic acid and ethanol are fed directly. DFAFC is advantageous due to its high catalytic activity, easier water management and minimal balance of plant. However, performance of the cell strongly depends on the feed concentration of formic acid due to mass transport limitations. Generally, high feed concentrations are needed. DEFC may be preferable due to the advantages of ethanol such as high energy density, safer to use and easy to store. However, in the electrochemical reactions a lot of acetaldehyde is produced which is a very flammable and harmful liquid. Additionally, reaction kinetics is very slow and ethanol crossover is a problem. BFC may be used in very low power applications. Mainly, there are two classes of BFC which are Microbial Fuel Cell and Enzymatic Fuel Cell. The first one has higher efficiency and complete oxidation of fuel; but lower power density. Hence, it is more applicable for larger scale applications such as generating power on the seafloor. The latter one has high power density but lower efficiency and incomplete oxidation of fuel. It may be used in small scale application such as implantable devices.

4. Applications

Niche applications are now becoming the main market area for PFCs, which include laptops, mobile phones, camcorders, digital cameras, portable generators for camping and other recreational activities, battery chargers, etc. In all of these applications, the consumer prefers small, lightweight and long operated devices, which may be provided by portable fuel cells. Additionally, batteries might not be able to supply the power needed for the new devices with a greater amount of functions. In this aspect, portable fuel cells should be preferred since they have a higher power density. For example, fuel cells can enable the universal connectivity of wireless devices, such as laptop computers and 3G phones. Currently, there are several companies developing portable fuel cells on DMFC technology.

The military defense research plays an important role in the development of PFCs since there is a big funding in this area. These fuel cells are important for military because the future soldiers are intended to have equipment needing high power such as night vision devices, global positioning systems, target designators, climate controlled body suits and digital communication systems. These should be light enough for soldiers to carry. They should also be able to operate for a long time. It is obvious that batteries cannot provide these energy needs at an acceptable weight. Therefore, PFCs are expected to be a must for the military for their future purposes. Another important point for military is the type of fuel used in these fuel cells since they prefer fuel that is available in the battle area in any part of the world such as diesel and JP-8. Hence, portable SOFC is the best option for the purposes where the fuel availability is the main criteria. However, PEMFC and DMFC may also be preferred depending on the size and purpose of a military application.

Some application examples of various PFCs as discussed in this paper are listed in Table 4 and ranked in regards to their market potential. It is obvious that portable DMFCs are ranked as the highest for market.

TABLE 4. Application examples and their rank of appearance in the market for PFCs.

Rank (1-highest, 7-lowest)	Fuel cell type	Application examples
1	DMFC	Laptop, cell phone, mp3 player, battery charger
2	PEMFC	Professional video camera, flashlight, bicycle light
3	SOFC	Camping devices, military applications
4	AFC	Portable charger, cell phone, PDA, digital camera
5	BFC	Microelectronics and biomedical applications
6	DFAFC	Cell phone
7	DEFC	Wearable military power packs

5. Conclusions

In this book contribution the portable fuel cells have been discussed as potential alternatives to replace batteries for portable applications where high power density, long operation time and lightweight are of great importance. There are various technologies available with each of them having advantages and disadvantages and operating with different fuel. Non-technical considerations such as safety, storage and distribution should be taken into account in choosing the fuel. The current trend shows that DMFCs are the leading fuel cell type for niche applications, whereas portable SOFCs are the most promising type for military applications where the fuel availability is the major concern.

Acknowledgements

The support of an Ontario Premier's Research Excellence Award and the Natural Sciences and Engineering Research Council of Canada is gratefully acknowledged.

References

1. NEC/TOKIN, 2006, *Lithium Ion Rechargeable Battery-General Catalog.*
2. Colpan, C.O., Dincer, I. and Hamdullahpur, F., 2007a. Thermodynamic modeling of direct internal reforming solid oxide fuel cells operating with syngas, *International Journal of Hydrogen Energy* 32(7), 787–795.
3. Colpan, C.O., Dincer, I. and Hamdullahpur, F., 2007b. A review on macro level modeling of planar solid oxide fuel cells, *International Journal of Energy Research* 32(4), 336–355 (2008).
4. Larminie J. and Dicks A., 2003, *Fuel Cell Systems Explained*, 2nd ed., Wiley, UK, 157–158.
5. JPS, 2004, *Fuel Cell Power System for Professional Video Camera Applications*, Jadoo Power Systems, Inc., White Paper, 1–8.
6. Cowey, K., Green, K.J., Mepsted, G.O. and Reeve, R., 2004, Portable and military fuel cells, *Current Opinion in Solid State & Materials Science* 8, 367–371.

AUTONOMOUS TEST UNITS FOR MINI MEMBRANE ELECTRODE ASSEMBLIES

EVGENI BUDEVSKI*, IVAN RADEV, AND EVELINA SLAVCHEVA
Institute of Electrochemistry and Energy Systems, Bulgarian Academy of Sciences, Sofia 1113, Bulgaria

1. Introduction

An autonomous device for the investigation and optimization of electrochemical energy converting cells (fuel cells, electrolyzer cells or bi-functional cells) and single electrodes operating with polymer electrolyte membranes and gaseous reactants and reaction products will be described. Two constructions using mini membrane electrode assemblies will be given offering a wide range of investigation opportunities for a deeper insight in the new high temperature gas phase electrochemistry of the system.

There were several beautiful lectures in this Institute[1] where we learned that there is no single approach to fuel cells even if we restrict our selves to PEM Fuel Cells and hydrogen as fuel. Fuel cells range from watts to megawatts with enormous application possibilities, giving to the engineer a vast field of imagination for design and innovations.

Nevertheless there is one simple element which is common to all kinds of electrochemical energy converters (EcEC) including:

- Fuel cells (FC) with chemical to electrical energy conversion,
- Electrolyzers (ELZ) with electrical to chemical energy conversion,
- Bi-functional EcEC (Bi-EcEC) with both ways on demand conversion, this element has got the name: *The membrane electrode assembly* or (MEA).

The MEA is the heart of an EcEC. It is the actual site of the electrochemical energy conversion. In a sense it is also a commercial end product. It is available on a square meter basis. If you would like to produce a fuel cell for your needs you buy the necessary amount of MEA, use

_____

* Evgeni Budevski, Institute of Electrochemistry and Energy Systems, Bulgarian Academy of Sciences, 10, Acad. G. Bonchev str., Bl.10, Sofia 1113, Bulgaria; e-mail: budevski@bas.bg

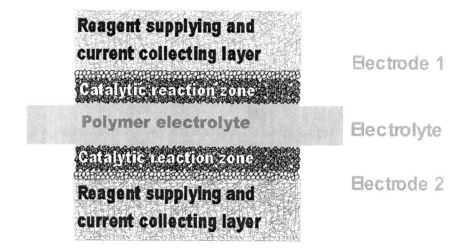

Figure 1. Main components of a membrane electrode assembly.

your creative skills and make the envelope. It is not and easy and inexpensive task.[2-4] The MEA may be even the cheapest item in your design. But the MEA dictates all. If you have a good MEA you may make a good EcEC.

The membrane electrode assembly is a sandwich of two electrode structures with a polymer electrolyte membrane in between, Figure 1.

The MEA is a typical electrochemical cell consisting of two electronic conductors in contact with an ionic conductor. The specific features of the MEA as an electrochemical cell are:

- The electrolyte is a solid ion conductive polymer
- The reagents are gases
- The electrodes are porous and gas permeative.

Before going into further details, however, I would like to list the major problems of the membrane electrode assemblies beginning with near term soluble and ending with the long term, eventually insoluble to our present knowledge, problems:

- The *membrane*: high selective ionic conductivity, chemical stability and operational simplicity (mainly humidity independence and lower operational temperatures)
- The *catalyst*: catalytic activity for the specific reactions, load and efficiency (mg/cm^2, or mg/W, respectively), price ($US per W)
- The *structure* of the gas diffusion electrode section serving as current collector and reagent transporting media and last

Figure 2. Commercially available framed MEA's (left) and single fuel cell connected to the conditioning control unit (right).

- The *highly irreversible oxygen reduction-evolution reaction* consuming almost half of the overall efficiency of the EcEC.

The development and optimization procedure needs an easy and fast preparation and screening technology. The present laboratory technologies do not offer an easy approach to the testing procedure. The common easily available investigation techniques such as scanning voltamograms, electrochemical impedance spectroscopy (EIS), and several other laboratory electrochemical techniques cannot always be clearly and unambiguously interpreted in terms of their significance for practical application. Keeping the real operation conditions we are bound to use fully developed cells (FC's or ELZ's) with the heavy periphery for control of the running conditions – reagents supply at definite pressure and composition, temperature, humidity, heat extraction, etc. Figure 2 depicts a standard test equipment. As seen it includes:

- The fuel cell stack
- The reagents (hydrogen and air oxygen) supply and conditioning unit
- The reaction product (water, electricity and heat) draining configurations
- The test procedure performing units (not shown on the figure).

Even the simplest quality assessment equipment (Figure 3) needs a periphery that is more complicated than the test object itself confining the experiments to single tests under constant surveillance.

In the fuel cell stack itself there are several components, forming only a very small part of the illustrated configuration that could be decisive for the final device performance. As already mentioned one of the essential components is: the intrinsic energy – converting device, the Membrane Electrode Assembly – the MEA.

Figure 3. Commercial fuel cell test equipment. Reagent conditioning and supplying unit (left), single fuel cell stack (center), testing unit (right).

2. The Membrane Electrode Assembly: Structure and Performance

2.1. STRUCTURE

The main components of a membrane electrode assembly are the polymer electrolyte, the catalytic reaction layer, and the reagent transporting and current collecting zones.

The most popular and easily available polymer electrolyte membrane (PEM) is Nafion, a DuPont product, with an operating temperature range between 0°C and 80–100°C. It has a specific conductivity of ca. 10^{-1} S cm^{-1} at 80°C, which is considered as acceptable for most of the applications. Nafion is sensitive to humidity, however, so that water vapor content in the gas phase is an essential prerequisite for an adequate electrochemical energy conversion. In an advanced state of development is a polybenzimidazole based polymer, a Celanese acetate Co. product. PBI is mainly designned for higher temperatures of operation of about 150°C.

The most popular catalysts are Platinum and Pt-Ruthenium alloys, and lately Iridium for the water splitting reaction.[5-13] To increase their efficiency the expensive catalysts are spread over the surface of catalyst carriers. The most popular catalyst carriers are carbon black particles of the size of less than one micron. The catalyst material is bonded with Teflon suspension and spread in form of a micro thin layer over the surface of the porous gas diffusion electro conductive layer, usually carbon fiber, woven or non-woven fabric. Both layers form the so called gas diffusion electrodes, GDE. Gas diffusion electrodes are commercially available on square meter basis.

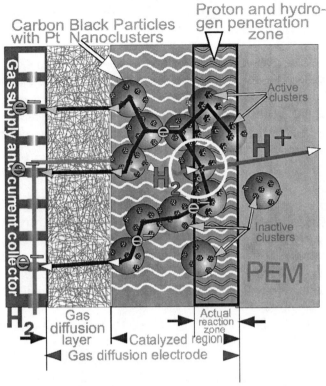

Figure 4. Structure of a gas diffusion layer. Note: the catalyzed region is over dimensioned. In the emphasized reaction site (encircled) H_2 is oxidized to protons and two electrons per H_2 atom are generated. The transport paths of H_2, electrons, and protons are indicated.

Two gas diffusion electrodes are pasted by hot pressing to the PEM using a solution of the membrane material to form the MEA. The structure of a MEA, as described, is represented in Figure 4.[14]

Due to a limited penetration debt of the proton conductive electrolyte the actual reaction zone is only a part of the catalyzed region. All catalyst particles which are outside the proton and hydrogen penetration zone or are not in electrical contact with the electro conductive gas diffusion layer are actually inactive, s. Figure 4. This affects significantly the efficient use of the catalyst material.

A new approach has been reported recently where the catalyst is deposited by dc magnetron sputtering.[9-12] The procedure involves the deposition of a thin adhesive layer of Ti of ca. 50 nm on a preferably flat surface of carbon fiber material (e.g. Toray paper) followed by the deposition of the catalyst (Pt or other catalyst metals or alloys) of a thickness ranging between 60 and 400 µg/cm^2. The catalyzed carbon paper is than soaked superficially with a diluted solution of the PEM material, dried, and hot pressed both sides on a PEM. The resulting structure is shown on Figure 1. As it will be seen later in this paper this procedure is not only very simple and reproducible, but it gives a higher efficiency of the catalyst materials. No matter what the production technology is, the modern concept is to locate the catalyst in a tiny layer between the current collector and the membrane using membrane solution as a bonding agent only.

2.2. PERFORMANCE

An electrochemical cell, e.g. a MEA, can be used to convert chemical to electrical energy and vice versa. Working in isothermal conditions the electrochemical energy conversion avoids the Carnot cycle efficiency limitations. The theoretical conversion efficiency is therefore close to unity. Due to kinetic limitations, however, the efficiency coefficient is rarely higher than 50–60%.

The electrochemical energy converter (EcEC) can be designed for *power generation*, e.g. as Fuel Cell (FC) with hydrogen as fuel, or for *energy storage*, e.g. as Electrolyzer (ELZ), with hydrogen for chemical energy storage. EcEC can be designed also as *bifunctional* electrochemical energy converters (Bi-EcEC), which depending on the direction of the current can be operated on demand in a FC mode or in ELZ mode.

In the hydrogen-oxygen EcEC the operative reaction

$$2H_2 + O_2 = 2H_2O + \Delta G \tag{1}$$

is split into two electrode reactions:

$$2H_2 = 4H^+ + 4e \tag{2}$$

and

$$4H^+ + O_2 + 4e = 2H_2O \tag{3}$$

Both reactions are reversible, in a directional sense. Hydrogen is oxidized in the forward direction of the first reaction (2) and evolved in the reverse direction. The forward, anodic, reaction is often described as the hydrogen oxidation reaction, or HOR, the reverse, cathodic, reaction is known as the hydrogen evolution reaction, or HER. The same applies to the second, oxygen-water reaction (3). In the forward, cathodic direction it is known as the oxygen reduction reaction, or ORR, in the opposite, anodic, direction it is called the oxygen evolution reaction, or OER.

The site of the electrochemical reactions is the place of change of the charge carriers – electrons (in the metallic phase of the electrodes) versus ions, or protons, respectively, (in the electrolytic phase), i.e. the electrode – electrolyte phase boundary, see Figure 4. The reaction step is known as the charge transfer reaction (CTR) giving rise to a shift of the electrode potential known as activation overpotential, or activation overvoltage, η_a. The kinetics of the CTR is given by the Butler-Volmer equation

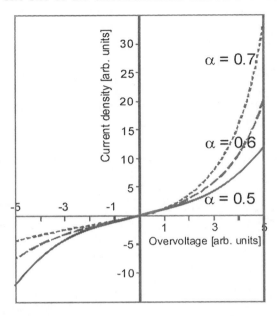

Figure 5. The Butler-Volmer current overvoltage relation. Depending on the charge transfer coefficient α the relation is symmetrical for α –0.5 and becomes asymmetrical for higher or lower α values.

$$j = j_0 \left[e^{\alpha z F \eta / nRT} - e^{-(1-\alpha) z F / nRT} \right] \qquad (4)$$

relating the current density j to the overvoltage η_a, with two kinetic coefficients: the exchange current density j_0 [A/cm^2] and the charge transfer coefficient α.

For small overpotentials, or current densities this relation gives a symmetrical linear j/η_a dependence both for positive and negative overpotential values. The slope $dj/d\eta_a$ has the dimension of Siemens/cm^2 and in the $j = 0$ region is proportional to the exchange current density j_0:

$$(dj / d\eta_a)_{j \to 0} = j_0 z F / nRT \qquad (5)$$

At higher overpotentials the j/η_a dependence is exponential both for positive and negative η_a values, and fairly symmetrical, depending on the charge transfer coefficient α, Figure 5.

The second rate determining step is the reagent transportation to an electrochemical reaction site, or the gas diffusion process, respectively. The arising overpotential is called the *concentration overpotential* η_c. The reaction rate – driving force relation, i.e. the j/η_c dependence, follows the Nernst equation characterized by a saturation current density j_d known as the limiting diffusion current density.

$$\eta_c = (RT/n_e F)\ln\left[(1 + j/j_{d,Ox})/(1 - j/j_{d,Red})\right] \quad (6)$$

The limiting current density reflects the depletion of the respective reagent at the phase boundary to zero values. j_d is proportional to the corresponding bulk concentration C_i and contains the diffusion coefficient D_i of the reaction species and a constant δ with dimension of distance:

$$j_{d,i} = \frac{n_e F}{v_i} \frac{D_i}{\delta} C_i \quad (7)$$

n_e and v_i are stoichiometric coefficients. n_e for the number of electrons and v_i for the number of molecules involved in the red-ox reaction.

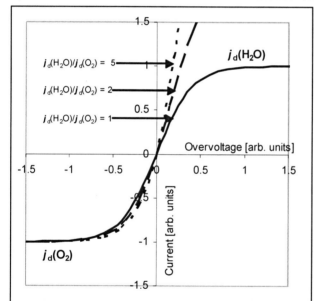

Figure 6. The polarization curve of a red-ox reaction according to the Nernst relation is symmetrical for equal current densities: $j_{d,Red} = j_{d,Ox}$. It becomes asymmetrical with growing $j_{d,Ox}$ values.

For red-ox reactions with two reaction species, Red and Ox, the relation has the form given by Eq. (6) with two limiting diffusion currents.

The oxygen reeduction – evolution reaction, Eq. (3), is a typical two species reaction with O_2 for Ox and H_2O for RED. It contains two diffusion limiting currents one for the anodic $j_d(H_2O)$ (oxidation of H_2O) and one for the cathodic $j_d(O_2)$ reactions (reduction of O_2) with limiting currents depending on the respective bulk concentrations. For this reaction (see Eq. (3)) n_e is 4 and $v(O_2) = 1$. The protons, with $v(H^+) = 4$ do not show diffusion limitations because they are transported by migration only.

The limiting current densities $j_{d,i}$ (Eq. (7)) contain essentially the distance parameter δ, which corresponds to a virtual distance from the electrode surface to the gas phase where the concentration becomes equal to the bulk concentration of the diffusing species. The distance, δ, defines the maximal (at $C_{x=0}$) surface concentration gradient: $(dC/dx)_{x=0} = C/\delta$, and hence, the limiting current.

δ is a complex parameter characterizing the specifics of the diffusion structure of the GDE depending on porosity, the tortuosity of the pores, and some other structural parameters. Next to the experimentally available limiting current concentration proportionality coefficient: $(dj_{d,i}/dCi) = n_e F D_i / v_i \delta$ (Eq. (7)) δ can be used as a constant characterizing the transportation qualities of the gas diffusion electrode.

The polarization curve of a red-ox reaction, Figure 6, is symmetrical for equal values of the limiting current densities: $j_d(H_2O)/j_d(O_2) = 1$. With increasing $j_d(H_2O)$ values the curve becomes asymmetrical with, under circumstances not recordable anodic limiting current. Even a five time higher value of $j_d(H_2O)$ renders a straight line currentvoltage dependence in the anodic, OER, experimentally available region.

The slope of the $j - \eta_c$ relation at zero current contains other information for the j_d values.

The third, rate determining process is the ionic migration of the protons under the influence of the potential gradient in the electrolyte. This process does not involve concentration gradients and diffusion because there is one charge transporting species only. The resulting current – potential shift relation is linear and adds linearly to the overall potential change ΔE. Note that a simple addition of the different overpotential contributions is only possible in the linear part of the current density – overvoltage relation.

Hence, the current – potential change relation, or as it is known the polarization curve, involves significant information concerning the rate determining steps and can be used for the evaluation of the structural determined performance characteristics of the electrodes.

3. The Reagent Recovery Principle and the Autonomous Test Cell Unit

The investigation of the performance characteristics of an EcEC needs a very precise and easy to maintain reagent composition in the bulk phase.

For reversible reactions, i.e. for reactions that can be run in the forward and backward directions, as the H_2-O_2 reaction, there is an easy way to secure this condition if the test cell (TC) is connected in series with a second reagent recovery (RRC) cell, operating in the reverse direction and

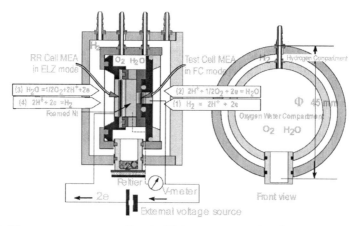

Figure 7. The autonomous test cell unit. Right: a cross section front view. Left: cross section side view. An exploded view showing constructional details is given in Figure 8.

the gas phase compartments of both cells are connected in a proper way, Figure 7.

With a 100% faradic efficiency the consumed or generated reagents of the test cell can be recovered quantitatively in the RR cell and returned back to the TC. The system can than be sealed from the environment and run theoretically indefinitely in time without a change of the operating conditions.[12] Figure 7 shows a compact constructional version of this principle.

The cell MEA's are enclosed in a cylindrical body consisting of two concentric tubes forming two compartments, right hand side of Figure 7. The inner tube houses the reagent recovery cell MEA and the test cell MEA (in a micro version in the case) separated by foamed metal (Ni). The open pore foamed metal gives a free diffusion transportation of the oxygen and water vapors (reactions 2 and 3) and a good electric serial connection of both cells. Hydrogen is transported freely in the outer compartment from the left hand cell to the test cell following reactions 1 and 4.

Water is not only a reagent but determines in an essential extent the conductivity of the membrane. The water vapor pressure is therefore strictly controlled by a water pool in the bottom part of the outer compartment at a temperature regulated by a Peltier element. This makes the water partial pressure a free variable, limited by the cell temperature as an upper value.

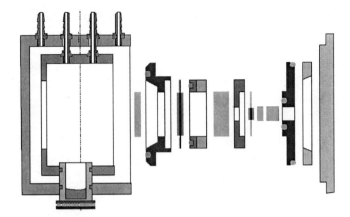

Figure 8. Exploded view of the test unit. Parts from left to right: foamed metal, RRC MEA frame, the RRC MEA, sealing flange, foamed metal, TC MEA frame, the TC MEA, foamed metal, two sealing flanges, end cover.

The test cell is subject of variations and is easily exchanged. The size of the MEA can be changed to micro dimensions, as depicted in Figure 7. The gas diffusion electrode structure is also a free parameter and may differ significantly from that of the RRC. Next to its simplicity of operation the main advantage of the unit is that the initial conditions are easily set and maintained precisely during operation.

The gas pressure and compositions (e.g. H_2, O_2, H_2O, additional gases like He, N_2, Ar, etc.), catalyst (e.g. kind and loading), gas diffusion structures, etc. can be varied to reveal different essential rate determining steps and find optimal configuration and structural solutions.

At the same time the test electrode is operating at 100% real conditions and therefore is extremely suitable for optimization, demonstration, and educational purposes.

The assumed 100% faradic efficiency is, however, only conditionally true, because hydrogen can diffuse trough the membrane of the TC giving rise to a non faradic reaction with consumption of the hydrogen of the fuel compartment and water generation on the left side catalytic layer of the TC on Figure 7. The easiest way of a complete control of this *cross-over effect* is to record the hydrogen pressure in the fuel (H_2) compartment and to add an additional current on the RRC keeping the hydrogen pressure level to a constant value. This current is than exactly equivalent to the H_2 cross-over.

4. Autonomous Test Unit for Single Electrode Optimization: The *EasyTest Cell*

It is a general practice in electrochemistry to test and optimize *single electrodes* before coupling them to a working cell. Consider a MEA enclosed in a H_2 environment, Figure 9.

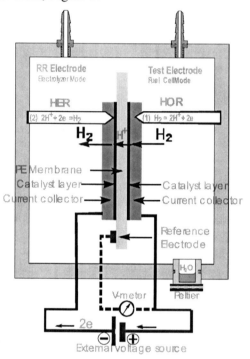

Figure 9. The H_2 test cell unit.

Using an external source we can pass a current across the MEA. With polarity as shown in the figure a HO Reaction (reaction 1) will proceed on the right hand side electrode with hydrogen consumption and proton and electron generation. On the left hand side electrode a HE Reaction will take place with conversion of the protons to hydrogen atoms and consumption of electrons supplied by the current connector. In this way the species consumed or produced on the right hand side electrode will be fully restored. The container can be closed and the reaction can proceed theoretically indefinitely under the set conditions, completely autonomously.[12]

The right hand electrode can be designed and used as a test electrode structure. Its performance characteristics can be measured in the traditional way using a reference electrode. The left hand side electrode serves as a counter electrode, only, with the function to restore the consumed or generated species on the test electrode. This *"EasyTest Cell"* is extremely simple for operation. The MEA is brought in contact with the connectors freely not needing any sealing. The initially set reagent conditions are maintained constant during stay or operation. Water content or humidity can be controlled by the presence of the liquid phase kept by a Peltier element at a freely selected temperature lower than the test cell temperature, as already described.

The ORR and OER reactions can be investigated in a similar way using an oxygen–water environment.

In an adequate construction the system is extremely easy to operate. The temperature can be increased to levels of the stability of the PEM, 150–180°C. The overall pressure is easily maintained up to 10 or 20 bars. The gas composition including water partial pressure (see the water pool and the Peltier element temperature regulator) can be set initially to any level and strictly maintained during operation. Partial pressure values as low as 0.001 bar can be maintained precisely.

A practical construction and experiments concerning some kinetic and mechanistic aspects of the reactions in H_2-O_2-H_2O-cells will be given in the second lecture.

References

1. *Mini-Micro Fuel Cells – fundamentals and applications*, edited by S. Kakac, L. Vasiliev and A. Pramuanjaroenkij (Springer Verlag, NATO science series, 2008).
2. S. Kim, Hydrogen Supply System for a Fuel Cell, *Pub. No.: US 2004/0115493 A1*, Pub. Date: Jun. 17 (2004).
3. R. Gomez, Proton Membrane Fuel Cells, *Pub. No.: US 2004/0096718 A1*, Pub. Date: May 20 (2004).
4. G. Wang and W. Wang, Miniature Fuel Cell system Having Integrated Fuel Processor and Electronic Devices, *Pub. No.: US 2003/0170515 A1*, Pub. Date: Sept. 11 (2003).
5. K. Taft, M. Kurano, and A. Kannan, Composite Polymer Electrolytes for Proton Exchange Membrane Fuel Cells, *Pub. No.: US 2004/0048129 A1*, Pub. Date: Mar. 11 (2004).
6. A. Pozio, M. De Francesco, A. Cemmi, F. Cardellini, and L. Giorgi, Comparison of High Surface Pt/C Catalysts by Cyclic Voltammetry, *Journal of Power Sources*, 105, 13–19 (2005).
7. B. Munoz, G. Kepner, M. Preveti, and X. Xi, Gas Diffusion Electrodes, *Pub. No.: US 2004/0086774 A1*, Pub. Date: May 6 (2004).
8. S. Smedley, Electrolyte-Particulate Fuel Cell Anode, *Pub. No.: US 2004/0053097 A1*, Pub. Date: Mar. 18 (2004).

9. S. Hirano, J. Kim, and S. Srinivasan, High Performance Proton Exchange Membrane Fuel Cells with Sputter-Deposited Pt Layer Electrodes, *Electrochimica Acta*, 42, 1587–1593 (1997).
10. E. Slavcheva, I. Radev, S. Bliznakov, G. Topalov, P. Andreev, and E. Budevski, Sputtered Iridium Oxide Films for Water Splitting via PEM Electrolysis, *Electrochimica Acta*, 52, 12, 3889–3894 (2007).
11. E. Slavcheva, I. Radev, S. Bliznakov, G. Topalov, P. Andreev, and E. Budevski, Sputtered Electrocatalysts for PEM Electrochemical Energy Converters, *Electrochimica Acta*, corr. proof (2007).
12. I. Radev, E. Slavcheva, and E. Budevski, Investigation of Nanostructured Platinum Based Membrane Electrode Assemblies in "EasyTest" Cell, *International Journal of Hydrogen Energy*, 32, 872–877 (2007).
13. I. Radev, E. Slavcheva, S. Bliznakov, and E. Budevski, Nanocomposite Electrocatalysts and Test Cell for Hydrogen Generation in PEM Electrolyser, in: *Nanoscience&-Nanotechnoloigy: Nanostructured Materials, Application and Innovation Transfer*, edited by E. Balabanova and I. Dragieva (Heron Press Science Series 6, Sofia, 2006) pp. 206–209.
14. E. Budevski, Structural Aspects of Fuel Cell Electrodes, *Journal of Optoelectronics and Advanced Materials*, 5, 1319–1325 (2003).

HEAT PIPES IN FUEL CELL TECHNOLOGY

L. VASILIEV* AND L. VASILIEV JR.
Luikov Heat and Mass Transfer Institute, National Academy of Sciences of Belarus, P. Brovka 15, 220072, Minsk, Belarus

1. Introduction

The big market now is available in portable electronics such as mobile phones, PDAs and laptop computers. Several companies such as Samsung, Toshiba and Panasonic are developing portable direct methanol fuel cells, as one of the best approaches to small fuel cell systems and a considerable amount of research is being conducted in this area.

The goal of this work is to design and develop a new fuel cell cooling system based on heat pipe concept [1–5]. New and original heat pipe approach is done based on the Luikov Institute heat pipe activity. The suggested cooling system based on heat pipe phenomena could be considered as a new approach to solve the above mentioned problem.

Heat pipes for fuel cell thermal management ought to have high effective thermal conductivity and be insensitive to the gravity forces. The vacant porous media for micro/mini heat pipes is a metal sintered powder wick or a silicon/carbon porous wafer with biporous (micro/macro pores) composition, saturated with working fluid. Heat pipes fuel cell management can be performed in different ways:

1. Micro/mini heat pipes for fuel cells thermal management (<10 W)
2. Heat pipes for medium fuel cells (10–100 W)
3. Heat pipes for portable fuel cells (>100 W)
4. Heat pipes systems for stationary fuel cells (stationary electricity generation)

---

*Leonard Vasiliev. Porous media laboratory, Luikov Heat and Mass Transfer Institute, National Academy of Sciences of Belarus, P. Brovka 15, 220072, Minsk, Belarus, e-mail: Leonard_Vasiliev@rambler.ru

2. Different Designs of Heat Pipes for FC Thermal Management

2.1. MICRO/MINI HEAT PIPES FOR FUEL CELL THERMAL MANAGEMENT (<10 W)

Micro/mini heat pipes have the cylindrical, flattened or flat shape and can be embedded bipolar plate for fuel cell stacks.

Micro/mini heat pipes for PEM fuel cells operate at relatively low temperatures, around 80–100°C, Figure 1. Micro (MHP) and miniature heat pipes (mHP) are small scale devices that are used to cool micro/mini fuel cells. Microchannels in MHP are fluid flow channels with small hydraulic diameters. The hydraulic diameter of MHPs is on the order of 10–500 μm, the hydraulic diameter of mHPs is on the order 2–4 mm. Smaller channels application is desirable because of two reasons: (i) higher heat transfer coefficient, and (ii) higher heat transfer surface area per unit flow volume.

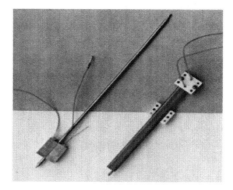

Figure 1. Cylindrical and flattened mini heat pipe for fuel cells thermal management developed in the Luikov Institute.

The most efficient thermal management of micro/mini fuel cells can be performed, if the two phase thermal control typical for the animal body or plant leaves is realized. So called micro heat pipe open type thermal control is feasible, if the heat loaded structures are covered with the mini porous layer, saturated with liquid [6–8]. The gas diffusion layer and gas channels of PEMFC ought to be done from the porous structure including the macro and mini pores. In heat pipes basic phenomena and equations are related with liquid-vapour interface, heat transport between the outside and the interface ("radial" heat transfer), vapour flow and liquid flow. There is a strong interaction between basic phenomena in heat pipes. Feedbacks may cause instabilities, waves, flooding, and performance jump. Basic equations are related to vapour flow in the MHP channel, liquid flow in the capillary structure, interface position between the vapour and liquid (mechanical equilibrium yields interface curvature K), radial heat transfer, vapour flow limit, capillary limit. Optimisation of the new copper sintered powder wick in miniature heat pipes with outer diameter 3–4 mm and length up to 200 mm was carried out in the Luikov Institute, Minsk since 1997. The maximum heat transfer rate for these HPs is almost 50 W [5]. Software was developed and used for prediction of round and flat heat pipes (including mHP) characteristics [8]. Heat pipe family qualified geometry is: circular tube diameter 3–25 mm, flat heat pipe thickness 2–20 mm, length 0.1–0.8 m, wall thickness 0.2–1.0 mm. Pipe material – copper 99.95% purity, wick – copper sintered powder with thickness 0.2–0.8 mm. Transport capacity 10–500 W. Water, methanol and propane are used as working fluids.

2.2. LOOP HEAT PIPE FOR MEDIUM FUEL CELLS (10–100 W)

Loop heat pipes (LHP), Figure 2 are an attractive alternative for heat regulation. The performance of the evaporator depends on the transport properties of the wick, i.e. permeability, thermal conductivity as well as structural characteristics of the wick: homogeneous or heterogeneous porous system, narrow or wide size distribution of the pores.

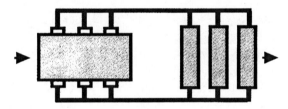

Figure 2. Loop heat pipe with some evaporators embedded in the FC stack and condensers.

A new type of miniature loop heat pipe (LHP) was investigated by the NASA Glenn Research Center, USA [9]. The principle application is electronic cooling at the chip level, but it is also very promising for the PEMFC stack cooling and thermal control. The heat pipe evaporator is constructed of silicon, such that there will be little thermal interface resistance between the source of the heat generation, the computer chip junction, and the working fluid. The device utilizes a Coherent Porous Silicon (CPS) wick that provides small effective pore radii, Figure 3.

This new technology is a type of Micro Electro Mechanical Systems (MEMS) process that allows one to "drill" a pattern of micron-sized holes in a silicon wafer. In fuel cells and heat pipes, the flow characteristics in the porous media (gas diffusion layers or capillary wick) are useful in modeling performance.

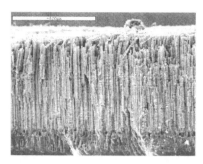

Figure 3. Silicon made porous structure for the LHP evaporator [9].

Permeability is a parameter that describes the relationship between pressure drop and mass transport through porous media.

In heat pipes effective pore radius is a parameter used to describe the available pressure rise for liquid pumping [5, 9]. In the LHP there is a possibility to use an evaporator above the condenser, the vapour flows through the vapour channels towards the condenser and the liquid goes back the evaporator due to the capillary pressure head of the porous wick.

2.3. LOOP THERMOSYPHONS FOR PORTABLE FUEL CELLS (>100 W)

Since the loop thermosyphon has larger critical heat flux than conventional thermosyphon it is convenient to use it in many different applications, for example, for Fuel Cells thermal control. The loop thermosyphon evaporator need to have a good thermal contact with the stack, the condenser ought to be cooled by air, or water. The loop thermosyphon transports thermal energy from a heat source to a sink by natural convective circulation without any external power supply such as a pump. Thermosyphon evaporator and a condenser are installed separately, but connected to each other by small diameter bendable pipes. Due to the coupling between momentum and energy transport theoretical analysis of the loop performance is very complicate, therefore it is necessary that these problems be solved by experimental investigation before applying the loop thermosyphon to heat exchanger design.

In the loop thermosyphon the heat transfer is considered to be affected by many factors, such as type and quantity of working fluid, pipe diameter, pipe length, and ratio of cooled surface to heated surface, the length of the adiabatic zone between heated and cooled sections, heat flux and operating temperature.

The evaporator and condenser of the loop thermosyphon can be made of carbon-steel, cupper, or aluminum. Propane, R 134a, R 600, ammonia or water can be used as working fluid. When the copper is allowed to be applied water is the best working fluid. In order to establish heat transfer correlations for the application in the design program for the loop thermosyphon heat exchanger, regression analysis could be applied to experimental data for heat transfer coefficients in evaporator and condenser.

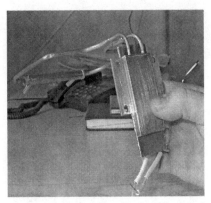

Figure 4. Copper/water loop thermosyphon for the FC stack thermal management.

Typical loop thermosyphon with the flat evaporator is shown on Figure 4. This thermosyphon is capable to transport Q_{max} near 100 W at the temperature of the adiabatic zone 100°C. The condenser is cooled by water circulation. The thermal resistance of thermosyphon R is 0.03 K/W.

2.4. LOOP HEAT PIPE WITH NONINVERTED MENISCUS OF THE EVAPORATION

Loop heat pipes are more flexible to compare with loop thermosyphons, due to its insincerity to the gravity field. Typical LHP for PEMFC with optimal heat flow rate 800 W at the working temperature near 80°C is shown on Figure 5.

LHP with the non inverted meniscus of the evaporation designed and tested in the Luikov Institute is made from copper and has the wick performed from copper sintered powder. The working fluid is water. The typical maximum heat flow rate of such evaporator is near 1,500 W, the thermal resistance of the evaporator R, = 0.06 K/W. The length of the evaporator is 70 mm, width –60 mm and thickness 12 mm. The wick porosity is >45% and the effective thermal conductivity of the wick 40 W/m K.

Another alternative to the conventional heat pipe is an aluminium (multi-channel) heat pipe panel (Figure 6) with propane as a working fluid to cool FC stack [10]. Pulsating heat pipe (PHP) is one of several oscillatory thermal transport cycles under development that are receiving attention as a potential semi-passive, high-power, high flux heat transport device.

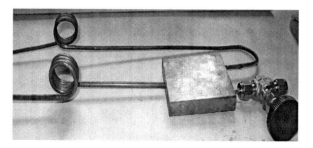

Figure 5. LHP flat evaporator made from copper with sintered powder porous structure and mini grooves for liquid suction.

2.5. PULSATING HEAT PIPE PANELS

The PHP is unique in that it is capable of generating driving pressures in excess of many mechanically pumped loops. Capillary forces do not limit the PHP and it is capable of transferring high heat loads over long distances and against significant resistance (i.e. gravity, small tube diameters, etc.).

The main parameters of flat heat pipe panels developed in the Luikov Institute, Minsk are: HP width –70 mm, HP height –7 mm, HP length –700 mm, evaporator length –98 mm, condenser length –500 mm, mass –0, 43 kg. HP thermal resistance R = 0.05 K/W, evaporator heat transfer coefficient α = 8,500 W/m² K, condenser heat transfer coefficient α = 2,500 W/m² K [11]. The working fluid (hydrocarbons) dynamic movement is stable with liquid filling ratio near 0.6 of the heat pipe volume.

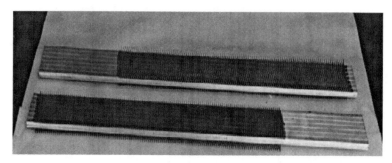

Figure 6. Aluminum pulsating heat pipe panel with mini channels inside and mini fins on the outer surface.

2.6. SORPTION HEAT PIPES

The sorption heat pipe (SHP) combines the enhanced heat and mass transfer in conventional heat pipes with sorption phenomena of a sorbent bed. Sorption heat pipe could be used as a sorption heat transfer element and be cooled and heated as a heat pipe [12, 13]. The original design of such a sorption heat pipe was patented in USSR in 1992 [12].

The sorption heat pipe (Figure 7) has a sorbent bed (adsorber/desorber and evaporator) at one end and a condenser and evaporator at the other end. Sorption heat pipe have a sorbent bed (adsorber/desorber and evaporator) at one end and a condenser + evaporator at the other end. Traditional two-phase thermal control system for FCs is sensitive to the vehicle acceleration and vibration. Sorption heat pipe thermal control is efficient for such cases.

The solid sorption cooler begins to function (to be switched on), when cooling possibilities of the conventional heat pipe are exhausted. Sorption heat pipe cooler is better than conventional loop heat pipe coolers in the cases, that:

- Require operation in large accelerations
- Require higher pumping capability, require more intense heat transfer in the evaporator
- Require operation of sorption heat pipe evaporator colder than the environmental temperature.

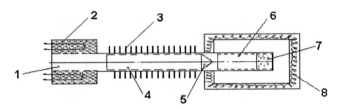

Figure 7. Sorption heat pipe 1 – vapor channel; 2 – sorption structure; 3 – finned surface of heat pipe evaporator/condenser; 4 – porous wick; 5 – porous valve; 6 – low temperature evaporator with porous wick; 7 – working fluid; 8 – cold box with thermal insulation.

The heat output of SHP developed in the Luikov Institute is 1,000 W (water), the thermal resistance of the evaporator – 0.03 K/W, pressure drop $\Delta P_{cap} = 200$ mbar.

3. Conclusions

Heat pipe concept as a thermal control system for Mini/Micro Fuel Cell is a powerful tool to increase FC efficiency.

Mini/Micro heat pipes are considered as advanced thermal control for Mini Fuel Cells with power generation 10–100 W, Loop heat pipes, pulsating heat pipes and sorption heat pipes are suggested as an advanced thermal control system for portable Mini Fuel Cells with power generation > 100 W.

References

1. James H. Krallik, *US 6,355,368 B1 Mar.12, 2002*, "Cooling Method and Apparatus for Use a Fuel Cell Stack"
2. Oh, Se Min and Vasiliev Leonard, *US 20030141045 A1, July 31, 2003*, "Miniature Heat Pipe and method of manufacturing the same"
3. David B. Saraff and Joel T. Schwendemann, *US 6,817,097 B2, Nov. 16, 2004*, "Flat Fuel Cell Cooler"
4. Amir Faghri, Mansfilled, *US Patent 2005/0026015 A1, Feb. 3, 2005* CT (US), "Micro heat pipe embedded bipolar plate for fuel cell stacks"
5. L.L. Vasiliev, A. Antukh, V. Maziuk, A. Kulakov, M. Rabetsky, and L.Vasiliev Jr., Oh Se MiN, Miniature Heat Pipes Experimental Analysis and Software Development. in: *Proceedings of the 12th International Heat Pipe Conference "Heat Pipes Science, Technology, Application"*, edited by Yu. Maidanik (Russian Academy of Sciences, Moscow, Russia, 2002), pp. 329–335.
6. P. Dunn and D.A. Reay, *Heat Pipes* (Pergamon Press, Great Britain, 1976), pp. 21–25.
7. V.G. Reutsky and L.L. Vasiliev, *Papers of the Academy of Science, Belarus*, 24, No. 11, 1033–1036 (1981).
8. V. Maziuk, A. Kulakov, M. Rabetsky, L. Vasiliev, and M. Vukovic, in: Miniature heat-pipe thermal performance prediction tool – software development, *Apllied. Thermal Engineering*, 21, 559–571 (2001).
9. M.M. Weislogel, E. Golliher, J. McQuillen, M.A. Bacich, J.J. Davidson, and M.A. Sala, Recent results of the micro scale pulse thermal loop in: *International two-phase Thermal Control Technology Workshop*, Newton White Mansion, Mitchellville (2002).
10. L. Vasiliev, A. Zhuravlyov, A. Shapovalov, and V. Litvinenko, Vaporization Heat Transfer in Porous Wicks of Evaporators, *Archives of Thermodynamics*, 25, No. 3, 47–59 (2004).
11. L.L. Vasiliev, Sorption machines with a heat pipe thermal control, *Proceedings international sorption heat pump conference*, September 24–27 (2002), Shanghai, China, pp. 408–413.
12. L.L. Vasiliev and V.M. Bogdanov (1992), *USSR patent 174411* "Heat pipe", B.I. No. 24, 30.06.1992.
13. L. Vasiliev and L. Vasiliev Jr. Sorption heat pipe – a new thermal control device for space and ground application, *International Journal of Heat and Mass Transfer*, 48, 2464–247 (2005).

HEAT TRANSFER ENHANCEMENT IN CONFINED SPACES OF MINI-MICRO FUEL CELLS

L. VASILIEV* AND L. VASILIEV JR.
Luikov Heat and Mass Transfer Institute, National Academy of Sciences of Belarus, P. Brovka 15, 220072, Minsk, Belarus

1. Introduction

The general goal of this paper is to develop a new option of mini/micro fuel cells cooling technology. Usually PEM fuel cells operate at relatively low temperatures, around 80–100°C, but need to be functional at the low temperature (–30°C). Therefore some hydrocarbon fluids can be welcomed for FC thermal management. The special porous coating of fuel cell stack (FCS) cooling system (mini channels) ensures the heat transfer enhancement, decreases and equalizes the temperature field on all FCS volume. Current PEMFC cooling system research may be divided into the following areas: thermosyphon coolers and two-phase circuits with a mechanical pump. Here we consider a new miniature passive DMFC (based on mini/micro heat pipe concepts) cooling system that is including in a FCS design. It is important due to 3/4 of the fuel cell thickness now belongs to the cooling system.

2. Different Designs of FC Thermal Management

2.1. TWO PHASE EVAPORATIVE COOLING OF FC BASED ON THERMOSYPHON AND MECHANICAL PUMP FLUID CIRCULATION

Recently some experiments were carried out in the Luikov Institute (Minsk), Belarus [1, 2] aimed for investigation of mini/micro scale heat transfer in long mini channels (100–300 mm) with its thickness 1.5–2 mm and heat flux ranges 10–300 W/cm^2 (Figure 1). To visualize the two-phase heat transfer the heat loaded tube with porous coating was structure. A

---

*Leonard Vasiliev. Porous media laboratory, Luikov Heat and Mass Transfer Institute, National Academy of Sciences of Belarus, P. Brovka 15, 220072, Minsk, Belarus, e-mail: Leonard_Vasiliev@rambler.ru

micro scale heat transfer effect was observed inside the porous body and a miniscale effect was typical in mini channels of the cooling device. These mini channels, or mini thermosyphon evaporators are cylindrical of flat, has a small diameter, or thickness less than capillary constant of the working fluid. Visual analysis and experimental results show, that such combination is favorable for the enhancement of the evaporation and two-phase convection heat transfer. For the convenience of the experiments propane was used as a working fluid. The availability of annular mini channel significantly promotes to intense heat transfer (up to 2.5–3 times) at heat fluxes <50 kW/m^2, as compared with the same heat transfer of the FCS element disposed in the liquid pool. The suggested cooling system based on heat pipe phenomena could be considered as a new approach to solve the above mentioned problem (Figure 2).

Since metal sintered powder wicks applied in such evaporators are generally near 50% porous, there is a large surface area available for liquid films evaporation. Typical copper sintered powder wicks handle 50–70 W/cm^2, and have been tested up to 250 W/cm^2. Besides copper sintered powder wicks, Al$_2$O$_3$ nano particles were used as heat transfer intensifier with boiling. The typical schema of such mini channel or thermosyphon

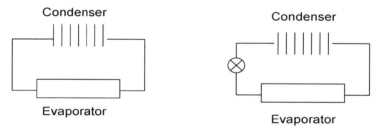

Figure 1. Different modes of mini thermosyphon (left) or two-phase forced convection FCS cooling system application with mechanical pump (right).

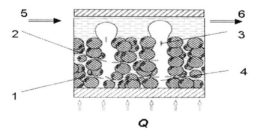

Figure 2. Element of the FCS mini channel with porous coating: 1 – micro-pore, 2 – meniscus of evaporation, 3 – macro-pore with vapor bubble, 4 – micro (nano) particle, 5, 6 – liquid flow input and output.

evaporator (Figure 2) shows us the presence of the liquid flow along the microstructured (porous) wall of the evaporator. The flow limitation is related to momentum conservation in the fluid flow along the mini channel, or thermosyphon loop. The static pressure of the liquid column in the liquid channel of thermosyphon is considered as the driving force at the evaporator entrance. For two-phase forced convection cooling system the driving force is ensured by mechanical mini-pump. The major pressure drop in the loop is related with the viscous pressure drop in the vapor and liquid channels. The maximum capillary pressure of the porous coating should exceed the pressure drop in micropores and macropores of capillary structure and viscous forces in the vapor channel under the nominal heat load.

The liquid flow is entering from condenser to the evaporator and move in mini channel along the tube. The bubbles increase the flow rate proportionally the heat flux. The vapor flow is available on the evaporator exit. Macro pores in the FCS are considered as vapor flow generators with small hydraulic diameters of 10–500 μm, the hydraulic diameter of mini channel is on the order 1–3 mm. Mini channels application is desirable because of two reasons: (i) higher heat transfer coefficient, and (ii) higher heat transfer surface area available per unit flow volume. In operation of the evaporator of the FCS cooling system the key components – vapor bubbles and liquid column static pressure work in concert and ensure a proper amount of liquid/vapor flow in the mini channel. New data were obtained for a flooded and partially flooded heat loaded tube at different heights of meniscus Z between liquid and vapor zones. Visual analysis (Figure 3) and experimental data (Figure 4) confirm the intensification of the evaporative heat transfer up to 8–10 times to compare with propane pool boiling on the smooth horizontal tube. Such design of the evaporator enhances heat

Figure 3. Fluid flow (propane) visualization in annular mini channel (thickness $\delta g = 0.3$ mm), and capillary-porous coating (thickness δ) over copper tube ($\delta = 0.3$ mm; $\delta_g = 3$ mm; $q = 34$ kW/m^2).

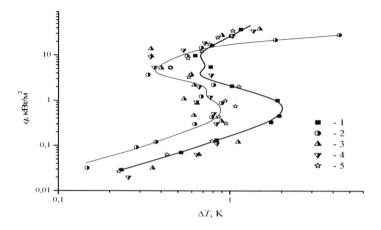

Figure 4. Heat transfer in annular mini channel as function of ΔT (porous coating thickness $\delta = 0.3$ mm) for different channel thickness δ_g (1 – liquid pool; 2 – $\delta_g = 0.5$ mm; 3 – $\delta_g = 1$ mm; 4 – $\delta_g = 2$ mm; 5 – $\delta_g = 3$ mm).

transfer up to 2.5–3 times (~40 kW/m² K) to compare with the same tube placed in the liquid pool due to two-phase forced convection heat transfer heat transfer inside the mini channel.

The uniform vapor streams are constantly ejected from macropores of the wick and act as a miniature pumps allowing the liquid to move along the channel. The part of the liquid is constantly sucked back into the porous coating by capillary forces of the wick. Visualization of the micro heat pipe effect inside the porous structure + two-phase forced convection in the annular mini-channel stimulated by vapor bubble generation testifies the availability of thermodynamically efficient mechanism of the mini/micro fuel cell cooling system.

3. Fuel Cartridge (FC) Design Based on H_2 Adsorption and Chemical Reactions for FCs Thermal Management

The activated carbon fiber "Busofit" and activated wood-based carbon particles were suggested in the Luikov Institute as perspective materials for gas storage systems in fuel cartridges [3, 4]. "Busofit" is a universal adsorbent, which is efficient to adsorb different gases (H_2, N_2, O_2, CH_4, and NH_3). Figure 5 shows the texture of "Busofit", which can be performed as a loose fibers bed or felt or as monolithic blocks with binder to have a good thermal conductivity along the filament.

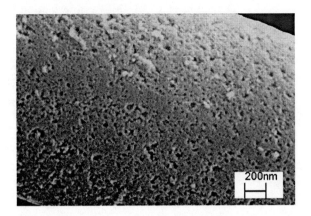

Figure 5. Photo of the active carbon filament "Busofit-M8", with a set of micropores on its surface, multiplied by 50,000.

"Busofit" has such advantages as high rate of adsorption and desorption; uniform surface pore distribution (0.6–1.6 nm); small number of macropores (100–200 nm) with its specific surface area 0.5 m²/g; small-number of mesopores with its specific surface area 50 m²/g. Hydrogen sorption capacity is near 253 ml/g (2.23 wt%) at 77 K and 0.1 MPa. The total volume V, associated with an active carbon adsorbent may be split up into its components:

$$V = V_c + V_v + V_{void} + V_\mu \qquad (1)$$

where V_c – the volume of the carbon atoms on which the adsorbent is composed; V_μ – micropores volume; V_v – meso- and macropores volume; V_{void} – the space inside the vessel free from adsorbent bed. This latter V_{void} can be eliminated by making the solid block of adsorbent. For carbon materials a linear relationship between BET surface area and volume sorption capacity of hydrogen at 77 K and 0.1 MPa was found as:

$$a_v = 0.0783\, S_{BET} + 84.02 \qquad (2)$$

Influence of micropore volume on sorption capacity of hydrogen for various carbon materials under the chosen conditions (77 K, 0.1 MPa) is described by the following linear correlation:

$$a_v = 119.12\, V_{DR} + 115.41 \qquad (3)$$

Complex compound sorbent bed was performed as a set of microcrystals of metal hydrides attached to the filament surface to increase the total sorption capacity of the sorbent bed. Due to the high density of metal hydrides the systems is very compact. Complex compound applications are best suited for systems, where power and volume are more critical than weight. Heat and mass transfer in the fuel cartridge (FC) of FCS ought to be high, which implies good porosity and high thermal conductivity of the sorbent bed. For this reason the metal hydride reaction beds inside the storage vessel are needed to be temperature controlled by heat pipe heat exchangers (Figure 7) to increase the total fuel cell efficiency and reliability including the FCS and FC.

During absorption/adsorption the FC operates as absorber in which hydrogen and metal hydride reacts with an exothermic reaction:

$$S_{(sol)} + n \cdot G_{(gas)} \rightarrow S \cdot G_{n(sol)}; \ \Delta H_{react} < 0 \qquad (4)$$

The enthalpy of reaction $\Delta H_{react.}$ of the metal hydride in Eq. (4) is about 50 kJ/NH$_3$ mol. and is a reason for sorbent bed heating. The low thermal conductivity of the metal hydride and the very high expansion factor S during its reaction with the gas G are two important coefficients being able to reduce the sorption capacity.

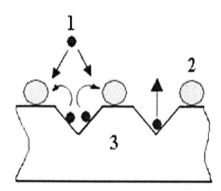

Figure 6. Activated carbon fibre (3) with metal hydride microcrystals on its surface (2) and hydrogen molecules (1) adsorbed.

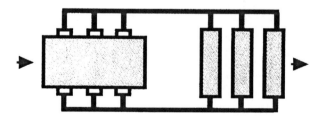

Figure 7. Loop heat pipe with some evaporators embedded in the FCS and in FC.

The use of additives (active carbon fiber) to metal hydride has two functions: the increase of the total sorption capacity (adsorption + absorption) and the maintenance of high porosity of the medium during the solid-gas reaction. The fast adsorption of hydrogen was checked for "Busofit" 5 minutes. The slow rate of absorption was determined for metal hydride $La_{0.5}Ni_5Ce_{0.5}$ – 35 minutes. The mean time of ad/absorption was noticed for the combination (40% "Busofit" + 50% $La_{0.5}Ni_5Ce_{0.5}$ + 10% binders) –15 minutes.

Heat flow has to be evacuated outside the vessel during the ad/absorption of complex compound, and gas diffusion has not to be slackened to reach the solid. The evacuated heat is vacant source of energy to preheat the FCS and to cool FC sorbent bed. Conversely, in desorption phase, the FC operates as regenerator in an endothermic reaction and can be used for the FCS cooling. It means, that permeability and kinetic of ad/absorption of "Busofit" and complex compound is higher to compare with $La_{0.5}Ni_5Ce_{0.5}$. The maximum temperature rise of the sorbent bed was determined for pure metal hydride $La_{0.5}Ni_5Ce_{0.5}$ – 370 K. It means the isosteric heat of hydrogen sorption for "Busofit" or combination of "Busofit" and $La_{0.5}Ni_5Ce_{0.5}$ is lower to compare with metal hydride. So, we obtained fast sorbent bed compound (40% "Busofit"+ 50% $La_{0.5}Ni_5Ce_{0.5}$+ 10% binder), which sorption capacity of of hydrogen is near the same as pure metal hydride, but the cycle of adsorption/desorption is three times less. The heat pipe based system of FCS thermal control (Figure 7) ensures the thermal stability and high thermal efficiency of the whole system (FC + FC).

4. Conclusions

Micro heat pipe effect inside the porous structure + two-phase forced convection in the annular mini channel is considered thermodynamically as an efficient evaporator and effective cooling device for micro/mini fuel cells.

Metal sintered powder structure with micro heat pipe phenomena stimulates the evaporative heat transfer in two-phase cooling system for fuel cells near 8–10 times to compare with pool boiling heat transfer inside thermosyphon smooth horizontal evaporator.

A new sorbent bed (complex compound "Busofit"+ metal hydride microcrystals on its surface) for hydrogen fuel cartridge was suggested. Experimental verification of the efficiency of the complex compound was validated.

Application of the heat pipe based fuel sell thermal management promotes to reduce the time of adsorption/desorption of the gas and increase the efficiency of FC + FCS. Some sorption capacity data for new sorbent materials were obtained.

References

1. L. Vasiliev, A. Zhuravlyov, A. Shapovalov, and V. Litvinenko, Vaporization heat transfer in porous wicks of evaporators, *Archives of Thermodynamics*, 25(3), 47–59 (2004).
2. L.L. Vasiliev, D.A. Mishkinis, A.A. Antukh, A.G. Kulakov, and L.L. Vasiliev Jr., in: *Proceedings International sorption heat pump conference*, edited by R. Wang, Z. Lu, W. Wang, and X. Huang (Science Press, Science Press New York Ltd. 2002), 408–413.
3. L.L. Vasiliev, L.E. Kanonvhik, and A.A. Antukh, Activated carbon fiber composites for ammonia, methane and hydrogen adsorption, *International Journal of Low Carbon Technologies*, 2/1, 95–111 (2006).
4. L.L. Vasiliev, L.E. Kanonchik, A.G. Kulakov, and V.A. Babenko, Hydrogen storage system based on novel carbon materials and heat pipe heat exchanger, *International Journal of Thermal Sciences*, 46, 914–925 (2007).

PERFORMANCE CHARACTERISTICS OF MEMBRANE ELECTRODE ASSEMBLIES USING THE *EASYTEST CELL*

EVGENI BUDEVSKI*, IVAN RADEV, AND EVELINA SLAVCHEVA
Institute of Electrochemistry and Energy Systems, Bulgarian Academy of Sciences, Sofia 1113, Bulgaria

1. Introduction

The autonomous test electrode assembly, the *EasyTest Cell*, will be described in a detailed constructional version. The testing opportunities offered by the system will be discussed. Some instructive examples will be given including: performance characteristics of the gas diffusion electrode and disclosure of transport limitations. Pt load optimization studies will be demonstrated.

Testing procedures, including performance characteristics and service life of electrochemical energy converting (EcEC) cells based on gas reactions can be very arduous and burdensome. The main reason for the problem is the need of constant reagent supply under strongly controlled conditions and composition, impeding greatly the experiment.

With reversible gas reactions, e.g. the hydrogen-oxygen-water electrochemical system, the consumed reagents in the cell under study, the test cell (TC), can be recovered quantitatively in a second, reagent recovery cell (RRC) connected in series with the first. If the reagent gas compartments of both cells are connected in a proper way the products of the RRC can be supplied to the reagent gas compartments of the TC, so that the system can be isolated from the environment and be run completely autonomously in a "close circuit material circulation", as already discussed in the first lecture.[1]

It is a general practice in electrochemistry to test and optimize single electrodes before coupling them to a working cell. A second electrode, known as "counter electrode", is used to pass a current through the test electrode (TE) and simulate its working conditions under real or close to real conditions. If the electrodes of a membrane electrode assembly operate

---

*Evgeni Budevski, Institute of Electrochemistry and Energy Systems, Bulgarian Academy of Sciences, 10, Acad.G. Bonchevstr., Bl.10, Sofia 1113, Bulgaria, e-mail: budevski@bas.bg

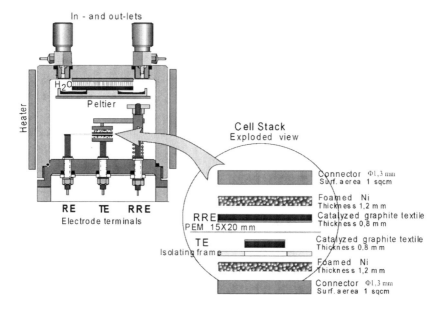

Figure 1. The *EasyTest Cell*. TE – test electrode; RRE – reagent recovery electrode; RE – reference electrode. On the right hand side an exploded view of the cell stack is given with dimensional details.

in the same gas environment, the reaction on the counter electrode will proceed in the opposite direction of that of the TE and with reversible reactions the current will restore the reagents involved in TE reaction. Under these conditions in a closed system the gas environment of the cell will remain constant during operation. Schematically this situation has been shown and discussed in the preceding lecture.[1] In Figure 1 a construction of an autonomous single electrode test system is represented.

Before starting the discussions, however, some definitions should be introduced:

An *electrochemical cell* can be used to convert chemical to electrical energy and vice versa. Working in *isothermal conditions* the electrochemical energy conversion avoids the Carnot cycle efficiency limitations. This, *electrochemical energy converter* (EcEC), can be designed for power generation, e.g. as *Fuel Cell* (FC) with hydrogen as fuel, or for energy storage, e.g. as *Electrolyzer* (ELZ), with hydrogen as storing reagent. The

EcEC can be designed also as *bi-functional electrochemical energy converters* (Bi-EcEC), which depending on the direction of the current can be operated in a FC mode or in ELZ mode.

The *polymer electrolyte membrane (PEM) cell* is an electrochemical cell based on a *solid, proton-conducting electrolyte* connecting the two electrodes of the cell. The electrodes are hot pressed directly both sides on the membrane forming the *membrane electrode assembly – MEA*. The MEA is the heart of the EcEC and can be seen as an *electrochemical processor* (ECP). ECP can be regarded as a functional description of the MEA which is the physical description of the device.

The structure and composition of a MEA, or an ECP, respectively, depend very little on their application in an EcEC. What really changes is generally the MEA size ranging form 1,000 cm^2 to mini dimensions of less than 1 cm^2.

The electrodes of a MEA have a *layered, porous, electro-conductive composite structure*. They are known as *gas diffusion electrodes* – GDE. Both sides, next to the PEM are the *reaction layers* containing a catalyst. To save the often expensive metal catalysts (Pt or Pt containing metals) the catalyst is distributed over the surface of a porous electro-conductive catalyst carrier in form of nano-particles. On top of the reaction layers are the *gas diffusion layers* serving as *current collectors* and *gas reagent transporting media*.

The essential characteristic of the electrode performance is the *current density – voltage curve*. In the case of single electrodes the voltage is given either as *electrode potential* or as *overvoltage*. The electrode potential is the *potential difference of the TE versus a reference electrode* (usually a hydrogen electrode). The overvoltage is the *deviation of the electrode potential under load versus the rest potential (i.e. at zero current) of the TE*.

2. The Autonomous Test Electrode Assembly (The *EasyTest Cell*)

Figure 1 shows a realization of the autonomous single electrode test system. The MEA, containing the test electrode (TE) is placed horizontally between the upper and lower terminals. RRE is a counter electrode serving as a reagent recovery electrode. The TE can be reduced to the desired minimized dimensions. The gas reagents are supplied laterally through the foamed metals serving at the same time as current collectors. The membrane is extended in one direction (see Figure 2) to make a contact with a reference electrode (RE). The starting gas reagents are supplied by the in- and out-let valves kept closed during operation.[2]

Temperature is controlled directly on the test electrode connector (Figure 1) using thermo resistor. A Peltier element is used to control the temperature of the water condenser and the water partial (vapor) pressure. On the right hand side the cell stack assembly is represented in an exploded view with dimensional details. Details can be read directly on the figure.

Figure 2. Laboratory prepared MEA's. From left: (a) solution casted 20 μm Nafion membrane; (b) PBI based membrane with pasted electrodes; (c) electrodes pasted on Nafion 117, and (d) a cut from commercial (Ion Power) MEA.[3]

Figure 2 shows MEA's prepared for use in the *EasyTest Cell*. In the figure tested electrodes based on different technologies are presented. The preparation technology and size of the electrodes used in this presentation are described in section 3.1 (Tables 1 and 2).

In Figure 3 the inside compartment of the *EasyTest Cell* is presented.

The *EasyTest Cell* offers the following investigation advantages allowing:

Figure 3. The inside section of the *EasyTest Cell*.

- To *avoid the use of reactants in pressurized* form or in larger difficult for management quantities
- To easily *minimize* the test electrodes down to mini dimensions and reduce the sometimes expensive materials to a minimum without affecting the practical significance of the results
- To run in parallel larger quantities of cells, for longer periods of time without surveillance, particularly for service life and degradation estimations
- To *set the cell working conditions* – temperature, total gas pressure, gas composition, reactant partial pressures including water, under a *close control*
- To *widen the working conditions* in non standard ranges for evaluation and assessment purposes.

3. Applications and Results

3.1. EXPERIMENTAL PARAMETERS AND CELL CONDITIONS

The experimental parameters and cell conditions in this section are given in Tables 1 and 2.

The *EasyTest Cell* mode of operation is given as an alphanumeric string, e.g.:

TABLE 1. Gas diffusion electrodes (description and data). *A* are laboratory made GDE, using dc magnetron sputtering. *E* is commercial ELAT product. Catalyst and loadings are given in mg/cm^2 omitting the point separator in the code name.

Origin	Technology	Structure	Code name	Catalyst and Loading (mg/cm^2)	TE Size cm^2	RRE Size cm^2
IEES-BAS	Dc magnetron sputtering (Laboratory product)	Toray paper, sublayer Ti (50 nm ca. 20 µg/cm^2), catalyst as indicated	A Pt040	Pt 0.40		
			A Pt030	Pt 0.30		
			A Pt012	Pt 0.12		
			A Pt006	Pt 0.06		
			A Pt/Ir	IrO$_x$ 0.08 Pt 0.06 IrO$_x$ 0.08	0.6	1.0
			A Ir040	Ir 0.40		
			A Ir030	Ir 0.30		
			A Ir020	Ir 0.20		
ELAT	Commercial ETEK product		E Pt050	Pt 0.5		

\<T80RH100TP125H0012Ar\> has the meaning:

- **T80** – cell temperature 80°C (default value. Can be omitted)
- **RH100** – liquid water is present in the cell at the cell temperature keeping a relative humidity of 100% and water partial pressure of 0.48 bar (default value. Can be omitted)
- **TP125** – total gas pressure 1.25 bar
- **H0012** – hydrogen partial pressure 0.012 bar
- **Ar** – argon has been added to balance the total gas pressure to 1.25 bar (omitted if not present).

TABLE 2. Cell conditions and gas composition code names. **T** is for temperature; **RH** for relative humidity; **TP** gives the total pressure in bar; **H**, **O** and **Ar** give the H_2, or O_2 partial pressure in bar, and Ar, if present, as balance to **TP**. Separator points are omitted in the code name. Ranges are given in parentheses. Items, if omitted, are in default values.

Cell temperature	Pressure, or partial pressure, respectively (bar)							
Value name	Total pressure and code name		Water		Hydrogen		Oxygen	Argon (balance to TP)
Default	Value	Name	Value	Name	Bar	Name	Bar	Name	Name
				Default		Name omitted if absent			
	6	TP600			5.52	H5520	5.52	O5520	
	5	TP500			4.52	H4520	4.52	O4520	
80°C	4	TP400	0.48	RH100	3.52	H3520	3.52	O3520	
T80	3	TP300	100		2.52	H2520	2.52	O2520	
	2	TP200	%		1.52	H1520	1.52	O1520	
	1.25	TP125			0.77–0.006	H(0770–0006)	0.77–0.006	O(0770–0006)	Ar

3.2. THE HYDROGEN ELECTRODE

The hydrogen electrode is the negative electrode of the EcEC based on the hydrogen-oxygen-water electrochemical system. The electrochemical reaction $H_2 = 2H^+ + 2e$ is reversible in direction. The electrode can thus operate as a *hydrogen oxidation reaction (HOR) electrode*, i.e. in a fuel cell (FC) mode, or as a *hydrogen evolution reaction (HER) electrode* in the opposite direction, i.e. in an electrolyzer (ELZ) mode.

The performance characteristics as evaluated from the current density – overvoltage relation, or, for short, the polarization curve, can be studied on a MEA enclosed in an *EasyTest Cell* as shown in Figure 1.

The membrane electrode assembly consists of three GDE: one serving as a test electrode (TE) – size 0.6 cm^2; the second is a counter electrode serving as reagent recovery electrode (RRE) – size 1 cm^2; and the third one is a reference electrode (RE) of smaller size cut from a commercial GDE (**E** Pt050).

The IEES-BAS laboratory made GDE, were prepared by dc magnetron sputtering of the catalyst (Pt, Ir, or a layered mixture of both) on a commercial (ELAT ETEK) graphite paper (Toray paper) over an adhesive layer of Ti (usually 50 nm).[2, 4, 5]

The reagent recovery electrodes (RRE) were prepared using commercial GDE (ELAT, ETEK).

The GDE were slightly surface wetted with diluted Nafion solution and hot pressed on a Nafion 117 membrane to form the MEA.[2, 4–6]

Polarization curves of the hydrogen electrode in an *EasyTest Cell* at a total cell gas pressure of 2.0 bar for different test electrodes are shown in Figure 4.

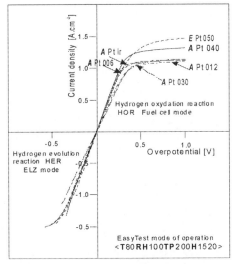

Figure 4. HOR-HER polarization curves of a hydrogen electrode in an EasyTest cell. TE and gas condition codes are given as defined in Table 1 and 2. RRE is **E** Pt050.

A first look on the figure confirms the reversibility of the hydrogen oxidation-evolution reaction. All polarization curves are symmetrical around the reversible hydrogen electrode potential and go smoothly in the reverse direction from the oxidation to the evolution region up to A/cm^2.

The slopes of the curves around the zero current potential are largely independent on type of electrode and catalyst loading.

The polarization curves in the HOR region show a tendency of saturation. In general, this seems to be is missing in the HER region.

Experimentally two important performance criteria can be evaluated from these curves.

1. The first one is the value of the initial current density/overpotential slope, $(dj/d\eta)_{j=0}$. Initial current density/overpotential slopes are summarized in Table 3 from *EasyTest Cell* experiments with six electrode types at 80°C and varying cell gas conditions.

TABLE 3. Initial slopes $(dj/d\eta_j)_{=0}$ and saturation current density coefficients (ventilation coefficient) $j_{sat}/p(H_2)$ of hydrogen electrode polarization curves in the *EasyTest Cell*. Cell gas conditions are <T80RH100TP(600–125)H(0006–5520)Ar> (see Table 2).

Test electrodes GDE in the HOR mode (fuel cell mode)				Initial slope	Ventilation coefficient
Origin	Catalytic layer	Name	Catalyst and load (mg cm^{-2})	$(dj/d\eta)_{j=0}$ (S cm^{-2})	$j_{sat}/p(H_2)$ (A cm^{-2} bar^{-1})
ELAT (ETEK)	30 % Pt on Vulcan XC	*E 050*	Pt (0.5)	4.8–3.6	9.3
IEES (BAS)	Dc magnetron sputtered catalyst on Toray paper with Ti (50 nm) adhesive sublayer	*A 040*	Pt (0.4)	4.2–3.3	2.4
		A 030	Pt (0.3)	3.5–2.6	2.4
		A 012	Pt (0.12)	3.6–2.7	1.6
		A 006	Pt (0.06)	3.9–3.0	1.4
		A Pt Ir	Ir(0.08)Pt(0.06) Ir (0.08)	4.2–3.8	1.9

The value of the slopes, of about 3–5 S/cm^2, is largely independent on the catalyst loading and the type *A* or *E* of the electrode. Unfortunately it is obviously overshadowed by the conductivity of the membrane to be used as a measure of the catalytic activity as discussed in lecture 1.[1] For an access to the exchange current density value j_o, impedance measurements would be needed to extract the *jR* potential drop from the initial slope value.

Nevertheless the test results show that there is a very high reserve in the Pt loading, if the structuring of the catalyst layer is made in a proper way.

2. The hydrogen oxidation reaction (HOR) in Figure 4 shows a tendency of saturation suggesting, transport limitations.

To identify this saturation values as diffusion limited currents, however, they should be proportional to the bulk concentration of the diffusing reagent, or, in this case, to the H$_2$ partial pressure.

Figure 5 shows the current density/overvoltage relation for different hydrogen partial pressures, $p(H_2)$, of a MEA in hydrogen atmosphere for the HO Reaction, as measured in an *EasyTest Cell*. The MEA has been prepared using Nafion 117 and GDE – *A* Pt Ir for the TE, and *E* 050 for the RRE. The overall pressure was varied between 6 and 1.25 bar. The hydrogen pp has been varied in the range of 0.006–5.52 bar. In the range below 1.25 bar Argon has been added to balance the overall pressure to a level of 1.25 bar (see Table 1).

Two regions can be distinguished in the course of the polarization curves for the different $p(H_2)$ values:

1. The higher pressure region (5.52–0.77 bar), where the polarization curves and the saturation values are largely independent on $p(H_2)$.
2. The lower pressure region (0.40–0.006 bar), where a proportionality between the current density saturation values and the hydrogen partial pressure can be estimated.

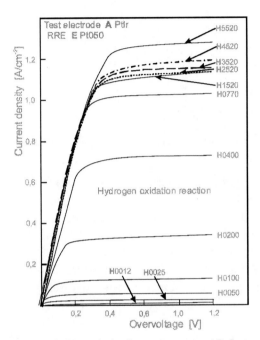

Figure 5. HOR polarization curves on A Pt Ir test eledtrodes at hydrogen pp's varying between 0.012 and 5.52 bar. RRE: E Pt050. Gas composition: <T80RH100TP(600–125)H(0.012–5.52)Ar> (see Tables 1 and 2).

Figure 6 presents this last region in a saturation current density $j_{sat} - p(H_2)$ relation. It clearly shows a well expressed proportionality between j_{sat} and $p(H_2)$ which allows assuming that the saturation currents are diffusion limited. Without any additional assumptions the constant relating the j_{sat} values to the partial pressure can be taken as a criterion for the diffusion, or reagent transportation, qualities of the GDE. This last coefficient can be understood also as ventilation coefficient for the GDE. The thus obtained value for the A Pt Ir GDE is 1.9 A cm^{-2} bar^{-1} at the given hydrogen partial pressure conditions.

The dominant diffusion limitation of the saturation current densities in the lower $p(H_2)$ regions is indisputable. The nature of the saturation of j in the higher pressure regions is, however, still unclear. At this stage we can only speculate that at these high currents a local increase of the electrode temperature can provoke a drying out of the membrane with a decrease of the membrane conductivity. In fact an increase of the electrode temperature of about 5°C has been detected on the T sensor of the TE terminal. This would undoubtedly result in a conductivity decrease and a current limitation as observed.

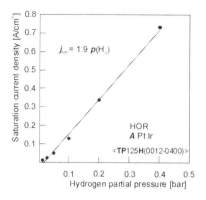

Figure 6. Saturation current density – hydrogen partial pressure dependence evaluated from Figure 5 for lower $p(H_2)$ values.

In the last column of Table 3 ventilation coefficients of six GD electrode versions included in MEA's are presented. The RRE was a standard GDE –

ELAT *E* Pt050. The gas composition was varied between 6 to 1.25 bar total pressure and 5.52–0.006 bar $p(H_2)$. The ventilation coefficient values are given in their range of variation.

The results again show an insignificant Pt load dependence. No matter what the reason for this independence is assumed it shows that the Pt load can be significantly reduced if current densities lower than 1.0 A/cm^2 (see Figure 4) are aimed.

3.3. THE OXYGEN ELECTRODE – THE OXYGEN REDUCTION REACTION

The oxygen electrode imposes the highest restrictions on the electrochemical energy conversion efficiency. On one hand the only reliable catalyst for the oxygen reduction reaction is still Pt. On the other Pt is not only expensive. Even in the fuel cell mode of operation its catalytic activity is largely insufficient giving rise to overpotentials of the order of hundreds of mV at minimal loads of mA/cm^2. Even more, Pt showed to have a low activity for the reverse oxygen evolution reaction (OER) of interest for the electrolyzer (ELZ) mode of operation. In this region Ir has recently been found to render a catalytic activity comparable to that of Pt in the OR reaction.[4-6]

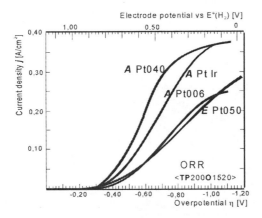

Figure 7. Polarization curves on 4 selected electrodes as indicated. Gas conditions: total pressure 2 bar, oxygen pp 1.52 bar: <**TP200O1520**>.

Nevertheless, the thermodynamic irreversibility of the ORR and OER reactions is responsible for the high inefficiency of all electrochemical energy converters based on oxygen electrodes. The real problem of the oxygen electrode is actually not to find a replacement of Pt or similar expensive catalyst materials but to find an even better catalyst. There is something that nature could tell us because oxygen is the main oxidation agent in the biological (human) fuel cell system.

Figure 7 shows polarization curves on four selected electrodes (*A* Pt040, *A* Pt006, *A* Pt Ir, and *E* Pt050) at cell gas conditions defined by <**T80RH100TP200O1520**> (see Tables 1 and 2). The counter, reagent recovery electrode (RRE), was a dc magnetron sputtered IrO_x 0.40 mg/cm^2 electrode (*A* Ir040). As catalyst carrier Toray paper with a Ti sub layer was as usually applied.

As seen, the current starts only after over potentials of 300 mV are exceeded and begins to increase exponentially with growing overpotentials. This behavior of the oxygen electrode, due to a very slow charge transfer reaction rate, is typical for the ORR and OER. This overpotential is connected with the high energy barrier of the charge (electron) transfer reaction. The transfer needs an activation energy. Therefore this potential shift is known also as *activation overpotential*.

In another aspect, the observed pronounced influence of the electrode type: *A* or *E* and catalyst structure is remarkable because it shows again a high reserve in Pt loading if a structure similar to that obtained by the dc

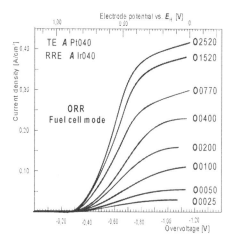

Figure 8. Polarization curves of a *A* Pt040 electrode at varying oxygen pp's as indicated, codes (see Tables 1 and 2).

magnetron sputtering technology is used. The *A* Pt Ir sandwich electrode with 0.06 mg/cm^2 Pt and 0.16 mg/cm^2 IrO$_x$ only is almost equivalent to the *A* Pt040 electrode with 0.4 mg/cm^2 Pt and much better than the *E* Pt050.

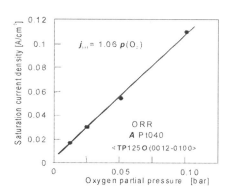

Figure 9. Saturation current density as function of oxygen pp as evaluated from Figure 8. TE: *A* Pt040. O$_2$ pp varying from 0.012 to 0.01 bar.

Figure 8 shows polarization curves of an *A* Pt040 electrode in the ORR mode at <T80RH100-TP(600-125)-O(2520-0012)Ar> cell conditions. The RRE was *A* Ir040, i.e. a dc magnetron sputtered IrO$_x$ 0.40 mg/cm^2 electrode. The oxygen pp level was varied between 0.012 and 2.52 bar (level 0.012 bar not shown). At partial pressure levels below 0.77 bar Ar was added to keep the balance of the TP to 1.25 bar.

At higher pp values the oxygen partial pressure does not affect significantly the course of the polarization curves. Below 0.77 bar, however, again a linear $j_{sat} - p(H_2)$ relation can be estimated.

Figure 9 shows the saturation current density on an *A* Pt040 electrode as function of oxygen pp. The ventilation coefficient for the *A* Pt040 electrode is 1.1 A cm^{-2} bar^{-1}. For the *A* Pt006 and the *A* Pt Ir electrodes 0.18 and 0.45

A cm^{-2}.bar^{-1}, were found, respectively. The ventilation coefficient of the GDE's for the ORR is several times lower than that of the HOR (see Table 3). This is easily understood, having in mind the more complicated diffusion conditions in the ORR with two diffusing species, O_2 and H_2O, and the lower diffusion coefficients of these species than that of H_2. The ventilation coefficients for the *A* Pt006 and the *A* PtIr electrodes are remarkably low.

Compared to the large activation overpotential the contribution of the reagent transportation limitations, however, is hardly the major problem, of the ORR, particularly at normal gas pressures, and lower current densities.

Before starting the discussions connected with the oxygen evolution reaction and more significant practical applications a change of the coordinate presentation should be discussed and introduced. Up to now we have used polarization curve presentations as current density *j* versus electrode overpotential η.

The overpotential η is the deviation of the electrode potential from its *thermodynamic reversible value* and is a measure for the reaction *driving force*. It is always the main parameter in theoretical relations giving the *rate of the electrochemical reaction, the current density,* as function of the driving force, the *overpotential*.

For practical uses, however, where we would be more interested in *cell voltages U* [V], a more suggestive presentation is the electrode potential *E* versus a reference electrode usually the hydrogen electrode – $E°(H)$ [V] as function of current density *j*. The upper scale of the polarization curve presentation in Figures 7 and 8 is given as $E°(H)$ value. As seen from the figures the scale is shifted by 1.18 V to the right – the shift potential being the theoretical value of the oxygen $E°(H)$ potential at 80°C.

In addition cell performance is given usually as the cell voltage *U* [V] as function of the cell load *j* [A/cm^2]. Therefore in the discussions to follow we will use polarization curves given as <u>*cell voltage U* (or *electrode potentials*</u> vs. H_2 reference electrode) [V] as function of the current density [A/cm^2].

3.4. THE OXYGEN ELECTRODE – THE OXYGEN EVOLUTION REACTION

The Oxygen Evolution Reaction (OER) is the electrolyzer mode of operation of the oxygen electrode in the H_2–O_2 electrochemical energy converter (EcEC).

The OER mode of operation has two major problems:

1. Pt shows a low activity for the OE reaction and can not be used as catalyst effectively.

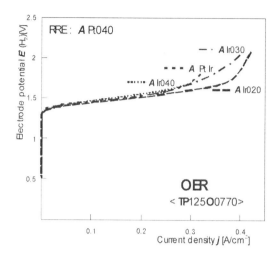

Figure 10. Oxygen evolution reaction (*EasyTest Cell*). TE: *A* Ir(0.2, 0.3, 0.4 mg/cm^2) and *A* Pt Ir. RRE: *A* Pt040. Gas composition <TP125O0770>. Curves for *A* Pt040 and *A* Pt Ir are partially overlapping. Note the change of coordinates!

2. Carbon is oxidized to carbon dioxide in contact with the nascent oxygen generated on the catalyst. Hence carbon can not be used as a catalyst carrier.

For the time being the best and apparently only acceptable catalyst for the OE reaction seems to be Ir, or IrO$_x$, respectively. IrO$_x$ is probably the state of the Ir surface at the oxidizing conditions of the OER.

Figure 10 shows the polarization curves of four GDE at OER mode of operation. The electrodes were prepared in the IEES using dc magnetron sputtered Ir, or Pt sandwiched between two Ir layers, respectively, on an adhesive sublayer of Ti (50 nm) on Toray paper (PEMEAS). The loads were 0.4 (*A* Ir040), 0.3 (*A* Ir030), and 0.2 (*A* Ir020) mg/cm^2 Ir, or 0.08, 0.06, and 0.08 mg/cm^2 Ir, Pt, and Ir for the *A* PtIr, in the sandwiched structure. The GDE codes are given in parenthesis. The sputtering was kindly done in IWE 1, RWTH Aachen University, Germany. For the RRE an *A* Pt040 GDE was used. The cell gas conditions were <T80RH100TP125-O0770> (see Table 2).

The polarization curves again show a large gap for the current start of more than 0.4 V with respect to the reversible oxygen electrode potential $E^o(H_2)$ of 1.18 V (at 80°C). The best performance shows the **A Ir020**

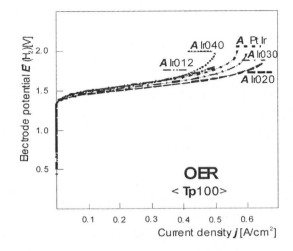

Figure 11. Oxygen evolution reaction (Electrolyzer). TE: *A* Ir(0.012, 0.2, 0.3, 0.4 mg/cm^2) and *A* Pt Ir. RRE: *A* Pt040. Gas composition <TP100>. *A* Pt040, *A* Pt012, and *A* PtIr are partially overlapping. Note the change of coordinates!

electrode with a reference value of 0.35 A/cm^2 at $E°(H_2)$ of 1.7 V. The *A* PtIr electrode, which will be used as a bi functional electrode in the Bi-EcEC development of sections 3.5 and 3.6, with a total loading of 60 microgram of Pt and of 160 µg of Ir per cm^2, only, is almost as good as the *A* Ir020 version for the OER. It has the same start but suffers obviously of worse gas reagent ventilation at higher current densities.

Figure 11 shows the same four electrodes operated in an electrolyzer cell with an *E* Pt050 electrode as the HER electrode. The polarization curves are given as $E°(H_2) - j$ dependence. The performance characteristics are close, or slightly better, than those of the same electrodes tested in an *EasyTest Cell*, Figure 10. The improvement could be due to the more precise water conditioning in the *EasyTest* case, and a not controllable possible water condensation on the oxygen electrode in the ELZ case.

It is interesting to note that the Ti sublayer, originally intended for improvement of the catalyst adhesion to the carbon paper substrate unexpectedly solved the problem with the carbon substrate oxidation. Experiments with mass spectrometry connected to an operating ELZ cell could not detect CO_2 contamination in the oxygen gas compartment of the ELZ. The reason for this protection is open for speculations, but it is easy to assume that before reaching the carbon material substrate the active oxygen radicals have sufficient time for deactivation.

3.5. THE BI – FUNCTIONAL ELECTRODES

The problem for a realization of a bi-functional electrochemical energy converter is to find catalysts or catalysts mixtures that are efficient both for the hydrogen oxidation and for the reverse, hydrogen evolution reaction. The same applies to the oxygen reduction-evolution reaction: the catalyst must be effective for the forward as well as for the backward reaction.

With Pt the hydrogen electrode does not impose any restrictions for a change of the reaction direction from a HOR to a HER (see Figure 4). The polarization curves go smoothly with the same slope in the opposite direction reaching high current densities.

The problems are connected again with the oxygen reduction-evolution reaction. Pt is an acceptable catalyst for ORR but largely insufficient for the opposite evolution reaction. Ir or IrO_x has shown a relatively good activity for the OER but is completely inactive for the opposite reaction.

It can easily be assumed that a mechanical mixture of both catalysts could produce a bi-functional catalytic layer. This seems to have been confirmed by the experiments discussed above. The layered structure Ir(0.08)Pt(0.06)Ir(0.08) (loadings are given in parenthesis in mg/cm^2) was found to be more than acceptable as a *first approach* to the solution of the problem:

The Pt-Ir electrode with 0.06 mg cm^{-2} Pt only shows a performance almost equivalent to that of the pure *A* Pt040 electrode with roughly seven times more Pt for the ORR (Figure 7). At the same time the Pt-Ir electrode with a total of 0.16 mg cm^{-2} of Ir is giving apparently a good compromise solution for the OER with 0.5 A/cm^2 at an electrode potential of $E°(H_2) = 1.75$ V, compared to the best electrode *A* Ir020 found in this series with 0.6 A/cm^2 at the same potential (see Figure 11).

The second problem is connected with the oxygen evolution reaction, OER. In general oxygen is not very active in its reaction with carbon materials. In the process of the water oxidation to oxygen, however, the oxidation process goes over the formation of very active oxidizing intermediates. These intermediates are short living and are deactivated in a very short time. They could be therefore completely ineffective for the oxidative process if the site of their formation is sufficiently distant from the carbon surface. According to the present experiments the adhesive sub layers Ti of about 50 nm showed to be sufficiently thick for this deactivation to occur. No degradation of the OER electrodes was observed, nor was CO_2 contamination found in the exhaust gases of the electrochemical reaction cell.

3.6. THE ELECTRO CHEMICAL ENERGY CONVERETER

The EcEC will be discussed in its three versions – the fuel cell, the electrolyzer and the bi-functional energy converter. Expected performance characteristics will be given as emulated polarization curves of the devices with reference to the presented electrode features.

Figure 12 shows the emulated curve for a fuel cell using the characteristics of the best electrodes presented in this series: A Pt006 for the hydrogen HOR electrode and A Pt040 for the oxygen ORR electrode. Data are taken from Figures 4 and 7 for gas conditions $<T80RH100TP200H1520O1520>$. The polarization curve of the fuel cell is calculated as the difference U between the ORR and HOR electrode potentials at the given current density.

Figure 13 shows the emulated polarization curve of a projected electrolyzer using the characteristics of the most suitable electrodes presented in this series: A Pt006 for the hydrogen HER electrode and A Ir020 for the oxygen OER electrode. Data are taken for the OER electrode from Figure 11 at gas conditions $<T80RH100TP100>$. Oxygen is evolved at atmospheric pressure.

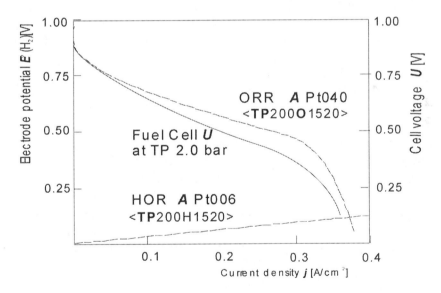

Figure 12. Emulated polarization curve of a fuel cell at 2 bar total pressure using electrode performance characteristics from *EasyTest Cell* measurements (Figures 4 and 7).

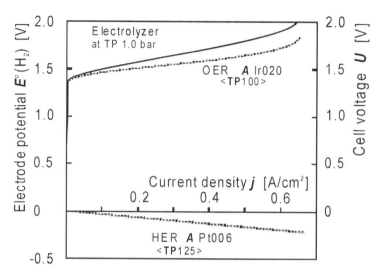

Figure 13. Emulated polarization curve of an electrolyzer at 1 bar total pressure using electrode performance characteristics from single electrode measurements (Figures 4 and 11).

Figure 14 shows emulated polarization curves for a bifunctional EcEC in both modes of operation. The Bi-EcEC consists of an *A* Pt Ir as oxygen and an *A* Pt006 as hydrogen electrode. The operating temperature is assumed as 80°C, and the total pressure is considered as 3 bar. The higher total pressure of 3 bar was selected because the Pt Ir electrode suffers from low ventilation and inferior performance at lower TP for the oxygen reduction, i.e. fuel cell mode of operation. The OER operation of the *A* Pt Ir electrode is taken from experiments with a real ELZ at 80°C and a TP of 1 bar assuming that the pressure difference is of minor influence on the electrode performance for the gas evolution reaction.

The upper, drawn curve is the emulated cell voltage of the device in an ELZ mode of operation. The curve is obtained as the difference between the *A* PtIr OER and the *A* Pt006 HER electrode potentials (negative for the HE reaction!). The lower drawn curve is the cell voltage of the Bi-EcEC in the FC mode of operation. The curve is obtained as the difference between the *A* PtIr ORR and the **A** Pt006 HOR electrode potentials.

The performance of the device is acceptable in the ELZ mode but obviously insufficient in the FC mode of operation. At 0.5 V cell voltage it gives some 150 mA cm^{-2}. The situation at 1 bar total pressure is even worse the Bi-EcEC giving less then 100 mA cm^{-2}, only, at the cell voltage of 0.5 V for the FC mode of operation.

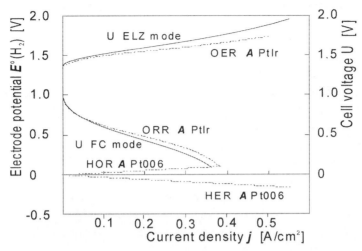

Figure 14. Emulated polarization curves (full lines) of bifunctional EcEC at 3 bar total pressure using electrode performance characteristics from single *A* PtIr and *A* Pt006 electrode measurements (dotted lines).

The main reason for the obtained low current densities lies in the very low ventilation of the *A* PtIr electrode, with a ventilation coefficient of 0.45 A cm^{-2} bar^{-1}. A better structuring of the *A* PtIr electrode can be recommended from this results.

4. Conclusions and Outlook

The presented results are very recent and have been selected to demonstrate the wide possibilities of an easy to handle experimental set-up concerning the development and optimization of electrochemical energy converters (fuel cells, electrolyzers or bifunctional units). The additional advantages of the new set-up are the very precise setting of the gas environmental conditions in a wide range of operation and their preservation during the experiment.

Some general conclusions of the set of experiments reported here can be made:

To save the often expensive catalyst material the catalyst should be dimensioned as a nano-partical layer located between the PEM solid electrolyte and the reagent transportation and current collecting layer. This makes the use of catalyst carriers fully expendable.

Carbon (graphite) fiber materials in form of woven or non woven fabrics can be used as gas transporting media if protected by a sub layer of Ti. A thickness of about 50 μm Ti has been found to protect the carbon material from oxidation even in the extremely aggressive environment of the oxygen evolution reaction.

The thickness of the catalyst layer can vary between 100 and 1,000 nm (60–400 μg cm^{-2}) if a good catalysts like Pt for the hydrogen and the oxygen ORR and IrOx for the oxygen OER are used.

A very essential finding is the fact that the transport limitations are located strongly around the catalytic layer.

Sputtering techniques can be used to produce the protective and catalytic layers with the offered high precision quantity and structure control and the wide range of modification possibilities.

The described reagent recovery principal allows not only the selection of the operating conditions widely outside the range of conventional investtigations giving a possibility to get an insight into kinetic details of the electrode reactions but offers also a very precise technique for standardized, particularly long term testing procedures.

References

1. E. Budevski, I. Radev, and E. Slavcheva, Autonomous Test Units for Mini Membrane Electrode Assemblies, in: *Micro-Mini Fuel Cells-Fundamentals and Applications*, edited by S. Kakac, L. Vasiliev, and A. Pramuanjaroenkij (Springer Verlag, 2008).
2. I. Radev, E. Slavcheva, and E. Budevski, Investigation of Nanostructured Platinum Based Membrane Electrode Assemblies in "EasyTest" Cell, *International Journal of Hydrogen Energy*, 32, 872 –877 (2007).
3. E. Budevski, I. Radev, and E. Slavcheva, The *EasyTest Cell* – an Enhanced MEA Investigation and Optimization Technique, in: *Proceedings of IHEC, Istanbul, Turkey, 13–15 July* (2005).
4. E. Slavcheva, I. Radev, S. Bliznakov, G. Topalov, P. Andreev, and E. Budevski, Sputtered Iridium Oxide Films for Water Splitting via PEM Electrolysis, *Electrochimica Acta*, 52, 12, 3889 – 3894 (2007).
5. E. Slavcheva, I. Radev, S. Bliznakov, G. Topalov, P. Andreev, and E. Budevski, Sputtered Electrocatalysts for PEM Electrochemical Energy Converters, *Electrochimica Acta*, corr. proof (2007).
6. I. Radev, E. Slavcheva, S. Bliznakov, and E. Budevski, Nanocomposite Electrocatalysts and Test Cell for Hydrogen Generation in PEM Electrolyser, in *Nanoscience & Nanotechnology: Nanostructured Materials, Application and Innovation Transfer*, edited by E. Balabanova and I. Dragieva (Heron Press Science Series 6, Sofia, 2006) 206– 209.

WATER TRANSPORT DYNAMICS IN FUEL CELL MICRO-CHANNELS

G. MINOR[1], X. ZHU[2], P. OSHKAI[3], P.C. SUI[3], AND N. DJILALI[3]*
[1]*Ballard Power Systems, Burnaby, BC, Canada, V5J 5J8*
[2]*Institute of Engineering Thermophysics, Chongqing University, Chongqing, 400030, China*
[3]*Department of Mechanical Engineering & Institute for Integrated Energy Systems, University of Victoria, Victoria, BC, V8W 3P6, Canada*

1. Introduction

The formation, phase change and transport of water have a major impact on the operation, performance and durability of PEM fuel cells, specifically in terms of start up, including freeze-start,[1] transient response and degradation.[2] Net water balance is primarily determined by the water production rate at the cathode, and transport across the membrane via diffusion and electro-osmotic drag. At higher currents, excessive water condensation can lead to 'flooding' of the cathode porous layers. Eventually, discrete water droplets can emerge through the pores of the gas diffusion layer (GDL) into the air distribution micro-channels (see Figure 1). Such droplets may grow and coalesce, partially or completely blocking the transport pathways for the reactant (particularly oxygen), and also causing a substantial increase in pressure drop. The resulting starvation of the reaction sites induces dips in current density[3,4] as well as flow maldistribution.

In order to promote water transport and mitigate flooding and the associated blockage of reactant transport, performance losses and degradation mechanisms, GDL's are commonly treated with PTFE (Teflon) to impart hydrophobicity. The mixed wetting properties of the surfaces combined with the small hydraulic diameters (~500 μm), and low Reynolds numbers (~10^3 or less), make the two-phase flows that prevail in PEM fuel cells quite different from the well documented two-phase flow processes found in more classical engineering applications. The micro-scale flows in PEM fuel cells are for the most part laminar and are characterized by much

---

*Ned Djilali, Dept. of Mechanical Engineering & Institute for Integrated Energy Systems, University of Victoria, Victoria, BC, V8W 3P6, Canada e-mail: ndjilali@uvic.ca

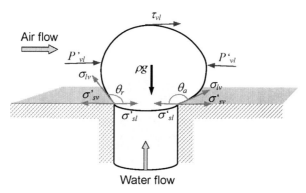

Figure 1. Schematic of droplet emerging from a pore on the surface of a gas diffusion layer, and forces acting on the droplet: subscripts v, l, and s represent gas, liquid, and solid, respectively, the superscript ' and " represent upstream and downstream sides of the water droplet; downstream and upstream contact angles θ_a and θ_r are the advancing and receding contact angles; the forces acting on the include shear stress (τ), surface tension (σ), pressure (P), and gravity (ρg).

larger ratios of surface to volume forces. In particular surface tension and viscous forces are predominant. Key question from the view point of design are understanding of the dynamics of liquid water droplets; of the detachment process, and of the ensuing entrainment and interaction with the channel walls and other droplet; and the characterization of the pressure drop.

The importance of two-phase flow in PEMFCs is underscored by the major thrust in the last few years in applying a variety of experimental techniques to investigations two-phase flow in the gas channels or gas diffusion layers of PEMFCs. This includes *in-situ* visualization in optically accessible fuel cells,[3,4,6-8] neutron scattering imaging,[9-12] IR-thermography,[13] NMR,[14] MRI,[15] and very recently employed synchrotron X-ray radiography.[16] *Ex-situ* observations have also been performed in model flow channels that isolate key flow features using fluorescence microscopy[17,18] and digital particle image velocimetry.[19]

In custom made cells with an optically accessible window, flow phenomena including droplet formation, flow regimes, and scenarios of droplet-droplet and droplet-wall interactions can be readily observed and some flow features can be quantified. Transparent windows, however, usually have different bulk properties (thermal and electrical) and surface properties (roughness and wettability) from those in actual fuel cells, and thus dictate a certain amount of caution in interpreting the observations. Techniques such as neutron or magnetic resonance imaging circumvent the need for a 'transparent' window but remain to date limited in their practicality, and in particular still lack temporal and spatial resolution.

The dynamics of a blob of water emerging from a pore at the bottom surface of a micro-channel (i.e. the GDL in a fuel cell) and subjected to a cross air stream is affected by factors including air velocity, water injection rate, wettability of the walls, surface tension of water, and channel dimensions. The forces acting on the droplet are depicted in Figure 1 and include shear stress (τ), pressure (P), and gravity (ρg), and the surface tension (σ) which contributes to pinning of the droplet.

On a flat surface, the contact angle (θ) defines geometrically the angle formed by a liquid at the boundary where liquid, gas and solid intersect as illustrated in Figure 2. A zero contact angle represents the completely hydrophilic case with total wetting and formation of liquid film on the surface, and conversely $\theta = 180°$ represents a completely hydrophobic (non wetting) case. Surface tension and wettability play a particularly important role, and in fuel cell micro-channels this is further complicated by the fact that three of the channel walls are hydrophilic, while the fourth (which corresponds to the GDL) is hydrophobic. The shape and motion of a droplet are determined by the combined effects of these forces, and some of the pertinent dimensionless numbers that represent the ratio of surface tension to the other forces are the capillary number ($Ca = \mu u/\sigma$), Weber number ($We = \rho u^2 d/\sigma$), and Bond number ($Bo = \rho g d^2/\sigma$). Because the processes are dynamic, with size, shape and location of any given water droplet changing with time, the characteristic dimensionless numbers as well as the flow regimes are also functions of time.

Kumbur et al.[20] presented a useful force balance model for a droplet in a fuel cell channel based on an idealized representation of the flow. The proposed approach assumes the droplet is rigid and a zero air velocity at the droplet-air interface (i.e. no-slip) and allows deduction of a droplet stability

Figure 2. Contact angle for hydrophilic (wetting) and hydrophobic (non-wetting) fluid/surface interface for fluid droplet on a flat surface (top) and fluid plug inside a channel (bottom).

map on the basis that when the drag force on the droplet due to the air flow exceeds the surface tension force, the droplet will shed down the channel with the air flow. A similar analysis was also performed independently by Chen et al.[21] This force balance model includes much of the pertinent physics and follows the experimental trends correctly, but because it neglects the deformation of the droplet and the three-dimensional interaction between the liquid and air phases, the model tends to underpredict droplet stability at low air flow rates, and overpredict them at higher air flow rates.

The difficult experimental environment of PEM fuel cells and the inherent limitations in current experimental techniques have prompted computational modelling work as well, but the focus has been largely on two phase flow in the porous components of the cell, and in particular on water saturation and capillary transport in the GDL.[22–25] These explicitly or implicitly assume that water appears as a homogeneous phase within the representative volume. This approach cannot be extended to the analysis of liquid water transport in a gas microchannel where large liquid interfaces are expected as a result of strong interactions between surface tension and the channel walls. Recently, CFD simulations based on the Volume of Fluid (VOF) method to track the gas-liquid have been used.[26–28] The thorough investigation of Theodorakakos et al.[28] combines VOF simulations with experimental validation. By implementing into the simulations measured critical values of the advancing and receding contact angles, Theodorakakos et al. were able to predict the critical droplet diameter corresponding to the shedding events. VOF simulations to date have typically considered a single droplet or an array of droplets to be initially either suspended in the channel, attached to the channel wall, or in the form of a water film. Zhu et al.[5, 29] have recently considered the emergence of a liquid droplet from a GDL into the gas channel, and found that both the emergence process as well as the detailed accounting of the pore geometry significantly impact detachment and subsequent dynamic evolution of the droplets.

In this Chapter, we discuss flow features and findings obtained from experimental and numerical investigations of water droplets in model flow channels reproducing scenarios relevant to PEM fuel cells.

2. Flow Field Inside a Water Droplet

In this section, visualization and velocity field measurements using microscopic digital particle image velocimetry (micro-DPIV) are presented and discussed to highlight the flow features *inside* a liquid droplet placed on a GDL in a micro-channel, and subjected to an air stream.

2.1. EXPERIMENTAL SET-UP AND VISUALIZATION

The experimental setup for droplet visualization is briefly described; details are provided by Minor.[19] The microchannel test sections were fabriccated using glass and polydimethylsiloxane (PDMS), a flexible, transparent polymer used in a variety of applications including microfluidic device fabrication by soft lithography. In order to simulate the fuel cell environment, one of the wall of the gas flow channel was bonded with a 300 μm thick, hydrophobically treated carbon paper layer having a 300 μm (E-TEK 40% wt. PTFE treated Toray carbon paper). The fabrication technique and materials of this test section allowed rapid and repeatable replacement of sections contaminated by the fluorescent particles required for visualization. The test section was placed in the stage of an inverted microscope, permitting a side view of the droplet, as shown in Figure 3. The experimental procedure consisted of introducing a 0.35 μl deionized water droplet on the carbon paper surface using a micro-pipette. The droplet was seeded with fluorescent tracer particles (0.0002% by volume).

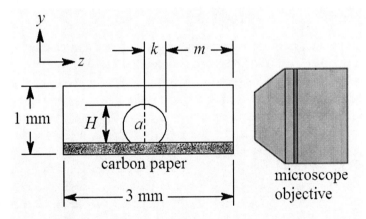

Figure 3. Cross sectional view of the gas channel with a droplet: m is the distance between the side wall and the point on the droplet closest to the objective; k and h are the half-width and height of the droplet; a is the central plane bisecting the droplet. (Reprinted from [19], Copyright Grant F. Minor, 2007. With permission.)

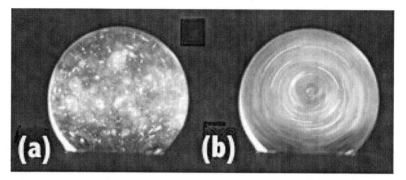

Figure 4. Visualization of a seeded droplet on a GDL (a) no airflow in microchannel; (b) air flow from right to left. Droplet height is approx. 700 μm. (Reprinted from [19], Copyright Grant F. Minor, 2007. With permission.)

Figure 4 shows visualizations of a droplet on the carbon paper GDL under quiescent conditions (no air flow in microchannel) and when air flows through the microchannel. The magnitude of the airflow (~2 m/s) is in this case not sufficient to deform the droplet but its effect on inducing internal flow within the droplet is clearly illustrated by the streaklines obtained using long time exposure.

2.2. VELOCITY FIELD MEASUREMENTS

Quantitative measurements of the velocity field within a droplet can be obtained by using micro-DPIV, a technique that has emerged as a powerful tool for investigating a range of microflow situations.[30] The main features of the technique are illustrated in Figure 5. Two successive digital images, separated by a time lapse Δt, are acquired by visualizing a laser illuminated region of the flow under a microscope. The images are then interrogated using a cross correlation algorithm, which based on the intensity of the scattered light allows identification of particle pairs on the two images. The displacement of each particle over the time lapse then yields the velocity, allowing reconstruction of the velocity field in the plane. Amongst the issues that need to be resolved to perform such measurements, one that is specific to this flow is optical distortion. Because the light from the fluorescent particles passes through a curved interface between the water droplet and the air, some optical distortion will occur due the different index of refraction between the two media and an optical distortion correction is required.[31]

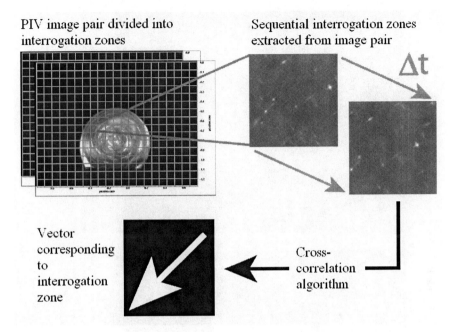

Figure 5. Digital particle image velocimetry: sequence of image acquisition and post-processing for calculating a velocity vector using cross-correlation.[32]

Example of velocity fields deduced from the micro-DPIV measurements are shown in Figure 6 for two air flow velocities (2.2 and 5.2 m/s) corresponding to Reynolds numbers of 212 and 500 respectively, which are representative of fuel cell operation. These images exhibit several features. First there is a very clear circulating flow pattern, with a centre of rotation that is offset and shifts downstream (left) with increasing air flow velocity. The magnitude of the velocity vectors tend to increase outwards from the centre of rotation, with the highest magnitude near the outer edge of the vector field, suggesting a non-zero velocity at the droplet-air interface. From the view point of a force balance analysis for the purpose of predicting detachment velocity, this would imply that the shear stress at the interface is lower than when a zero-velocity (no-slip) condition is assumed. The Figure also shows a substantial deformation of the droplet at the higher air velocity accompanied by a larger contact angle hysteresis (difference between the advancing and receding contact angles).

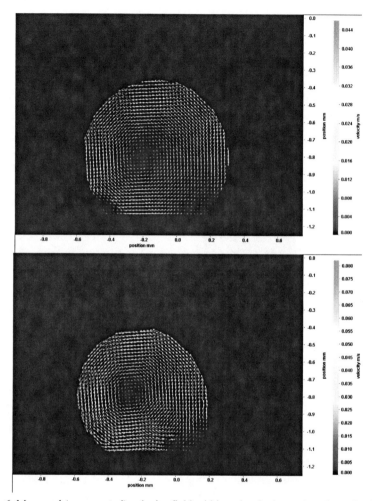

Figure 6. Measured (uncorrected) velocity field within a droplet in a microchannel subjected to an airflow of 2.2. m/s (top) and 5.2 m/s (bottom). (Reprinted from [19], Copyright Grant F. Minor, 2007. With permission.)

3. CFD Analysis of Droplet Dynamics

Characterization of the detachment process and of the subsequent dynamics of a droplet is of special interest, particularly in terms of the impact on water coverage (which inhibit mass transport from the cathode channel to the catalyst layer) and on the pressure drop (which induces parasitic losses as well as flow maldistribution). Detailed dynamic tracking of a droplet following its detachment is challenging experimentally, but considerable insight can be obtained using simulations with interface tracking capabilities. In this Section, simulations are presented and analyzed, focusing on the process of emergence and subsequent detachment of a water droplet in a simplified model of a microchannel with a GDL. The 3D discussed here simulations build on the 2D analysis presented by Zhu et al.[29] using the volume-of-fluid technique in conjunction with an interface reconstruction algorithm.

3.1. NUMERICAL MODEL AND VOF METHOD

The bulk flow Reynolds number in fuel cell microchannels is typically less than 1,000. In addition, the heat and mass transfer are not expected to have a significant direct effect on droplet dynamics. Thus unsteady, isothermal and laminar flow conditions are assumed in the numerical model.

In the VOF technique, a single set of momentum equations is shared by both fluid phases, with volume averaged properties calculated as a function of the volume fraction of each phase in each computational cell. The flow is then governed by the Navier-Stokes equation:

$$\frac{\partial}{\partial t}(\rho \vec{u}) + \nabla \cdot (\rho \vec{u} \vec{u}) = -\nabla p + \nabla \cdot \left[\mu (\nabla \vec{u} + \nabla \vec{u}^{\mathrm{T}}) \right] + \rho \vec{g} + \vec{F} \quad (1)$$

where p is the static pressure, $\vec{F}$ is a momentum source term related to surface tension, ρ and μ are the volume averaged density and dynamic viscosity. The interface between phases is tracked for each computational cell by computing the volume fraction for the fluid phase k:

$$C_k(x,y,z,t) = \begin{cases} 0 & \text{(outside } k^{\text{th}} \text{ fluid)} \\ 1 & \text{(inside } k^{\text{th}} \text{ fluid)} \\ 0 \sim 1 & \text{(at the } k^{\text{th}} \text{ fluid interface)} \end{cases} \quad (2)$$

where C_k is the volume fraction function of kth fluid. For all the fluids, the sum of the volume fraction function is equal to 1.

$$\sum_{k=1}^{n} C_k = 1 \quad (3)$$

The volume fraction function C_k is obtained by solving a volume fraction equation

$$\frac{\partial}{\partial t}(C_k \rho_k) + \nabla \cdot (C_k \rho_k \vec{u}_k) = 0 \qquad (4)$$

Surface tension is accounted for by using the continuum surface force (CSF) model,[33] and is expressed in terms of the pressure jump across the interface, which depends on the surface tension coefficient, and is implemented in the momentum equation as a body force. Once the volume fraction is computed for each cell in the domain, a piecewise linear interface calculation (PLIC) method [34] is used to geometrically reconstruct the interface between phases.

Figure 7 illustrates the three dimensional computation domain and typical surface mesh distributions. The simulation examples presented here are for a 1,000 µm long microchannel with a 250 × 250 µm square cross, a pore of diameter D = ϕ50 µm located on the bottom surface which simulates the GDL surface. The wetting properties of this surface can be prescribed independently of the other walls. This scenario simulates the transport into the channel of water that would have condensed and/or been transported in the GDL from the catalyst sites. The fixed location of the GDL pore is consistent with experimental observation that indicates water usually appears preferentially on certain site of the GDL surface.[18]

Uniform velocity profiles are specified at channel air inlet and pore water inlet, and a classical convective outflow condition is used at the microchannel outlet. No-slip boundary conditions are imposed on all solid

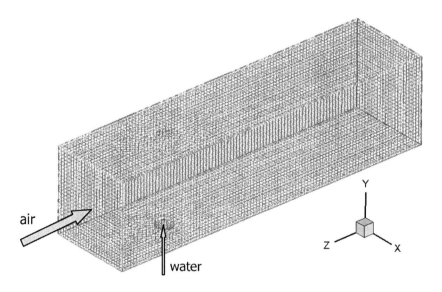

Figure 7. Computation domain and mesh for the three dimensional simulations.

walls, together with a specification of surface tension and contact angles. These are set to 45° for all walls, except for the bottom wall (GDL) for which the contact angle is varied in the parametric study.

3.2. DYNAMIC EVOLUTION OF EMERGING WATER DROPLET

The base case simulations discussed here correspond to an air velocity of 10 m/s (Re = 158) and a water is injection velocity of 1 m/s (Re_w = 50). This is of the same order as velocities used in *ex-situ* fuel cell experiments and corresponds to operation under high current density and water production rates.[34]

The dynamic motion and deformation of a blob of water emerging from a GDL pore into a gas channel are expected to be affected by factors including wettability of the walls, air velocity, water injection rate, and pore size; and the deformation and displacement of the droplet will depend on the balance of shear stress, surface tension, pressure, gravity, and inertial forces. In a microchannel, surface tension will play a particularly important role and for the baseline conditions considered here, the capillary, Weber and Bond numbers take the values $Ca = 1.4 \times 10^{-2}$, $We = 0.7$ and $Bo = 3.4 \times 10^{-4}$. The relatively small values of Ca and Bo are consistent with the expected dominance of surface tension over viscous and gravitational forces. Wettability of the wall, which is represented by the contact angle (θ), is a key parameter influencing not only the static shape of a droplet (*cf.* Figure 2) but also its dynamics evolution as will be shown.

The impact of wettability is clearly illustrated in Figures 8 and 9which show the time evolution of the droplet interface first for $\theta = 45°$, i.e. a hydrophilic surface, and for $\theta = 140°$, which corresponds to the typical Teflon treated GDL used in PEM fuel cell. In both cases the contact angle for the side (and top) walls of the microchannels, which correspond to the collector plate in a PEMFC, are fixed at $\theta = 45°$. In the hydrophilic surface case, water emerging from the GDL into the gas channel spreads forming a relatively thin layer on the surface. The water is inhibited from spreading upstream due to the air stream, but spreads downstream and towards both side walls of the microchannel. The water eventually reaches the side walls, and is subsequently and quickly dragged up and spreads onto the hydrophilic side. A concave gas/liquid interface results at the corners of the side walls and the bottom wall, inducing a surface tension gradient between the center and the corner of the channel. This gradient in turn drives water from the center to the sides, and with a recession of the downstream liquid front on the GDL surface can be observed. The overall liquid flow pattern results in significant coverage of the GDL surface. In the

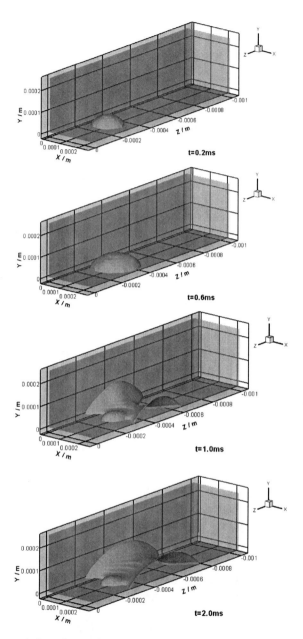

Figure 8. Time evolution of emerging water droplet into the microchannel for θ = 45°. Air velocity in microchannel is 10 m/s (Re = 158) and a water injection velocity through pore is 1 m/s (Re_w = 50).

WATER TRANSPORT IN FUEL CELL MICRO-CHANNELS 165

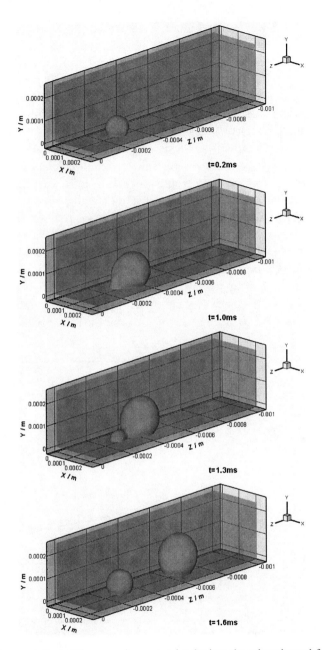

Figure 9. Time evolution of the emerging water droplet into the microchannel for θ = 140°. (Flow conditions as in Figure 8).

hydrophobic surface case in Figure 9, the emerging water initially forms into a quasi-spherical droplet with distinctly larger height. This shape has a smaller solid/liquid interface as well as a shorter triple-phase contact line. This combined with a larger pressure drag due to the protrusion of the droplet into the microchannel induces rapid downstream shedding. The water droplet detaches and exist the domain at a much earlier time.

The simulations clearly show the effectiveness of the hydrophobic treatment in water removal. Analysis of the simulations allows characterization of the critical diameter at which the droplet detaches as a function of Reynolds number (air flow rate) as shown in Figure 10. The detachment diameter as well as the microchannel friction factor, shown in Figure 11, decrease with increasing air inlet velocity as expected, and the computed values are a useful first step in providing data for specifying the required flow rate for proper water management as well as determining the wet pressure drop. An important point is that accounting for the emergence process through a *pore* appears to be essential to predicting the detachment diameter and critical velocity. Zhu et al.[29] showed that investigating the dynamics starting from droplet sitting on the surface yields substantially different results.

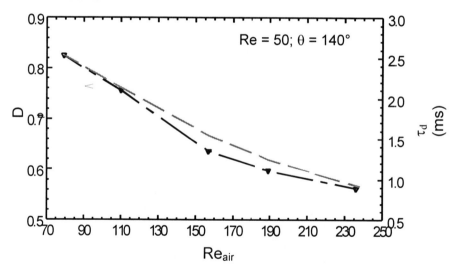

Figure 10. Variation of droplet detachment diameter and detachment time as a function of air flow Reynolds number; solid red line presents is a curve fitted correlation.

Another output of practical interest from the simulations is the tracking of the water coverage ratio, shown in Figure 12, and defined as the ratio of water covered area to total GDL surface. The water coverage ratio is plotted

WATER TRANSPORT IN FUEL CELL MICRO-CHANNELS 167

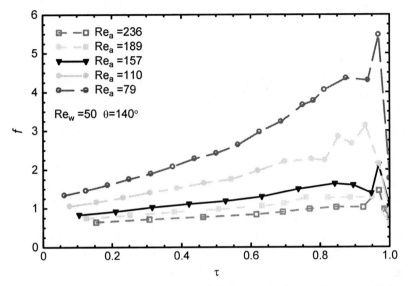

Figure 11. Effect of the Reynolds number of air flow on the flow resistance coefficient.

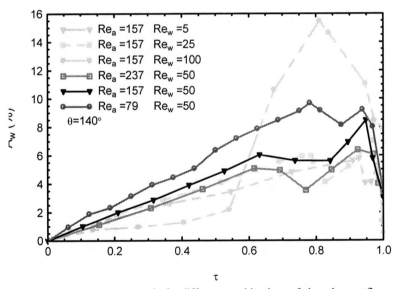

Figure 12. Water coverage ratio for different combinations of air and water flows.

with respect to dimensionless time for a number of combinations of water and air Reynolds numbers, and is a measure of flooding and mass transport resistance.

4. Conclusions

The cathode gas distribution micro-channels in PEM fuel cells are susceptible to partial flooding under some operating conditions. The resulting two-phase flow regimes differ significantly from the well documented two-phase flows encountered in more classical engineering applications. Some of the distinguishing features are the fibrous structure of the porous media, the important role of surface forces, and of hydrophobicity. The understanding, characterization and eventual control of the two-phase flow in PEM micro-channels is critical to improving performance as it impacts effective distribution of the oxidant, parasitic power losses incurred by the additional pressure drop, and lifetime. A number of emerging experimental and numerical techniques and tools have emerged in the last few years, and have started yielding insight into the two-phase flows pertinent to PEM fuel cells, but much remains to be done to satisfactorily characterize the various regimes and to develop models and correlations that can be used with some confidence in design.

Acknowledgements

This work was made possible with the support to the Natural Sciences and Engineering Research Council of Canada (NSERC), the MITACS Network of Centres of Excellence, and Ballard Power Systems.

References

1. J.J. Kowal, A. Turhan, K. Heller, J. Brenizer and M.M. Mench, Liquid Water Storage, Distribution, and Removal from Diffusion Media in PEFCS, *Journal of the Electrochemical Society*, 153(10), A1971–A1978 (2006).
2. A. Taniguchi, T. Akita, K. Yasuda and Y. Miyazaki, Analysis of Electrocatalyst Degradation in PEMFC Caused by Cell Reversal During Fuel Starvation, *Journal of Power Sources*, 130, 42–49 (2004).
3. X.G. Yang, F.Y. Zhang, A.L. Lubawy and C.Y. Wang, Visualization of Liquid Water Transport in a PEFC, *Electrochemical and Solid-State Letters*, 7, A408–A411 (2004).
4. K. Tüber, D. Pocza and C. Hebling, Visualization of Water Buildup in the Cathode of a Transparent PEM Fuel Cell, *Journal of Power Sources*, 124, 403–414 (2003).
5. X. Zhu, P. C. Sui and N. Djilali, Numerical Simulation of Emergence of a Water Droplet from a Pore into a Microchannel Gas Stream, *Microfluidics and Nanofluidics*, in press/online (2008).
6. M.M. Mench, Q.L. Dong and C.-Y. Wang, *In Situ* Water Distribution Measurements in a Polymer Electrolyte Fuel Cell, *Journal of Power Sources*, 124, 90–98 (2003).

7. K. Sugiura, M. Nakata, T. Yodo, Y. Nishiguchi, M. Yamauchi and Y. Itoh, Evaluation of a Cathode Gas Channel with a Water Absorption Layer/Waste Channel in a PEFC by using Visualization Technique, *Journal of Power Sources*, 145, 526–533 (2005).
8. F.-B. Weng, A. Su, C.-Y. Hsu and C.-Y. Lee. Study of Water-Flooding Behaviour in Cathode Channel of a Transparent Proton-Exchange Membrane Fuel Cell, *Journal of Power Sources*, 157, 674–680 (2006).
9. R. Satija, D.L. Jacobson, M. Arif and S.A. Werner, *In Situ* Neutron Imaging Technique for Evaluation of Water Management Systems in Operating PEM fuel cells, *Journal of Power Sources*, 129, 238–245 (2004).
10. D. Kramer, J. Zhang, R. Shimoi, E. Lehmann, A. Wokaun, K. Shinohara and G.G. Scherer, *In Situ* Diagnostic of Two-Phase Flow Phenomena in Polymer Electrolyte Fuel Cells by Neutron Imaging. Part A. Experimental, Data Treatment, and Quantification, *Electrochimica Acta*, 50, 2603–2614 (2005).
11. M.A. Hickner, N.P. Siegel, K.S. Chen, D.N. McBrayer, D.S. Hussey, D.L. Jacobson and M. Arif, Real-time Imaging of Liquid Water in an Operating Proton Exchange Membrane Fuel Cell, *Journal of the Electrochemical Society*, 153(5), A902–A908 (2006).
12. N. Pekula, K. Heller, P.A. Chuang, A. Turhan, M.M. Mench, J.S. Brenizer and K. Unlu, Study of Water Distribution and Transport in a Polymer Electrolyte Fuel Cell Using Neutron Imaging, *Nuclear Instruments and Methods in Physics Research A*, 542, 134–141 (2005).
13. A. Hakenjos, H. Muenter, U. Wittstadt and C. Hebling, A PEM Fuel Cell for Combined Measurement of Current and Temperature Distribution, and Flow Field Flooding, *Journal of Power Sources*, 131, 213–216 (2004).
14. K.W. Feindel, S.H. Bergens and R.H. Wasylishen, The Use of H NMR Microscopy to Study Proton-Exchange Membrane Fuel Cells, *ChemPhysChem*, 7, 67–75 (2006).
15. S. Tsushima, K. Teranishi, K. Nishida and S. Hirai, Water Content Distribution in a Polymer Electrolyte Membrane for Advanced Fuel Cell System with Liquid Water Supply, *Magnetic Resonance Imaging*, 23, 255–258 (2005).
16. I. Manke, Ch. Hartnig, M. Grunerbel, W. Lehnert, K. Kardjilov, A. Haibel, A. Hilger, J. Banhart and H. Riesemeier, Investigation of Water Evolution and Transport in Fuel Cells with High Resolution Synchrotron X-Ray Radiography, *Applied Physics Letters*, 90(17), Art. No. 174105 (2007).
17. S. Litster, D. Sinton and N. Djilali, *Ex Situ* Visualization of Liquid Water Transport in PEM Fuel Cell Gas Diffusion Layers, *Journal of Power Sources*, 154, 95–105 (2006).
18. A. Bazylak, D. Sinton, Z.S. Liu and N. Djilali, Effect of Compression on Liquid Water Transport and Microstructure of PEMFC Gas Diffusion Layers, *Journal of Power Sources*, 163(2), 784–792 (2007).
19. G. Minor, Experimental Study of Water Droplet Flows in a Model PEM Fuel Cell Gas Microchannel, MASc Thesis, University of Victoria, Canada (2007).
20. E.C. Kumbur, K.V. Sharp and M.M. Mench, Liquid Drop Behavior and Instability in a Polymer Electrolyte Fuel Cell Flow Channel, *Journal of Power Sources*, 161, 333–345 (2006).
21. K.S. Chen, M.A. Hickner and D.R. Noble, Simplified Models for Predicting the Onset of Liquid Water Droplet Instability at the Gas Diffusion Layer/Gas Flow Channel Interface, *International Journal Energy Research*, 29, 1113–1132 (2005).
22. Z.H. Wang, C.Y. Wang, and K.S. Chen, Two-Phase Flow and Transport in the Air Cathode af Proton Exchange Membrane Fuel Cells, *Journal of Power Sources*, 94(1), 40–50 (2001).

23. L.X. You and H.T. Liu, A Two-Phase Flow and Transport Model for PEM Fuel Cells, *Journal of Power Sources*, 155(2), 219–230 (2006).
24. T. Berning and N. Djilali, A 3D, Multiphase, Multicomponent Model of the Cathode and Anode of a PEM Fuel Cell, *Journal of the Electrochemical Society*, 150(12), A1589–A1598 (2003).
25. U. Pasaogullari and C.-Y. Wang, Liquid Water Transport in Gas Diffusion Layer of Polymer Electrolyte Fuel Cells, *Journal of the Electrochemical Society*, 151(3), A399–A406 (2004).
26. P. Quan, B. Zhou, A. Sobiesiak and Z.S. Liu, Water Behavior in Serpentine Micro-Channel for Proton Exchange Membrane Fuel Cell Cathode, *Journal of Power Sources*, 152(1), 131–145 (2005).
27. K. Jiao, B. Zhou and P. Quan, Liquid Water Transport in Parallel Serpentine Channels with Manifolds on Cathode Side of a PEM Fuel Cell Stack, *Journal of Power Sources*, 157, 124–137 (2006).
28. A. Theodorakakos, T. Ous, A. Gavaises, J.M. Nouri, N. Nikolopoulos and H. Yanagihara, Dynamics of Water Droplets Detached from Porous Surfaces of Relevance to PEM Fuel Cells, *Journal of Colloid and Interface Science*, 300, 673–687 (2006).
29. X. Zhu, P.C. Sui and N. Djilali, Dynamic Behaviour of Liquid Water Emerging from a GDL Pore into a PEMFC Gas Flow Channel, *Journal of Power Sources*, 172(1), 287–295 (2007).
30. D. Sinton, Microscale Flow Visualization, *Microfluidics and Nanofluidics*, 1, 2–21 (2004)
31. G. Minor, P. Oshkai and N. Djilali, Optical Distortion Correction for Liquid Droplet Visualization using the Ray Tracing Method: Further Considerations, *Measurement Science and Technology*, 18(11), L23–L28 (2007).
32. LaVision, *Device-Manual for Davis 7.0: Micro PIV*. Göttingen, Germany (2004).
33. J.U. Brackbill, D.B. Kothe and C. Zemach, A Continuum Method for Modeling Surface-Tension, *Journal of Computational Physics*, 100, 335–354 (1992).
34. D.L. Youngs, *Time-Dependent Multi-Material Flow with Large Fluid Distortion*, Academic Press, New York, 273–285 (1982).
35. A. Bazylak, D. Sinton and N. Djilali, Dynamic Water Transport and Droplet Emergence In PEMFC Gas Diffusion Layers, *Journal of Power Sources*, 176(1), 240–246, (2008).

POLYMER ELECTROLYTE FUEL CELL SYSTEMS FOR SPECIAL APPLICATIONS

U. PASAOGULLARI[*]
Department of Mechanical Engineering, and Connecticut Global Fuel Cell Center, University of Connecticut, Storrs, CT 06269

1. Introduction

Polymer electrolyte fuel cells (PEFC) have experienced significant interest for many applications over the last few decades, ranging from portable power generation in sub-watt to several hundred watts range, to automotive propulsion and stationary power generation. Due to lower noise and heat signature, portable PEFCs became the center of attention in mission critical systems, e.g. as a portable battery recharging system. PEFC systems present a unique alternative for portable power generation at around 200–500 W. Due to intrinsic compact design requirement, however, there are several challenges remaining to be tackled in the system level as well as the stack design and operation. Particularly, fuel storage for these systems still remains a challenge, as compressed or liquid hydrogen do not meet volume, weight and safety requirements. In addition, water management presents another issue since the amount of water that can be carried for humidification of the polymer electrolyte (i.e. polymer electrolyte membrane, PEM) is limited.

The operation of a PEFC is based on the electrochemical oxidation of hydrogen on the anode side and electrochemical reduction of oxygen on the cathode side (Eqs. (1) and (2)). PEFC Electrochemical Reactions:

Anode: $\quad H_2 \rightarrow 2 H^+ + 2 e^-$ (1)

Cathode: $\quad \frac{1}{2} O_2 + 2H^+ + 2e^- \rightarrow H_2O$ (2)

Overall: $\quad H_2 + \frac{1}{2} O_2 \rightarrow H_2O$ (3)

Figure 1 shows a schematic of a PEFC system where inlet gases are humidified by use of external humidifiers with water storage. As seen, the system consists of: (i) the fuel cell stack; (ii) air handling system with compressor and filter; (iii) hydrogen storage and supply system; (iv)

---

[*]Ugur Pasaogullari, Connecticut Global Fuel Cell Center, 44 Weaver Rd U-5233, Storrs, CT 06269, USA, e-mail: ugur.pasaogullari@uconn.edu

humidification and water recuperation system; and (iv) auxiliary components that perform the thermal management of the fuel cell stack and power conditioning (not shown) of the system.

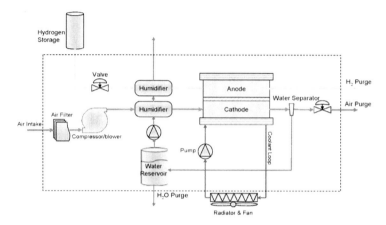

Figure 1. Schematics of a portable PEFC system.

Majority of the available space in portable PEFC systems is shared by fuel storage and the fuel cell stack; hence they compete for the available space. Larger fuel storage increases the runtime of the system before refilling; however an increased stack increases the total power and/or efficiency of the system, as PEFCs run more efficiently at low power densities as shown in Figure 2. Therefore a larger fuel cell stack operating at lower power densities offers higher efficiency, decreasing the fuel consumption. This clearly shows that there exists an optimum between the size of the fuel storage and the fuel cell stack.

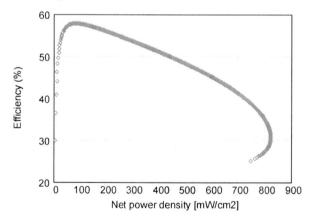

Figure 2. Efficiency vs. power density of a typical PEFC.

2. Hydrogen Storage and Generation on Board

Figure 3 shows weight and volume storage capacities of several hydrogen storage options, including metal hydrides, chemical hydrides and molecular (liquid and compressed) hydrogen.[1] Although the targets shown are for vehicle propulsion systems, similar, albeit not as aggressive, targets also exist for portable fuel cell systems. As seen, several technologies, including metal and chemical hydrides compete as an option for portable hydrogen storage systems.

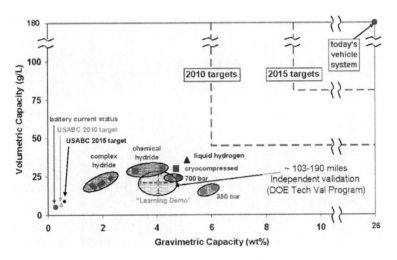

Figure 3. Existing hydrogen storage technologies.[1]

2.1. CHEMICAL HYDRIDES

Chemical hydrides are among the most viable options for portable hydrogen storage due to their compactness and safety. The term "chemical hydrogen storage" also commonly used for chemical hydrides and it refers to storage technologies, in which hydrogen is generated through a chemical reaction. Although, many different chemical hydrides exist today, either in research, prototype or commercial level, many applications focus on boron based chemical hydrides.

Figure 4 summarizes the chemical hydrogen storage technologies, and compares the energy efficiency and volumetric hydrogen storage capacity of different technologies.[2] Chemical hydrides typically provide higher volumetric capacity then liquid or compressed hydrogen, whereas compressed hydrogen provides in excess if 80% ideal energy conversion efficiency.

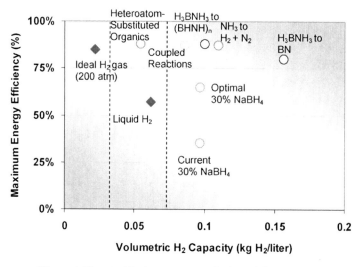

Figure 4. Chemical hydrogen storage [Adopted from Ref. 2].

2.1.1. *Sodium Borohydride*

Perhaps the best known chemical hydride for hydrogen storage is the sodium borohydride ($NaBH_4$). Sodium borohydride is not freely available; therefore it has to be synthesized. Brown-Schlesinger process[3] is one of the many routes for synthesis of sodium borohydride. The process is outlined in Figure 5. As seen from the process, hydrogen is needed to make sodium borohydride; therefore it becomes a chemical hydrogen storage method, rather than a hydrogen generation device.

The decomposition of sodium borohydride follows the catalytic reaction below, and has a theoretical hydrogen storage of 10.918 wt% and around 0.152 kg H_2/l^2.

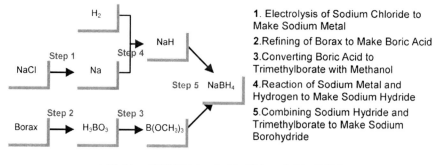

$$4\ NaH + B(OCH_3)_3 \rightarrow NaBH_4 + 3\ NaOCH_3$$

Figure 5. Brown-Schlesinger synthesis of $NaBH_4$ [Adapted from Ref. 3].

$$NaBH_4 + 2\ H_2O \rightarrow 4\ H_2 + NaBO_{2(aq)} + 300\ kJ \quad (4)$$

In addition to indirect use of $NaBH_4$ by decomposition into H_2, fuel cells based on electrochemical oxidation of $NaBH_4$ are also proposed.[4-6] These fuel cells can be based on cationic or anionic exchange electrolytes. In anionic exchange electrolyte types, the following reactions occur:

Anode: $\quad BH_4^- + 8\ OH^- \rightarrow BO_2^- + 6\ H_2O + 8\ e- \quad (5)$

Cathode: $\quad 2\ O_2 + 4\ H_2O + 8e^- \rightarrow 8OH- \quad (6)$

where OH^- is transported across the electrolyte from cathode to anode. In cationic exchange electrolyte based systems, the electrochemical oxidation and reduction reactions are, respectively:

Anode: $\quad BH_4^- + 8\ OH^- \rightarrow BO_2^- + 6\ H_2O + 8\ e- \quad (5)$

Cathode: $\quad 2\ O_2 + 4\ H_2O + 8e^- \rightarrow 8OH- \quad (6)$

where Na^+ is transported across the electrolyte from anode to cathode. As seen, both anionic and cationic forms have the same half cell reactions, and therefore they both have EMF (electro-motive force) of 1.64 V, compared to 1.23 V of H_2 based systems. The difference between the anionic and cationic systems is the ion that is exchanged across the electrolyte. For more compact systems, hydrogen peroxide can replace air as the oxidizer.[6]

2.1.2. Amine Boranes

In addition to sodium borohydride, amine boranes such as ammonia borane (AB, BH_3NH_3) have recently started receiving significant attention. In addition to hydrolysis routes, AB can be decomposed through pyrolysis and releases up to 2 moles of H_2 per mole of AB at temperatures around 150°C, giving a theoretical hydrogen storage of up to 13 wt%. With temperatures around 450°C, AB releases the third mole of H_2, bringing the theoretical storage to 19.6 wt%.[7]

$$BH_3NH_{3\ (l)} \rightarrow H_2BNH_{2\ (s)} + H_{2\ (g)}\ \sim 137°C\ \Delta H_r = -(21.7 \pm 1.2)\ kJ/mol \quad (7)$$

$$x(H_2BNH_2)_{(s)} \rightarrow (H_2BNH_2)_{x\ (s)}\ \sim 125°C \quad (8)$$

$$(H2BNH2)_{x\ (s)} \rightarrow (HBNH)_{x\ (s)} + x\ H_{2\ (g)}\ \sim 155°C \quad (9)$$

$$x(H_3BNH_3)_{(l)} \rightarrow (HBNH)_{x\ (s)} + 2x\ H_{2\ (g)} \quad (10)$$

In addition to higher H_2 content, pyrolysis based ammonia borane hydrogen storage occurs through thermally controlled reactions, and requires no noble catalyst. The reaction is typically controlled by controlling the temperature of ammonia borane. Recent studies on ammonia borane[8-9] show that they are quite competitive with other chemical hydrides.

3. Fuel Cell Stack

As shown in Figure 1, the fuel cell system consist of many components, arguably the most important being the fuel cell stack. PEFC stacks can provide high efficiencies and longer runtimes when coupled with an efficient hydrogen storage system; however their performance and durability heavily rely on optimization of the material properties and operating conditions. A fuel cell stack consists of repeating units called "single cells", each of which generate around 0.4–0.8 V during operation. The cells are stacked in series to generate the required operating voltage. A single cell consists of an anode, a cathode and an electrolyte. Both anode and cathode are comprised of a gas channel, maintaining reactant and product flow; a porous gas diffusion media (GDM), typically carbon paper or carbon cloth and maintains the transport of reactants and products between the reaction sites and gas channels, as well as electron and heat transfer between the reaction sites (catalyst layers) and bipolar plates. Catalyst layers typically consist of a composition of open pore spaces, ionomer phase and catalyst particles. The catalyst particles are typically carbon supported platinum, which is made by dispersing 2–5 nm Pt crystals on to 30–70 nm Carbon particles. The anode and cathode are separated by the electrolyte, which is typically a perfluorosulfonic acid (PFSA) based polymer electrolyte membrane (PEM), such as Nafion by DuPont. The catalyst layer can either be coated on gas diffusion media (i.e. gas diffusion electrodes, GDE) or on membrane (i.e. catalyst coated membrane, CCM). The operation of a PEFC is strongly dependent on the water balance in the cell, which requires rigorous water management.

3.1. WATER MANAGEMENT

Water management is defined as the optimization of operating conditions and material characteristics to maintain the hydration of PEM to provide required ionic conductivity, while eliminating the flooding (i.e. blocking of gas transport pathways by liquid water). Water management in portable PEFCs is especially challenging, due to requirements of decreased weight, which limits the amount of water that can be carried with the fuel cell system, as well as the auxiliary equipment that handles the water management.

As shown in Figure 6, water management is complicated with multi-modal water transport. In addition to diffusion of water vapor in open pore spaces; there is electro-osmotic drag of water molecules across the membrane as well as permeation due to hydraulic pressure gradient. In two-phase conditions, additional modes of water transport due to capillary flow of liquid water in porous gas diffusion media becomes pertinent, as well as the removal of liquid water droplets from the hydrophobic surface of the GDL and liquid film flow on hydrophilic channel walls.

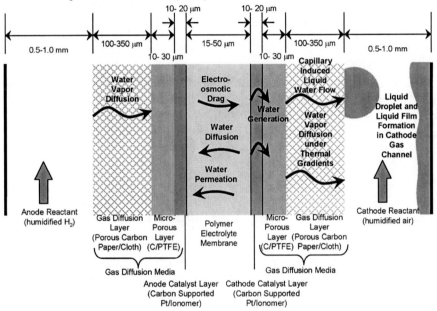

Figure 6. Schematics of water transport mechanisms.

Water management issues vary significantly depending on the cell design. Air breathing systems, which are typically used for low power units, do not have flow cathodes. In these systems, flooding of the cathode becomes a significant issue, especially with humidified hydrogen, since there is no forced air flow to remove the droplets that form on the surface of the cathode. Air-breathing PEFC systems are optimal for lower power (<20 W) applications.

When the cell is operated on stoichiometric anodes, which is typically called "*dead-ended anode*" since anode outlet is blocked, water accumulation in the anode channel and gas diffusion media may hinder H_2 transport, resulting in a phenomenon called "fuel starvation". In addition to lower performance, fuel starvation may cause accelerated degradation of the cell through corrosion of carbon support in catalyst layers, which results in platinum oxidation and dissolution, significantly reducing the life of a PEFC system. To eliminate this, anode side is periodically purged with high hydrogen flow rates, which decreases overall system efficiency due to wasted fuel.

As mentioned above, water management deals with hydration of the polymer electrolyte as well as the elimination of the flooding. These two phenomena are described in the next sections.

3.2. DYNAMICS OF HYDRATION OF POLYMER ELECTROLYTE

Fu and Pasaogullari[10] have modeled the dynamics of membrane hydration during the start-up of a fuel cell, focusing on the net exchange of water between anode and cathode. The net water transport coefficient, defined as the net number of water molecules carried from anode to cathode per number of protons, and is a effective result of electro-osmotic drag and back-diffusion.

$$j_{H_2O} = \alpha\, j_{H^+}\qquad(11)$$

$$\alpha = \frac{\dfrac{\rho_{dry}}{EW} D_\lambda \nabla \lambda + n_d\, I/F}{I/F}\qquad(12)$$

where I is the current density, F is the Faraday's constant, n_d is the electro-osmotic drag coefficient, λ is the membrane water content (defined as the number of H_2O per SO_3^- in the PEM), ρ_{dry} is the dry density and EW is the equivalent weight of the membrane.

Figure 7 shows the change in net water transport coefficient, α during a isothermal start up at a constant voltage of 0.6 V. Initially α is larger than electro-osmotic drag coefficient due to a positive water content gradient from anode to cathode as a result of higher humidifier temperature on the anode side. As the cell operates, due to water production, the water content on the cathode side of the membrane rise, which results in increase in the current density. This increases the electro-osmotic drag; therefore an initial peak in net water transport coefficient is seen. As the water content differential across the membrane builds-up, back-diffusion increases which decreases the net water coefficient. It is also seen that the micro-porous layer (MPL), which is a fine mixture of carbon particles and PTFE binder, coated on the gas diffusion layer (GDL) further decreases the net water transport coefficient due to increased resistance to water removal from MEA.

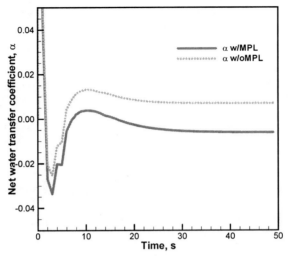

Figure 7. Average net water transport coefficient during an isothermal start-up at $V_{cell}=0.6$ V. Cell temperature is 80°C, anode and cathode humidifier temperatures are set as 60°C, and 20°C, respectively. Flow rates on both sides correspond to 1 stoichiometry at a reference current density of 1 A/cm^2.[10]

During this start-up operation, the water content, λ in the cathode side of the membrane near the cathode outlet increases from 3 to almost 9, which corresponds to a swelling strain of 0.06 mm/mm in the membrane, according to measurements by Kusoglu et al.[11] Since the membrane is partially constrained, membrane cannot swell but rather hygral residual stresses develop in the membrane, which could be as high as 2–3 MPa, which comparable to yield strength of Nafion. This example, by itself, shows the significance of the hydration dynamics on the durability of polymer electrolytes.

Modeling of PEFCs provides very useful information about the operation; however it requires extensive validation before they can be utilized in design. In such an attempt, Fu and Pasaogullari[12] have compared the experimental data of Lu et al.[13] with the predictions of computational model. As seen from Figure 8, reasonable agreement in current density distribution is achieved; however match in net water transport data is far from satisfactory. More accurate material properties are required for validation in net water transport coefficient.

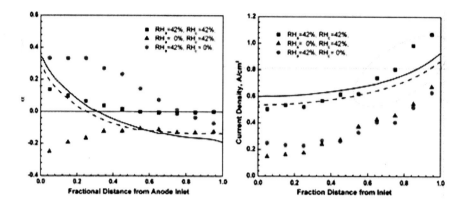

Figure 8. Validation against distributions of current density and net water transport coefficient.[12]

3.3. FLOODING AND TWO-PHASE TRANSPORT IN PEFCS

The other aspect of the water management deals with formation and transport of liquid water in PEFCs. Handling of liquid water in PEFCs remains as a challenge, since too much liquid water blocks the open pores of the catalyst layers, gas diffusion media, and gas channels, and hinders the reactant transport; adversely affecting the performance of the PEFC. In addition, flooding by liquid water may cause fuel starvation especially in portable fuel cells, where the PEFC is operated on close to stoichiometric ratios on fuel side, accelerating the degradation through carbon corrosion. The liquid water transport in PEFCs consists of four physical phenomena:

(i) Liquid water formation and transport in catalyst layers;
(ii) Liquid water flow across hydrophobic GDM;
(iii) Droplet dynamics at the GDM surface;
(iv) Two-phase flow in gas channels.

Pasaogullari and Wang[14] have proposed a theory of liquid water flow in hydrophobic GDM of PEFCs. According to this theory, liquid water is generated in the cathode catalyst layer due to production and drag from anode, and forms a tree-root like percolation network inside the GDM, as shown in Figure 9. Once the water reaches the GDM surface, it either forms large cross-section area droplets (on hydrophobic surfaces) or spreads on the surface (on hydrophilic surfaces).

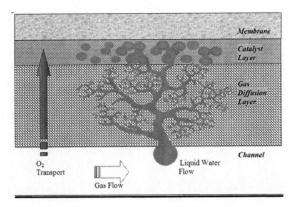

Figure 9. Liquid water transport in hydrophobic GDM of a PEFC.[14]

Liquid water flow in GDM is governed by capillary action. The capillary pressure is defined as the difference in wetting and non-wetting phase pressures. In a hydrophobic GDM, the wetting phase is the gas phase therefore the liquid pressure is higher than the non-wetting phase pressure. In either of the hydrophobic and hydrophilic GDM, the liquid pressure increases with increasing liquid saturation (i.e. the void fraction occupied by liquid water). This results in a liquid pressure gradient across the GDM, since the liquid saturation is higher on the catalyst layer side of the GDM compared to the channel side. This pressure gradient (formed due to capillarity) becomes the driving force for liquid water flow. The capillary pressure is typically written by use of a Leverette function, $J(s)$:

$$p_c = p_g - p_l = \sigma_{lv} \cos(\theta_c)(\varepsilon/K)^{0.5} J(s) \qquad (13)$$

Currently, there are many groups developing the appropriate Leverette functions for PEFC diffusion media, both computationally[15–16] and experimentally.[17] The capillary flux then is written as:

$$\mathbf{j}_l = \frac{\lambda_l \lambda_g}{\nu} K \nabla p_c \qquad (14)$$

where λ_l and λ_g are the mobility of liquid and gas phases, respectively, ν is the two-phase viscosity and K is the absolute permeability of the porous media. Phase mobility λ_l and λ_g are functions of individual phase viscosities as well as the relative permeabilities.

Utilizing the above theory, Pasaogullari et al.[18–19] developed a computational model and analyzed the effects of the construction of the diffusion media. They have analyzed the bi-layer construction, which consists of a macro-porous substrate and a micro-porous layer coating. The micro-porous coating has a much smaller mean pore size and higher Teflon content, consequently higher contact angle. With the distribution of material properties in addition to liquid saturation, the capillary flux becomes:

$$\mathbf{j}_l = \frac{\lambda_l \lambda_g}{\nu} K \left[\frac{\partial p_c}{\partial s} \nabla s + \frac{\partial p_c}{\partial d_p} \nabla d_p + \frac{\partial p_c}{\partial \theta_c} \nabla \theta_c \right] \qquad (15)$$

Figure 10 shows the effect of the micro-porous coating on the macro-porous GDL. As seen, the MPL becomes the dominant component governing the liquid saturation near the catalyst layer-GDM interface.

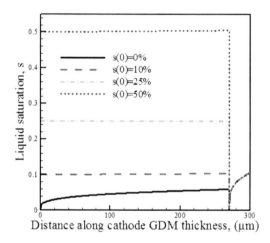

Figure 10. Liquid saturation distribution across the GDM.[19]

The effect of MPL was also investigated on the net water transport coefficient in two-phase conditions. In two-phase conditions, portion of the polymer electrolyte becomes saturated with liquid water, as shown in Figure 11, therefore permeation of water due to hydraulic pressure gradient becomes effective through expansion of nano-channels in the membrane in which liquid water can penetrate through.

Due to smaller pore size, the liquid pressure is higher in the MP, which results in a higher hydraulic pressure gradient across the PEM. As seen in Figure 12, especially at low current densities, the larger hydraulic pressure

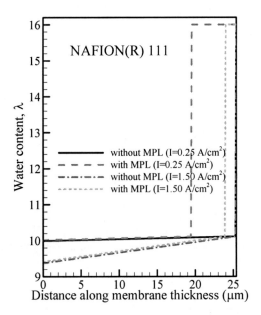

Figure 11. Effect of MPL on water content distribution across the PEM.[19]

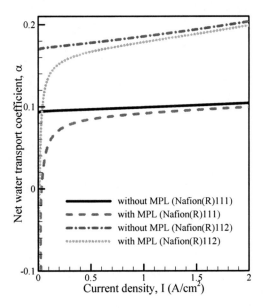

Figure 12. Effect of MPL on net water transport coefficient.[19]

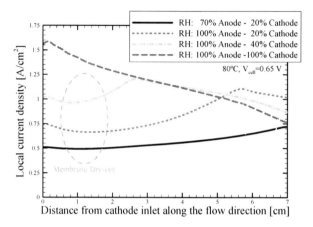

Figure 13. Current density distribution along the flow direction.[21]

gradient increases the back-flux of water from cathode to anode, which results in lower net water transport coefficient. The net water transport coefficient, α can be calculated as:

$$\alpha \frac{I}{F} = n_d \frac{I}{F} - D_\lambda \frac{\rho_{dry}}{EW} \frac{\partial \lambda}{\partial x} - K_{mem} C_{H_2O}^{liquid} \frac{\partial p_h}{\partial x} \qquad (16)$$

where x direction is from anode to cathode, K_{mem} is the hydraulic permeability of the polymer electrolyte, $C_{H_2O}^{liquid}$ is the molar concentration of liquid water and p_h is the hydraulic pressure.

The same theory has been applied to a multi-dimensional PEFC modeling framework originally developed by Um et al.[20] for single-phase transport conditions. The developed model,[21] as shown in Figure 13 predicts that all there aspects of water balance may occur in a PEFC simultaneously: (*i*) Membrane dry-out near the inlet; (*ii*) Well hydrated PEM with no adverse effects of liquid water around the middle of the cell; (*iii*) Flooded cathode towards the outlet.[21]

3.4. INNOVATIVE WATER MANAGEMENT SCHEMES

As mentioned above, due to volume and weight restrictions in portable systems, instead of carrying additional water for PEM hydration, it is desirable to recuperate the product water of cathode oxygen reduction reaction, and utilize it for humidification of gas streams. Several techniques are being researched for this, and examples to these are enthalpy wheel by Emprise Corporation,[22] and Nafion membrane shell-and-tube type exchange

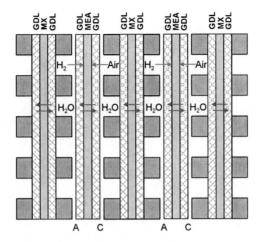

Figure 14. Internal humidification by composite bipolar plate concept.[24-25]

based humidifiers by PermaPure LLC.[23] However, these external components compete for available volume and weight with the rest of the system.

Fu[24-25] proposed a composite bipolar plate, which performs internal humidification by balancing water contents in the anode and cathode sides. The composite bipolar plate design contains a water permeable membrane that is embedded within the bipolar plate as shown in Figure 14. The membrane selectively allows water transport by diffusion from the cathode channel of one cell with a higher water concentration to the anode channel of the adjacent cell with a lower water concentration while effectively remaining impermeable to hydrogen and oxygen gases. The concept provides additional means to re-distribute water in the cell to improve the overall humidification. Thus, in case of ambient or dry operation, the ORR water product in the adjacent cathode is transported to the dry anode through the mass exchanger (MX) in addition to water transport through the PEM, which is typically towards cathode side due to the electro-osmotic drag.

Figure 15 shows the performance the proposed concept. It is seen that once the relative humidity of the cathode channel increases due to product water, the composite bipolar plate enables the hydration of the anode side due to water transport from cathode channel to anode channel.

The proposed concept appears to improve the water balance in the fuel cell, by providing an additional means of water transport between anode and cathode. In conventional PEFCs, the electro-osmotic drag dominates over the

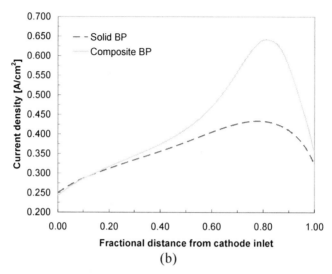

Figure 15. Comparison of local current density distribution between the conventional solid bipolar plate and the proposed composite bipolar plate.[25]

back-diffusion flux even with thin membranes; therefore dehydration of anode occurs. With this concept, the exchange of water between anode and cathode is improved by a water permeable bipolar plate.

4. Conclusions

Polymer electrolyte fuel cells are promising candidates for portable applications in the range of 200–500 W, and they provide several advantages over the other candidates such as direct methanol fuel cells (DMFC), batteries and internal combustion engine generators. They are more efficient than DMFCs, and efficient hydrogen storage techniques are being developed to used in these systems. They are lighter and provide longer runtimes compared to batteries, and they do not have any heat and noise signature, which internal combustion generators suffer from.

However, many obstacles in fuel storage and stack performance need to be overcome before the technology wins the competition. Higher volumetric storage capacity hydrogen storage technologies are needed for, and chemical hydrides, particularly borohydrides, provide a great potential in that. Fuel cell stack performance and durability needs to be improved and water management is among the major issues that affects these two key parameters.

References

1. S. Satyapal, G. Ordaz, C. Read, N. Stetson, J. Adams, J. Alkire, P. Bekke, and G. Thomas, Hydrogen Storage, 2007 DoE Hydrogen Program Merit Review and Peer Evaluation Meeting, May 15 2007.
2. B. Thomas, DoE Chemical Hydrogen Storage Center of Excellence, 2006 DoE Hydrogen Program Merit Review and Evaluation Meeting, May 16, 2006.
3. Y. Wu, M.T. Kelly, and J.V. Ortega, Review of Chemical Processes for the Synthesis of Sodium Borohydride, DoE Report, August 2004.
4. J.H. Wee, K.Y. Lee, and S.H. Kim, Sodium borohydride as the hydrogen supplier for proton exchange membrane fuel cell systems, *Fuel Processing*, 87(9), 811–819 (2006).
5. D. Browning, B, Lakeman, K. Pointon, and A. Rose, The Direct Sodium Borohydride Fuel Cell for Unmanned Underwater Vehicle Application, 2005 Fuel Cell Seminar, November 15–18, 2005, Palm Springs, CA.
6. G.H. Miley, N. Luo, J. Mather, R. Burton, G. Hawkins, L. Gu, E. Byrd, R. Gimlin, P.J. Shrestha, G. Benavides, J. Laystrom, and D. Carroll, Direct $NaBH_4/H_2O_2$ fuel cells, *Journal of Power Sources*, 165, 509–516 (2007).
7. A. Raissi, Hydrogen Storage in Ammonia and Aminoborane Complexes, 2002 DoE Hydrogen Progrem Review Meeting, May 21, 2002.
8. R. Kellett, S. Nosheen, S. Frueh, W, Willis, S. Suib, D. Irwin, C. Giersz, and C. Mallery, Studies on Ammonia Borane Pyrolysis for Portable Fuel Cell Hydrogen Generation, 5th ASME International Conference on Fuel Cell Science, Engineering and Technology, June 18–20, 2007, New York.
9. L. Li, B. Schmid, R.S. Smith, B.D. Kay, J. Linehan, W. Shaw, N. Hess, A. Stows, C. Brown, L. Daemen, M. Gutowski, and T. Autrey, Hydrogen Storage in HNBH Systems, Materials Research Society Meeting, San Francisco, April 2006.
10. R.S. Fu and U. Pasaogullari, Dynamics of Water Transport in Polymer Electrolyte Fuel Cells during Start-up in Under-Humidified Conditions, 210th Electrochemical Society Meeting, October 29-November 2, 2006, Cancun, Mexico.
11. A. Kusoglu, A.M. Karlsson, M.H. Santare, S. Cleghorn, and W.B. Johnson, Mechanical response of fuel cell membranes subjected to a hygro-thermal cycle, *Journal of Power Sources*, 161, 987–996 (2006).
12. R.S. Fu and U. Pasaogullari, Net Water Transfer Coefficient as Validation Data for Computational Modeling of Polymer Electrolyte Fuel Cells, 5th ASME International Conference on Fuel Cell Science, Engineering and Technology, June 18–20, 2007, New York.
13. G.Q. Lu, F.Q. Liu, and C.-Y. Wang, An approach to measuring spatially resolved water crossover coefficient in a polymer electrolyte fuel cell, *Journal of Power Sources*, 164, 134–140 (2007).
14. U. Pasaogullari and C.Y. Wang, Liquid water transport in gas diffusion layer of polymer electrolyte fuel cells, *Journal of Electrochemical Society*, 151(3), A399–A406 (2004).
15. P.K. Sinha and C.Y. Wang, Pore-network modeling of liquid water transport in gas diffusion layer of a polymer electrolyte fuel cell, *Electrochimica Acta*, 52, 7936–7945 (2007).
16. J.T. Gostick, M.A. Ioannidis, M.W. Fowler, M.D. Pritzker, Pore network modeling of fibrous gas diffusion layers for polymer electrolyte membrane fuel cells, *Journal of Power Sources*, 173, 277–290 (2007).

17. E.C. Kumbur, K.V. Sharp, and M.M. Mench, A validated leverett approach for multiphase transport in PEFC media: Part 1: hydrophobicity effect, *Journal of Electrochemical Society*, 154(12), B1295–B1304 (2007).
18. U. Pasaogullari and C.Y. Wang, Two-phase transport and the role of micro-porous layer in polymer electrolyte fuel cells, *Electrochimica Acta*, 49, 4359–4369, (2004).
19. U. Pasaogullari, C.Y. Wang, and K.S. Chen, Two-phase transport in polymer electrolyte fuel cells with bi-layer cathode gas diffusion media, *Journal of Electrochemical Society*, 152(8), A1574–A1582 (2005).
20. S. Um, C.Y. Wang, and K.S. Chen, Computational fluid dynamics modeling of proton exchange membrane fuel cells, *Journal of Electrochemical Society*, 147(12), 4485–4493 (2000).
21. U. Pasaogullari and C.Y. Wang, Two-phase modeling and flooding prediction of polymer electrolyte fuel cells, *Journal of Electrochemical Society*, 152(2), A380–A390 (2005).
22. Humidicore, Emprise Corporation, http://www.humidicore.com/
23. Perma Pure FC-Series Humidifiers, http://www.permapure.com
24. R.S. Fu and U. Pasaogullari, "An Internal Water Management Scheme for Portable Polymer Electrolyte Fuel Cells", Proceedings of Fuel Cell 2006, 4th International Conference on Fuel Cell Science, Engineering and Technology, June 19–21, 2006, Irvine, CA.
25. R.S. Fu, M.S. Thesis, University of Connecticut, 2006.

SORPTION AND DIFFUSION SELECTIVITY OF METHANOL/WATER MIXTURES IN NAFION

DANIEL T. HALLINAN JR. AND YOSSEF A. ELABD[*]
Drexel University
Department of Chemical and Biological Engineering
Philadelphia, Pennsylvania 19104

1. Introduction

During the past 40 years membranes have gained large importance, particularly in separations, actuators, and fuel cells. Their advantage being the ability to control solute flux through the membrane, particularly rapid transport of one component with the exclusion of other components.[1] In multicomponent transport through a polymer membrane, the ratio of fluxes is termed selectivity and can be caused by differences in sorption and diffusion. The former is attributed to thermodynamic interactions, while the latter is considered transport effects. This solution-diffusion mechanism was developed in the 1940s. Around 1950, ion-exchange resins were developed, allowing far greater selectivity than is possible solely from chemical interactions.[2] Ion-exchange resins are the precursors of polymer electrolyte membranes (PEMs) used in fuel cells.

Nafion is a PEM made by DuPont, consisting of a perfluorinated backbone with perfluoroether side chains that terminate in a sulfonic acid group. Nafion is widely used as a PEM in fuel cells. This work will focus on its use in the direct methanol fuel cell (DMFC), which has several benefits, such as the ability to produce energy with a renewable fuel. Also, because the DMFC is simple, converting chemical energy directly into electricity, it has the potential for high efficiency and high power density, which is ideal for portable power applications.[3] However, an important problem called fuel crossover hinders current DMFCs running with Nafion as the PEM. Fuel crossover occurs because methanol is able to swell and pass through Nafion. In swelling the membrane, the mechanical properties are decreased and the crossover rate (methanol flux) is increased. Methanol not only

---

[*]E-mail: elabd@drexel.edu

reacts at the anode, where it contributes to power production, but it also passes through the membrane and reacts at the cathode producing a counter current or mixed potential, actually decreasing the power available from the fuel cell.[4] In addition, some methanol may leave the cathode outlet as lost fuel. These effects all contribute to decreased efficiency and performance of DMFCs and require operation at low methanol feed concentrations.[5] Unfortunately, low methanol feed concentrations limit the anode reaction rate and therefore maximum achievable power.

It is widely accepted that a new PEM for the DMFC is needed. In fact, many groups are focusing on that goal.[3] Others are studying sorption and transport in Nafion using techniques like gravimetric sorption, NMR spectroscopy, pulsed field gradient (PFG) NMR, and electrochemical methods.[6-15] These techniques provide overall sorption and diffusion of a multicomponent system or self diffusion coefficients. More useful are mutual, binary diffusion coefficients, which account for the concentration gradient of each diffusant.[16] The approach herein is to thoroughly examine each component distinctly, methanol and water, in terms of their sorption and mutual diffusion in the Nafion membrane, in order to fully understand the complex transport mechanisms and therefore identify an informed direction for membrane development. There are limited studies on fundamental transport properties of methanol in Nafion and no clear consensus on transport property trends among various studies. Specifically, it is not clear what the main contributing factors behind increased methanol flux with increasing methanol solution concentration are – methanol sorption or diffusion or both. Therefore, more fundamental investigations and new experimental techniques in this field would be of significant interest.

In this study, the diffusion and sorption of methanol and water in Nafion were measured using time-resolved Fourier transform infrared – attenuated total reflectance (FTIR-ATR) spectroscopy. This technique has been used by numerous investigators to measure diffusion in polymers,[17-19] but has not been used to measure diffusion in Nafion. Many of these studies have compared their results to more conventional transport experiments (e.g. permeation cell, dynamic gravimetric sorption) and report excellent agreement between the techniques. However, unlike conventional transport measurements, time-resolved FTIR-ATR spectroscopy provides molecular-level contrast between diffusants and the polymer based on bond vibrations absorbing light at different wavelengths. In other words, changes to the polymer and the diffusant(s) in the polymer can be measured in real time on a molecular level during the diffusion process. This molecular contrast allows for the measurement of multicomponent diffusion and sorption in polymers. In this study, the effective mutual methanol and water diffusion coefficients and concentrations in Nafion were measured with time-resolved FTIR-ATR spectroscopy as a function of methanol/water solution concentration, and the effect of these values on methanol flux and water to methanol selectivity were determined.

2. Experimental

2.1. MATERIALS

Nafion 117(1,100 g dry polymer/moles sulfonic acid, 178 μm dry thickness) was purchased from Aldrich. Hydrogen peroxide (Aldrich, 30–32 wt%), reverse osmosis (RO) water (resistivity ~16 MΩ cm), and sulfuric acid (Aldrich, 99.999% purity, A.C.S. reagent) were used to purify Nafion. Virgin, electrical-grade polytetrafluoroethylene (PTFE 1016 μm and 254 μm thick) was purchased from McMaster-Carr for gravimetric sorption experiments. Methanol (≥99.8%, Aldrich A.C.S. reagent) and RO water were used in sorption and diffusion experiments.

2.2. MEMBRANE PREPARATION

Nafion and PTFE membranes for gravimetric sorption were cut into approximately 3 and 5 cm^2, respectively. Nafion membranes for FTIR-ATR experiments were trimmed to the size of the long reflecting face of the zinc selenide (ZnSe) crystal, 6 × 1 cm. Samples for conductivity were punched into circles ~1.6 cm in diameter. All Nafion samples were subsequently purified, similar to a procedure reported elsewhere,[20] by refluxing in 3 wt% hydrogen peroxide, then in RO water, next in 0.5 M sulfuric acid, and finally in RO water again. Membranes were rinsed thoroughly with RO water after every one hour step.

2.3. GRAVIMETRIC SORPTION

For gravimetric sorption (weight uptake) experiments, all Nafion membranes, weighing approximately 100 mg, were purified as described above, then soaked in RO water for at least one day prior to immersing in solutions. PTFE membranes, weighing approximately 5 and 1 g for the 1,016 and 254 μm thicknesses, respectively, were used as received. Membranes were immersed in a large excess of water, methanol, or a methanol/water mixture (3, 6, 13, 27, or 57 wt% methanol). Wet weights were measured 2–3 times over the course of several days, for Nafion, or several weeks, for PTFE, to ensure equilibrium sorption. Equilibration time was longer for samples in methanol solutions than for those in water. The weighing process involved removing the membrane from liquid, carefully patting its surface to remove excess liquid, immediately placing on a Mettler Toledo AB54-S balance (0.1 mg accuracy), and finally returning it to its respective liquid. After reaching wet equilibrium, the Nafion membranes were dried for 3–5 days at ambient conditions and dry weights recorded. Laboratory conditions during these experiments ranged from 18–24°C and 10–20% relative humidity. Dry weights of the PTFE membranes were measured before immersion in solution. PTFE dry weights were invariant between vacuum drying and ambient drying. The weight uptake or sorption of liquid was determined by:

$$C_T = \frac{m_{wet} - m_{dry}}{m_{wet}} \qquad (1)$$

where C_T is total solute concentration by weight, m_{wet} is wet membrane weight, and m_{dry} is dry membrane weight. At least three samples were studied at each concentration. A minimum of three expe riments were conducted on each sample and the values reported are the average of these experiments. The average standard deviation was 7% of the average C_T values.

Dimensions: length, width, and thickness of samples used for gravimetric sorption were also measured. These measurements provide volume of the samples as a function of methanol concentration, which, with weight uptake, can be used to calculate swollen Nafion densities. Lateral dimensions of extruded Nafion 117 were measured with digital calipers (VWR) and found to be slightly anisotropic (17–36%). The membrane in each solution was measured at 5 different locations. These measurements were repeated three times over the time frame of the gravimetric experiments, resulting in an average standard deviation that is 1% of each average lateral dimension. Membrane thickness was measured with a digital micrometer (Mitutoyo) with 1 μm accuracy. Each thickness measurement was the average of 5–10 readings at different positions on the membrane and was repeated at least twice on each sample (at each solution concentration). The values reported are the average of those experiments. The average standard deviation was 3% of the average membrane thickness.

2.4. FOURIER TRANSFORM INFRARED – ATTENUATED TOTAL REFLECTANCE (FTIR-ATR) SPECTROSCOPY

Infrared spectra were collected using an FTIR spectrometer (Nicolet 6700 Series; Thermo Electron). Each spectrum was the average of 32 scans, had a resolution of 4 cm^{-1}, and was corrected by a background spectrum of the lank ATR element. The spectrometer was equipped with a temperature-controlled ATR cell (Specac, Inc.) containing a multiple-reflection, ZnSe crystal (Specac, Inc.). The crystal was 0.6 cm thick, 1 cm wide, 6 cm long, and had entrance and exit faces beveled at 45°. The mercury-cadmium-telluride (MCT) detector was liquid nitrogen-cooled.

For each experiment, a background spectrum of the ATR crystal was recorded and saved. Then the flow-through ATR cell was opened, the ZnSe crystal was removed, and a hydrated Nafion sample was placed on the crystal. The membrane-covered crystal was returned to the cell, which was then closed. Dry air was flowed through the ATR cell for four hours in order to dry the membrane. The hydroxyl stretching and bending vibrations of water in Nafion were monitored during approach to dry steady state. After reaching a dry steady state, the flow-through cell was filled with RO water in order to re-hydrate the membrane. Re-hydration required two hours, and steady state was determined by monitoring the same hydroxyl vibrations. The drying step was necessary to drive-off any bulk water between the membrane and crystal. At the end of each experiment the membrane-covered crystal was examined for complete adhesion, which was achieved. Hydrated Nafion was the starting condition chosen for these experiments, hence the re-hydration step. Other work, in preparation and not presented here, begins with a dry steady state.

2.5. MULTICOMPONENT SORPTION (STEADY-STATE FTIR-ATR SPECTROSCOPY)

Steady-state FTIR-ATR spectra of Nafion equilibrated with aqueous solutions of 0, 3, 6, 13, 27, and 57 wt% methanol provide the relative amounts water and methanol sorbed in the membrane. If calibrated with gravimetric sorption, FTIR-ATR can provide multicomponent sorption data (i.e. the concentration of water and methanol in the membrane). The peak heights of the absorption bands associated with H-O-H bending of water, A_W, and C-O stretching of methanol, A_M, were measured at each solution concentration. A mass balance on the solutes gives:

$$C_T = C_M + C_W \qquad (2)$$

where C_T, C_M, and C_W are, respectively, the total solute concentration, the methanol concentration, and the water concentration in the membrane.

For weak to moderate IR absorption, concentration can be related to absorbance through a differential Beer-Lambert law that incorporates the evanescent decay of the ATR infrared absorption:

$$A(t) = \int_0^\lambda \varepsilon C(t) \exp\left(\frac{-2z}{d_p}\right) dz \qquad (3)$$

where d_p is the depth of penetration of the evanescent wave into the sample. For a film that is 10 times thicker than the depth of penetration ($\ell/d_p > 10$), Eq. (3) simplifies to:

$$A(t) = \left(\frac{\varepsilon d_p}{2}\right) C(t) \qquad (4)$$

Combining constants and substituting Eq. (4) into Eq. (2) gives:

$$C_T = \frac{A_M}{\varepsilon_M d_{pM}} + \frac{A_W}{\varepsilon_W d_{pW}} \qquad (5)$$

where A_i, ε_i, d_{pi} represent the absorbance, extinction coefficient, and depth of penetration for species i, where M and W correspond to methanol and water, respectively. Dividing Eq. (5) by A_W gives:

$$\frac{C_T}{A_W} = \frac{1}{\varepsilon_M d_{pM}} \frac{A_M}{A_W} + \frac{1}{\varepsilon_W d_{pW}} \qquad (6)$$

Plotting C_T/A_W versus A_M/A_W yields the calibration constants for methanol (slope) and water (y-intercept).[21]

2.6. DIFFUSION (TIME-RESOLVED FTIR-ATR SPECTROSCOPY)

Time-resolved spectra for diffusion experiments were collected every 12.75 seconds. To begin each diffusion experiment, a well-stirred, methanol/water solution of known concentration was pumped at 5 ml/min into the ATR cell (over the hydrated Nafion 117 membrane). This flow rate was chosen to avoid any mass transfer resistance at the liquid/polymer interface and to maintain a constant source concentration, while not producing excessive amounts of waste. The ATR outlet was not recycled. With this flow rate, the flow-through cell (V = 550 µl) was completely replenished with a fresh methanol/water solution twice per data point (spectra). All diffusion experiments were performed at 25°C.

The "effective" diffusion coefficient (neglecting multicomponent effects) of methanol or water in Nafion for this ATR system can be descrybed by one-dimensional, binary, Fickian diffusion and has been solved analytically:

$$\frac{C-C_0}{C_{eq}-C_0} = 1 - \frac{4}{\pi} \times \sum_{n=0}^{\infty} \frac{(-1)^n}{2n+1} \exp(-Df^2 t)\cos(fz) \quad (7)$$

where

$$f = \frac{(2n+1)\pi}{2} \quad (8)$$

Substitution of Eq. (7) into Eq. (3) and integration yields[17]:

$$\frac{A(t)-A_0}{A_{eq}-A_0} = 1 - \frac{8}{\pi d_p [1-\exp(-2/d_p)]} \times \sum_{n=0}^{\infty} \frac{1}{2n+1} \left[\frac{\exp(-f^2 Dt)[f\exp(-2/d_p)+(-1)^n 2/d_p]}{(2/d_p)^2 + f^2} \right] \quad (9)$$

where $A(t)$ is the ATR absorbance at time t and A_{eq} is the absorbance at equilibrium. D is the only adjustable parameter in this model, which can be regressed to determine effective diffusion coefficients for methanol and water in Nafion. At least two diffusion experiments were conducted at each solution concentration (3, 6, 13, 27, 57 wt% methanol in water).

2.7. ELECTROCHEMICAL IMPEDANCE SPECTROSCOPY (EIS)

Proton conductivity was calculated from membrane thickness and resistance. Membrane resistance was measured using EIS (Solartron SI 1287 Electrochemical Interface and SI 1260 Impedance/Gain-Phase Analyzer) at alternating current (AC) frequencies ranging from 100 Hz to 1 MHz. Two 1.22 cm^2 stainless steel blocking electrodes were used to measure resistance perpendicular to the plane of the membrane. All membranes were immersed in RO water, methanol, or methanol solutions (3, 6, 13, 27, or 57 wt%) for one week and then quickly removed from solution and placed between the electrodes in a sealed cell. Membrane resistance was taken as the real part of the impedance data. Membrane thickness was measured after re-immersing each membrane in its respective solution for one day after each conductivity experiment. Conductivity values for each concentration are an average of multiple experiments, where the average standard deviation was

6% of those values. Samples were allowed to re-equilibrate in their solutions for three days between repeat experiments. A schematic diagram of the apparatus and more details can be found elsewhere.[3, 22]

3. Results

3.1. GRAVIMETRIC SORPTION

Figure 1 depicts the total solute concentration sorbed by Nafion (◇) and PTFE (△) from methanol/water solutions ranging from 0 wt% methanol to 100 wt% methanol. The amount of solvent sorbed into Nafion increases with increasing methanol concentration to ~45 wt%, where it seems to plateau. Contrast this with the lack of solvent taken up by PTFE. The chemical structure of Nafion consists of a completely fluorinated backbone, identical to PTFE, off of which are perfluoroether side chains that terminate

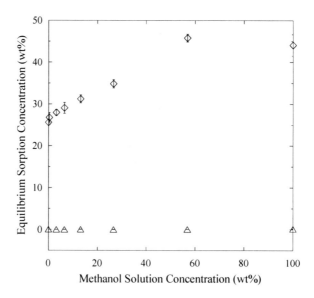

Figure 1. Total equilibrium sorption, by weight, of water and methanol in Nafion (◇) and PTFE (△) plotted as a function of methanol solution concentration. Extreme points show gravimetric uptake from pure solvents, water and methanol. Note: methanol solution concentration is grams of methanol per total grams of aqueous methanol solution.

in a sulfonic acid moiety. The side chains are both ionic and hydrophilic and therefore phase separate from the highly hydrophobic backbone. Contrasting the lack of uptake by PTFE with the large uptake by Nafion of water (25.6 wt%) and methanol (44.1 wt%) is interesting because it suggests that all the solute exists in the hydrophilic, ionic regions of the polymer.

As discussed in the experimental section, multicomponent sorption can be measured with FTIR-ATR, by calibrating with total sorption from gravimetric experiments. This was done for all concentrations being considered, using the steady-state absorbances of the H-O-H bending of water and the C-O stretching of methanol, which are shown in Figure 2a. The calibration yields component concentrations of methanol, C_M (□),and water, C_W (○) in Nafion (Figure 2b). The plateau in total solute concentration appears to be caused by methanol, the concentration of which reaches a maximum around 33 wt%. Water concentration in Nafion decreases linearly with increase of methanol concentration in the equilibrating solution.

In Figure 3, the density of the Nafion/solvent system is plotted as a function of the methanol solution concentration. The density of dry Nafion, not shown in Figure 3, was measured as 1.94 g/ml, which compares with that reported by others, 2.05 g/ml.[23] There is a large decrease in density when Nafion is water hydrated to 1.58 g/ml, which agrees well with other findings.[24] The density decreases further with increasing methanl concentration, eventually reaching 1.16 g/ml in pure methanol. The solid line in Figure 3 is a calculated density based on volume additivity between the polymer and solvents. This calculation was performed by first finding the methanol/water concentration in the membrane from multicomponent sorption. The density of methanol/water mixtures from established thermodynamic data was used, which accounts for the volume change on mixing between methanol and water.[25] Unless dissolution takes place, which was not observed, the mass of polymer is unchanged and the dry density is known.

The density of the polymer/solvent system was calculated as the concentration weighted sum of the mixture density and the polymer density. As can be seen is Figure 3, volume additivity does not strictly hold in this system. This is not surprising because there are strong ionic interactions between the solvents and polymer. In other words, the difference between measured and calculated densities should correspond to the volume change upon mixing between methanol/water and Nafion.

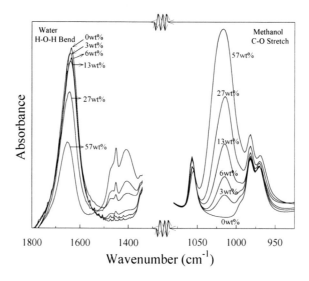
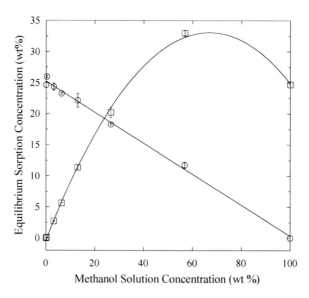

Figure 2. (a) Steady-state infrared spectra of Nafion equilibrated in designated methanol/water solutions. Note: 0 wt% corresponds to hydrated Nafion. (b) Component concentrations in Nafion as a function of equilibrating solution concentration for methanol, C_M (□), and water, C_W (○).

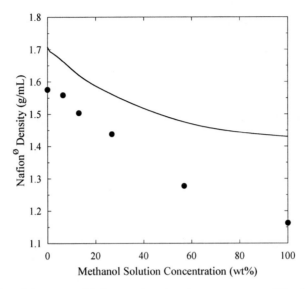

Figure 3. Density of the system: Nafion, methanol, and water versus equilibrating methanol solution concentration. Solid line depicts density calculated from component concentrations assuming volume additivity (without accounting for volume change upon mixing) between solutes and polymer.

3.2. MULTICOMPONENT SORPTION

One of the reasons FTIR-ATR is unique is because it provides multicomponent sorption data, which allows for the calculation of partition coefficients. Figure 4 displays the partition coefficients of methanol, $K_M = C_M/C_{Bm}$ (□), and water, $K_W = C_W/C_{Bw}$ (○), versus equilibrating solution concentration. C_{Bm} and C_{Bw} are the grams of methanol and water, respectively, in solution per total grams of solution. It should be noted that the concentration units used in the ratio of internal concentration to external concentration affect the resulting values of the partition coefficients.

In this work, partition coefficients were calculated on a weight percent basis. This was for simplicity since the concentrations in the membrane were measured in weight percent. For instance, the partition coefficient of water, which is relatively constant, averages 0.26 ± 0.03. If concentrations are converted to a volume basis, then the water partition coefficient in Nafion averages 0.15. If excess volume between water and methanol is ignored, the partition coefficient becomes 0.38. Clearly, avoiding volume-based conversions and thereby extra variability is desirable demonstrated by the difference in measured and calculated densities in Figure 3. The partition coefficient of methanol in Nafion ranges from 0.30 at low concentration to a maximum of 0.87 around 15 wt% methanol, decreasing to 0.44 when Nafion is equilibrated in pure methanol. This maximum is interesting and may suggest cooperative swelling of water and methanol in Nafion. The idea of cooperative swelling is further supported by consideration of the Hildebrand solubility parameters. One of the solubility parameters of Nafion, 16.7 $(cal/cm^3)^{1/2}$, falls in between those of methanol, 14.5 $(cal/cm^3)^{1/2}$, and water, 23.4 $(cal/cm^3)^{1/2}$.[26]

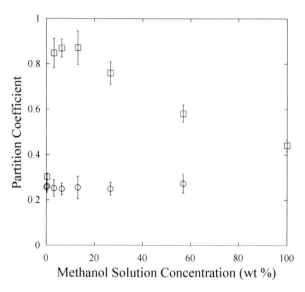

Figure 4. Partition coefficients for methanol, $K_M = C_M/C_{Bm}$ (□), and water, $K_W = C_W/C_{Bw}$ (○), vs. methanol solution concentration.

3.3. MULTICOMPONENT TRANSPORT

In addition to being able to measure multicomponent sorption, FTIR-ATR can measure the sorption kinetics or diffusion of multiple components. Figure 5 shows time-resolved spectra of 6 wt% methanol diffusing into hydrated Nafion 117. The insets show the absorbance decrease of water H-O-H bending and the absorbance increase of methanol C-O stretching.

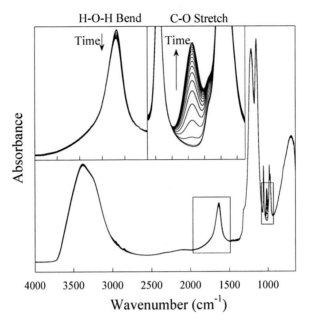

Figure 5. Infrared spectra of 6 wt% methanol diffusion in hydrated Nafion at selected time points. Insets show decrease of water H-O-H bending band and increase of the methanol C-O stretching band as a function of time.

The absorbance of each component can be measured at each time point and plotted versus time to yield diffusion curves seen in Figure 6a. Equation (9) has been regressed to these curves to yield effective diffusion coefficients. The effective methanol diffusion coefficient in Nafion for this experiment is 2.75×10^{-6} cm^2/s, and the effective water counter-diffusion coefficient is 4.44×10^{-6} cm^2/s. Repeating this process multiple times for each methanol concentration considered and averaging the results at each concentration yields Figure 6b. The effective methanol diffusion coefficients (□) are plotted against the methanol concentration in the membrane and increase from $2.61 \pm 0.05 \times 10^{-6}$ cm^2/s at 2.7 ± 0.1 wt% to $5.84 \pm 0.06 \times 10^{-6}$ cm^2/s at 33.0 ± 0.4 wt%. The water diffusion coefficients (○) are plotted versus the water concentration in the membrane. Both methanol and water diffusion coefficients are not strong functions of concentration.

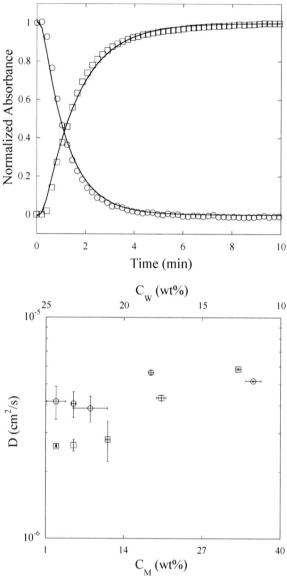

Figure 6. (a) Time-resolved, normalized absorbances for the methanol C-O stretching vibration (□) and water H-O-H bending vibration (○) in Nafion. Solid lines are regressions to the ATR solution of binary, Fickian diffusion, Eq. (9). The regressions determine the effective methanol diffusion coefficient in Nafion ($D_M = 2.75 \times 10^{-6}$ cm^2/s; 6 wt%, 25°C) and the effective water counter-diffusion coefficient ($D_W = 4.44 \times 10^{-6}$ cm^2/s). (b) Semilog plot of effective methanol diffusion coefficients (□) and water counter-diffusion coefficients ○) in Nafion versus their respective concentrations within the membrane.

Fluxes of methanol and water in Nafion can be calculated by combining the steady-state and transient FTIR-ATR data. As Figure 6b shows, the diffusion coefficient and concentration in the membrane of each component is known for each experiment. Methanol flux is then:

$$J_M = \frac{D_M \Delta C_M}{\lambda} \qquad (10)$$

and water flux:

$$J_W = \frac{D_W \Delta C_W}{\lambda} \qquad (11)$$

where λ is thickness of the membrane.

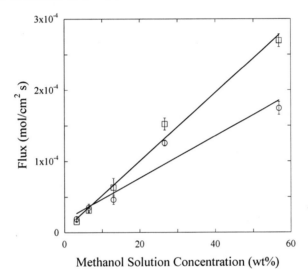

Figure 7. Methanol flux, $J_M = D_M \Delta C_M/\ell$ (□), and water flux, $J_W = D_W \Delta C_W/\ell$ (○), plotted versus equilibrating methanol solution concentration. Solid lines are linear fits to the data.

During a transient experiment, initially the entire membrane is hydrated with water, having a concentration, C_o, of 25.6 wt% water and 0.0 wt% methanol. When the experiment begins, the concentration in the membrane at the interface with the solution equals the concentration that the entire membrane will reach at steady state, C_{eq}. So the concentration gradient used to calculate the initial, maximum flux is $\Delta C = C_{eq} - C_o$. The resulting flux has been converted to conventional units (mol/cm² s) with the system density from Figure 3 and the molecular weight of each solute. As shown in Figure 7, the methanol flux increases with increasing concentration of the solution from $1.51 \pm 0.07 \times 10^{-5}$ mol/cm² s at 3 wt% methanol to 27.0 ± 0.9

× 10^{-5} mol/cm² s at 57 wt% methanol. The water flux also increases, from 1.92 ± 0.33 × 10^{-5} mol/cm² s at 3 wt% methanol to 17.4 ± 0.9 × 10^{-5} mol/cm² s at 57 wt% methanol. The methanol flux increases more steeply than the flux of water, which is undesirable.

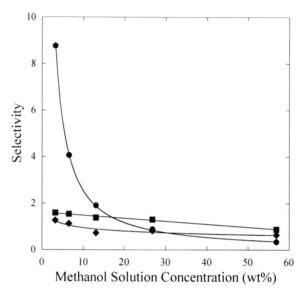

Figure 8. Sorption selectivity (●), diffusion selectivity (■), and flux selectivity (♦) of Nafion, where selectivity is the ratio of the water concentration in the membrane, diffusion coefficient, or flux, respectively, to that of methanol. Solid lines are trend lines.

A better DMFC membrane would demonstrate a shallower slope of methanol flux. To consider the issue from the perspective of selectivity, each of the water coefficients can be ratioed to the corresponding methanol coefficient. A higher selectivity equates to a better performing DMFC membrane. This is shown in Figure 8. Interestingly, a DMFC using Nafion performs best with less than 10 wt% methanol feed, which corresponds to a high sorption selectivity in Figure 8. Although the diffusion and flux selectivities of Nafion decrease with increasing methanol concentration, it is slight compared to the order of magnitude decrease in sorption selectivity (8.8–0.35 from 3 wt% to 57 wt% methanol). In order to operate DMFCs at higher methanol concentrations, the sorption selectivity for water over methanol of the PEM must be improved.

TABLE 1. Water and methanol concentrations in the membrane, effective diffusion coefficients, and fluxes.

C_{Bm} (wt%)	C_M (wt%)	C_W (wt%)	$D_M \times 10^{-6}$ (cm²/s)	$D_W \times 10^{-6}$ (cm²/s)	$J_M \times 10^{-5}$ (mol/cm²s)	$J_W \times 10^{-5}$ (mol/cm²s)
0	0.0	25.6 ± 0.2				
3	2.7 ± 0.1	24.4 ± 0.6	2.61 ± 0.03	4.15 ± 0.71	1.51 ± 0.07	1.92 ± 0.33
6	5.6 ± 0.01	23.2 ± 0.1	2.64 ± 0.11	4.06 ± 0.55	3.09 ± 0.30	3.49 ± 0.54
13	11.3 ± 0.3	22.2 ± 1.1	2.80 ± 0.57	4.07 ± 0.56	6.34 ± 1.29	4.63 ± 0.67
27	20.2 ± 0.8	18.3 ± 0.1	4.32 ± 0.09	5.63 ± 0.12	15.2 ± 0.8	12.5 ± 0.4
57	33.0 ± 0.4	11.7 ± 0.5	5.84 ± 0.04	5.16 ± 0.02	27.0 ± 0.90	17.4 ± 0.9
100	44.1 ± 2.6	0.0				

3.4. CONDUCTIVITY

Table 2 shows proton conductivity of Nafion membranes equilibrated in methanol/water solutions. Nafion proton conductivity decreases with increasing methanol concentration (decreasing water content in the membrane). Proton conductivity decreasing with decreasing water content has been demonstrated for polyelectrolyte membranes equilibrated with activities of water vapor.[24] Interestingly, this study shows that a similar relationship holds for fully liquid swollen membranes. The decrease in proton conductivity with increasing methanol concentration from 31 mS/cm in pure water to 17 mS/cm in pure methanol is slight compared to the large increase in methanol flux, and therefore should contribute little to DMFC performance decrease. In fact, the significant conduction ability of Nafion swollen purely with methanol suggests methanol can solvate the sulfonic acid sites of Nafion.

Continuing the selectivity analysis requires converting proton conductivity to proton flux. Proton conductivity of a membrane with immobilized anions, can be defined in terms of flux of protons, J_{H+}, and the gradient of the electric field potential, $\nabla \phi$, which, if expressed in volts, requires the use of Faraday's constant, $\Im = 9.652 \times 10^4$ C/mol.[27]

TABLE 2. Hydrated Nafion 117 proton conductivity as a function of methanol solution concentration.

C_{Bm} (wt%)	0	3	6	13	27	57	100
σ (mS/cm)	31	30	27	24	19	17	17

$$\sigma = \frac{J_{H^+}\Im}{\nabla\phi} \quad (12)$$

For conductivity experiments, the membrane is in equilibrium with the solution in which it is immersed, where there are no concentration gradients, and the only driving force is the electric potential. Therefore, the flux of protons can be expressed as:

$$J_{H^+} = \sigma\frac{\Im}{RT}\nabla\phi \quad (13)$$

where the gradient of the electrostatic potential is 10 mV divided by the thickness of the membrane, which is known at each concentration. Figure 9 depicts proton/methanol selectivity expressed as proton flux divided by methanol flux as a function of proton flux.

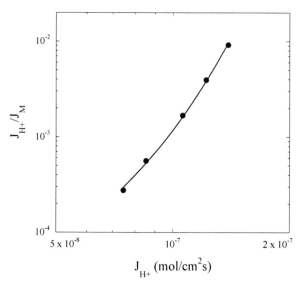

Figure 9. Proton flux to methanol flux in Nafion equilibrated with methanol/water solutions versus proton flux. Solid line is an exponential trend line.

The highest selectivity, 9.2×10^{-3}, and proton flux, 1.4×10^{-7} mol/cm^2 s, are at the lowest methanol concentration, 3 wt%. The lowest selectivity, 2.8×10^{-4}, and proton flux, 7.5×10^{-8} mol/cm^{2}s, are at the highest methanol concentration, 57 wt%. In an ideal DMFC membrane, the proton/ methanol selectivity and proton flux would remain high as the methanol solution concentration is increased. Unfortunately, for Nafion as with water/ methanol sorption selectivity, the proton/methanol selectivity decreases almost two orders of magnitude. This reiterates the need to exclude methaol from DMFC membranes.

4. Conclusions

FTIR-ATR is a powerful tool for studying selectivity in polymers, because it is able to measure not only multicomponent sorption, but also multicomponent diffusion with a single experiment. This ability allows the relation between diffusion and sorption to be examined directly. Since the concentration in the membrane and the diffusion coefficient for methanol and water can now be measured, the concentration dependence of the fluxes can be effectively examined, and are seen to increase with increasing methanol concentration. Furthermore, the sorption, diffusion, and flux selectivities were examined and decreased with increasing methanol concentration, sorption selectivity most significantly. This work shows that improving PEMs for the DMFC requires reducing the concentration dependence of the methanol flux and improving the sorption selectivity of the membrane for water over methanol.

Acknowledgments

The authors acknowledge the financial support of the National Science Foundation (CAREER 0644593; IGERT 0221664), the US Army Research Office (W911NF-05-1-0036), the U.S. Department of Education through the Graduate Assistance in Areas of National Need (GAANN) Fellowship, and the George Hill, Jr. Fellowship.

References

1. R.W. Baker, *Membrane Technology and Applications* (McGraw-Hill, New York, 2000), p. 1.
2. *Membrane Separation Processes*, edited by P. Meares (Elsevier, Amsterdam, 1976), Preface.
3. N.W. DeLuca and Y.A. Elabd, *Journal of Membrane Science*, 282, 217–224 (2006).
4. J. Larminie and A. Dicks, *Fuel Cell Systems Explained*; 2nd edn (Wiley, New York, 2003) pp. 141–161.
5. N.W. DeLuca and Y.A. Elabd, *Journal of Power Sources*, 163, 386–391 (2006).
6. M.W. Verburgge, *Journal of the Electrochemical Society*, 136, 417–423 (1989).
7. D. Nandan, H. Mohan, and R.M. Iyer, *Journal of Membrane Science*, 71, 69–80 (1992).
8. E. Skou, P. Kauranen, and J. Hentschel, *Solid State Ionics*, 97, 333–337 (1997).
9. C.M. Gates and J. Newman, *AIChE Journal*, 46, 2076–2085 (2000).
10. Z. Ren, T.E. Springer, T.A. Zawodzinski, and S.J. Gottesfeld, *Journal of the Electrochemical Society*, 147, 466–474 (2000).
11. S. Hietala, S.L. Maunu, and F. Sundholm, *Journal of Polymer Science Part B Polymer Physics*, 38, 3277–3284 (2000).
12. M. Saito, S. Tsuzuki, K. Hayamizu, and T. Okada, *Journal of Physical Chemistry B*, 110, 24410–24417 (2006).
13. D. Rivin, C.E. Kendrick, P.W. Gibson, and N.S. Schneider, *Polymer*, 42, 623–635 (2001).
14. H.A. Every, M.A. Hickner, J.E. McGrath, and T.A. Zawodzinski, *Journal of Membrane Science*, 250, 183–188 (2005).
15. J.R.P. Jayakody, A. Khalfan, E.S. Mananga, S.G. Greenbaum, T.D. Dang, and R. Mantz, *Journal of Power Sources*, 156, 195–199 (2006).
16. J.L. Duda and J.M. Zielinski, in: *Diffusion in Polymers*. edited by P. Neogi. (Marcel Kekker, New York, 1996), 143–172.
17. Y.A. Elabd, M.G. Baschetti, and T.A. Barbari, *Journal of Polymer Science Part B: Polymer Physics*, 41, 2794–2807 (2003).
18. Y.A. Elabd, and T.A. Barbari, *AIChE Journal*, 48, 1610–1620 (2002).
19. Y.A. Elabd, and T.A. Barbari, *AIChE Journal*, 47, 1255–1262 (2001).
20. T.A. Zawodzinski Jr., M. Neeman, L.O. Sillerud, and S. Gottesfeld, *Journal of Physical Chemistry*, 95, 6040–6044 (1991).
21. D.T. Hallinan Jr. and Y.A. Elabd, *Journal of Physical Chemistry B*, 111(46), 13221–13230, (2007).
22. Y.A. Elabd, C.W. Walker, F.L. Beyer, *Journal of Membrane Science*, 231, 181–188 (2004).
23. D.R. Morris and X. Sun, *Journal of Applied Polymer Science*, 50, 1445–1452 (2003).
24. R.F. Silva, M. De Francesco, and A. Pozio, *Journal of Power Sources*, 134, 18–26 (2004).
25. *Perry's Chemical Engineers' Handbook*; 7th edn, edited by R.H. Perry and D.W. Green (McGraw-Hill, New York, 1997), p. 111.
26. H. Chen, J.D. Snyder, and Y.A. Elabd, *Macromolecules*, 41(1), 128–135 (2008).
27. E.N. Lightfoot, *Transport Phenomena and Living Systems – Biomedical Aspects of Momentum and Mass Transport* (Wiley, New York, 1974), 154–194.

KINETICS AND KINETICALLY LIMITED PERFORMANCE IN PEMFCs AND DMFCs WITH STATE-OF-THE-ART CATALYSTS

H.A. GASTEIGER* Y. LIU, D. BAKER, AND W. GU
*Acta S.p.A., Via di Lavoria 56/G, 56040 Crespina, Italy
General Motors Corporation – Fuel Cell Activities
Honeoye Falls, 14472 NY, USA*

1. Introduction

Over the past 10 years, extensive R&D efforts to optimize H_2/air-fed proton exchange membrane fuel cell (PEMFC) performance resulted in power densities near 1 W/cm^2 at ≈0.6 V and much reduced MEA (membrane electrode assembly) platinum loadings of ≈0.45 mg$_{Pt}$/cm$^2_{MEA}$[1]. These accomplishments were largely driven by the implementation of thin membranes (≈25 μm), highly conductive bipolar plate materials/coatings, and empirical improvements in electrode design. In order to close the remaining performance gap for automotive applications, it is critical to quantify the various voltage losses in state-of-the-art H_2/air MEAs, as this will enable more targeted future MEA materials and design development.

Similarly, much progress has been made in materials and engineering design of direct methanol fuel cells (DMFCs),[2] and it is instrumental to deconvolute materials related losses (catalyst activity and ohmic resistances) from mass-transport related losses. This analysis will again enable the determination of performance gains which can be made by either MEA materials or MEA design improvements.

Thus, this paper will review the activity of currently known anode and cathode catalysts in H_2/air PEMFCs and DMFCS. Using these well-known kinetics, an analysis of the various voltage loss contributions will be conducted in order to determine the impact of mass transport losses and proton conduction losses, with the hope of being able to provide a clear focus on future development needs with regards to materials development and MEA engineering optimization.

---

*Hubert Gasteiger, Acta S.p.A., Via di Lavoria 56/G, 56040 Crespina, Italy, e-mail: hubert.gasteiger@gmail.com

2. H2/Air PEMFC Catalysis and Performance

Current PEFMC power densities and Pt loadings equate to a Pt-specific power density of ≈0.5 g_{Pt}/kW, compared to automotive targets of <0.2 g_{Pt}/kW dictated by both platinum cost and supply constraints.[1] In addition, MEA power densities must be maintained at the current levels of ≈0.9 W/cm^2 in order to minimize the cost of other MEA components (membranes, diffusion media, bipolar plates). At the same time, cell voltages at maximum power density must be >0.6 V to achieve high voltage efficiency (note that the commonly quoted maximum power density at cell voltages near 0.4–0.5 V is not relevant for automotive applications). In order to determine the most efficient approach toward reaching the above targets, i.e. to determine whether new MEA materials are required or whether improved engineering design of MEAs is sufficient, it is necessary to quantify the various voltage loss terms contributing to the measured cell voltage, E_{cell}:

$$E_{cell} = E_{rev} - iR_\Omega - \eta_{HOR} - |\eta_{ORR}| - i \cdot \left(R^{eff}_{H^+,ca} - R^{eff}_{H^+,an}\right) - \eta_{tx,O2(dry)} - \eta_{tx,O2(wet)} \quad (1)$$

where i is the current density, R_Ω is the sum of the Ohmic resistances of proton conduction through the membrane and of electron conduction which is primarily due to the electric contact resistance between the flow-field plates and the diffusion media.[3] The overpotentials for the hydrogen oxidation reaction (HOR) and the oxygen reduction reaction (ORR) are described by η_{HOR} and η_{ORR}, respectively. Contrary to the purely Ohmic resistance term, R_Ω, the effective resistances to proton conduction in the electrodes, $R^{eff}_{H^+,ca}$ and $R^{eff}_{H^+,an}$, are not simple Ohmic resistances as they depend on current density.[4] Finally, the gas-transport related losses for fuel cell operation with pure H_2 and air are mainly due to the O_2 transport resistance from the cathode flow-field through the diffusion medium (DM) and into the cathode electrode. Conceptually, these can be separated into transport losses which would occur if the DM and the electrode were entirely free of liquid water, $\eta_{tx,O2(dry)}$, and another term which is the additional O_2 transport loss in the presence of liquid water in the DM and the electrode, $\eta_{tx,O2(wet)}$. In the following, we will quantify the different voltage loss terms in Eq. (1).

2.1. OHMIC RESISTANCE LOSSES

The Ohmic resistance, R_Ω, consists of two components: (i) the proton conduction resistance of the membrane, R_{memb}, which strongly depends on relative humidity (RH) and is only a very weak function of temperature, and (ii) the electronic resistance, R_{elec}, which mainly depends on the compressive force applied to the cell and the surface properties of the flow-field plates. For graphite-based flow-fields under a typical compressive stress of 250 psig, the latter has a value of $R_{elec} \approx 0.025$ $\Omega \cdot cm^2$.[3] The RH-dependence of the resistance of a 25 μm thick Nafion211 membrane (1050 equivalent weight (EW) in units of gpolymer/mol H$^+$) was measured by AC-impedance[5] and is shown in Figure 1, illustrating the very strong RH-dependence of typical perfluoro-sulfonicacid (PFSA) membranes.

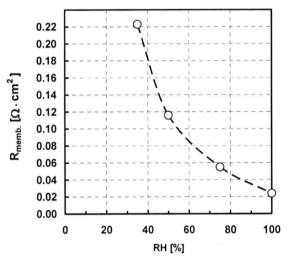

Figure 1. Resistance of a 25 μm thick Nafion211 membrane (1050 EW) as a function of RH at 80°C, measured by AC-impedance.[5]

2.2. HOR KINETIC LOSSES

The kinetic losses for the oxidation of pure H_2 are very small, owing to the fast H_2 oxidation kinetics on the currently used carbon-supported platinum catalysts (Pt/C). While the H_2 oxidation and evolution kinetics in a PEMFC were shown to be consistent with a Butler-Volmer equation, its Tafel-slope, b, and exchange current density, i_0, could only be determined within a factor of two (80°C and 100 kPa$_{abs}$ H_2)[6]: $b = 2.303 \cdot RT/F$ to $2.303 \cdot 2RT/F$ (where R = 8.314 J/mol/K and F = 96,485 As/mol) and $i_0 = 0.24$ to 0.60 A/cm$^2_{Pt}$.

In view of the very high exchange current density, the overpotential losses of a Pt/C anode electrode should be negligible, even for the very low anode platinum loadings of 0.05 mg$_{Pt}$/cm^2 which are needed to meet automotive cost targets. This is illustrated in Figure 2, which shows the anode overpotential, η_{HOR}, as a function of Pt-mass normalized current density, also referred to as Pt *mass activity* (in units of A/mg$_{Pt}$). Quite clearly, the HOR kinetics are linear up to ≈ 500 A/mg$_{Pt}$, and under automotive requirements of

≈30 A/mg$_{Pt}$ (0.05 mg$_{Pt}$/cm^2 at 1.5 A/cm^2) the voltage losses are only ≈2 mV. Thus, the η_{HOR}-term in Eq. (1) is negligible over the entire current density range.

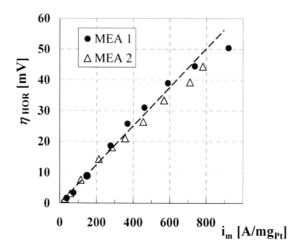

Figure 2. H$_2$ overpotential as a function of Pt mass activity, obtained on a 5% Pt/Vulcan-carbon (ETEK, USA) anode catalyst at 80°C and 100 kPa$_{abs}$ H$_2$.[6] The two different symbols refer to repeat measurements on different MEAs.

2.3. ORR KINETIC LOSSES

In PEMFCs, the overpotential for the ORR, η_{ORR}, versus current density, i, was shown to be consistent with simple Tafel kinetics, with a single Tafel-slope in the entire relevant cathode potential range:[7]

$$\eta_{ORR} = \frac{2.303 \cdot RT}{\alpha_c F} \cdot \log\left[\frac{i}{10 \cdot L_{ca} \cdot A_{Pt,el} \cdot i_{o,s(T,p_{O_2})}}\right] \quad (2)$$

where α_c is the cathodic transfer coefficient, L_{ca} (mg$_{Pt}$/cm^2) is the cathode Pt loading, $A_{Pt,el}$ (m^2$_{Pt}$/g$_{Pt}$) is the electrochemically available Pt surface area in the MEA, and $i_{o,s(T,p_{O_2})}$ (A/cm^2$_{Pt}$) is the catalyst-specific exchange current density for the ORR which depends on temperature and O$_2$ partial pressure.

Conceptually, the exchange current density at any oxygen partial pressure and temperature can be related to an exchange current density at some arbitrary reference conditions as displayed in Eq. (3):[8]

$$i_{o,s(T,p_{O_2})} = i_{o,s}^* \cdot \left(\frac{p_{O_2}}{p_{O_2}^*}\right)^\gamma \cdot \exp\left[\frac{-E_c^{rev}}{RT} \cdot \left(1 - \frac{T}{T^*}\right)\right] \quad (3)$$

Where $i^*_{o,s}$ is the catalyst specific exchange current density normalized to the reference oxygen partial pressure ($p^*_{O_2}$) of 101.3 kPa and a reference temperature (T^*) of 80°C (353 K), γ is the kinetic reaction order with respect to oxygen partial pressure and E^c_{rev} is the activation energy of the ORR at the reversible cell potential (zero overpotential).

The ORR kinetics over a wide range of current density (0.01–0.5 A/cm^2), temperature (35–80°C) and O$_2$ partial pressure (40–400 kPa), could be described quantitatively by Eqs. (2) and (3) using a cathodic transfer coefficient of $\alpha_c = 1$ together with the fitting parameters listed in Table 1.[7]

TABLE 1. ORR kinetic parameters for 50% Pt/C in a PEMFC, based on a fit to Eqs. (2) and (3) using a value of $\alpha_c = 1$. Reference conditions: $T^* = 80°C$ and $p^*_{O2} = p^*_{H2} = 101.3$ kPa$_{abs.}$[7]

Parameter	Fitted value
$i^*_{o,s}$ [A/cm$^2_{Pt}$]	$2.5 \pm 0.3 \times 10^{-8}$
γ	0.54 ± 0.02
E_c^{rev} [kJ/mol]	67 ± 1

Quite obviously, the exchange current density for the ORR is approximately seven orders of magnitude lower than that for the HOR (see 2.2). Therefore, the voltage penalty for lowering the current cathode loadings of ≈0.3–0.4 mg$_{Pt}$/cm^2 to the target of ≈0.15 mg$_{Pt}$/cm^2 is on the order of 25–35 mV, and improved catalysts need to be developed to allow for reduced MEA Pt loadings without loss in cell voltage efficiency and power density. The automotive requirement for the activity of novel cathode catalysts vs. that of currently used Pt/C is '4x',[1] and a factor of ≈2x can be obtained using Pt-alloy catalysts (e.g. PtCo/C[1, 9, 10]). The ORR kinetics of PtCo/C, the most commonly used Pt-alloy catalyst, follow the same simple Tafel behavior as pure Pt, and have very similar kinetic parameters:[9]

Thus, PtCo/C alloys may be a stepping stone toward the 4x mass activity requirement, but their long-term durability still needs to be determined. Other

TABLE 2. ORR kinetic parameters for 30% PtCo/C in a PEMFC, based on a fit to Eqs. (2) and (3) using a value of $\alpha_c = 1$. Reference cond.: $T^* = 80°C$ and $p^*_{O2} = p^*_{H2} = 101.3$ kPa$_{abs.}$[9]

Parameter	Fitted value
$i^*_{o,s}$ [A/cm$^2_{Pt}$]	4.8×10^{-8}
γ	0.50
E_c^{rev} [kJ/mol]	79

promising 4x catalyst concepts are: (i) 10 × higher activity of the (111)-face of Pt3Ni;[11] and (ii) de-alloyed Pt-alloy/C catalysts.[12]

2.4. PROTON TRANSPORT LOSSES IN THE MEA

Next we will discuss the proton conduction loss terms in the electrodes, $R^{eff}_{H+,ca}$ and $R^{eff}_{H+,an}$ (see Eq. (1)), which are related to the electrodes' sheet resistance, R_{sheet}. In principle, R_{sheet} can be obtained from the RH-dependent conductivity of the ionomer, $\sigma_{(RH,T)}$:

$$R_{sheet} = \frac{t_{electrode}}{\sigma_{(RH,T)} \cdot \varepsilon_i / \tau} \quad (4)$$

where $t_{electrode}$ is the thickness of the electrode, ε_i is the ionomer volume fraction in the electrode, and τ is an effective tortuosity. The ionomer volume fraction depends on the ionomer/carbon weight-ratio (I/C-ratio) in the electrode; for Vulcan or Ketjen carbon-supports, $\varepsilon_i \approx 0.2 \times$ (I/C) [1]. For I/C-ratios of $\geq 1/1$, τ was observed to be ≈ 1,[13] so that R_{sheet} can principally be calculated based on the generally known values of $\sigma_{(RH,T)}$. However, recent studies showed that τ can be a strong function of the I/C-ratio for the frequently used lower I/C-ratios,[5] and it is therefore critical to directly measure R_{sheet} for a specific electrode design.

The basic AC-impedance approach to measure R_{sheet} in PEMFCs was developed several years ago,[14] but careful quantification of R_{sheet} vs. I/C-ratio and RH using this approach was not done until recently.[5] Figure 3 illustrates the strong dependence of R_{sheet} on the I/C-ratio over the entire range of practical interest for electrodes using 50% Pt/Vulcan catalyst (0.4 mg_{Pt}/cm^2) and 1050 EW Nafion ionomer on a 25μm thick Nafion211 1050 EW membrane,[5] More detailed analysis showed that τ (see Eq. (4)) strongly deviates from unity, while the RH-dependence of R_{sheet} was reasonably consistent with the RH-variation of the membrane (viz., $\approx 10 \times$ higher resistance at 35% compared to 100% RH as shown in Figure 1).

In summary, R_{sheet} is an intrinsic electrode property which must be measured for each specific electrode design. Once the RH and temperature dependent R_{sheet} values are known, the effective resistance from proton conduction losses in the electrodes, $R^{eff}_{H+,ca}$ and $R^{eff}_{H+,an}$, can be determined using a current distribution model for either Tafel kinetics (ORR) or linear kinetics (HOR). However, owing to the fast kinetics of the HOR with pure H_2, the current distribution is very much skewed toward the anode/membrane interface, such that the proton transport resistance losses

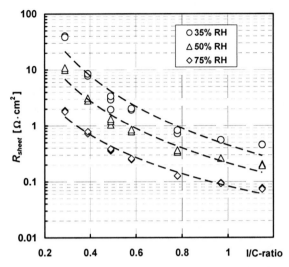

Figure 3. Cathode electrode R_{sheet} vs. I/C-ratio at 35%, 50%, and 75% RH in H2/N2 at 80°C. Data points represent independent measurement on different MEAs.[5] 50%Pt/Vulcan (Tanaka, Japan) at 0.4 mg$_{Pt}$/cm^2 with 1050 EW Nafion ionomer.

in the anode electrode are usually negligible, except under very low RH conditions.[15] Therefore, under most conditions $R^{eff}_{H^+,an}$ is essentially negligible compared to $R^{eff}_{H^+,ca}$, so that it will suffice here to only focus on the proton conduction losses in the cathode electrode.

A cathode electrode current distribution model for simple Tafel kinetics in the absence of O$_2$ mass transport losses in the electrode shows that the effective proton conduction resistance, $R^{eff}_{H^+,ca}$, can be obtained via:[6]

$$R^{eff}_{H^+,ca} = \frac{R_{sheet}}{3+\zeta} \quad (5)$$

where ζ is a current density dependent correction factor, which only depends on the ratio of proton conduction resistance, R_{sheet}, over the charge transfer resistance, b/i, (here b is the Tafel slope corresponding to 2.303·RT/F (see 2.3)[6]). Thus, at very low current densities or at very low R_{sheet} (high I/C-ratio and/or high RH, see Figure 3), the current distribution in the cathode is controlled by the charge transfer resistance (low x-axis values in Figure 4) and $\zeta \approx 0$. This result is consistent with one would expect based on a simple transmission-line model.[14] At higher current densities and/or higher R_{sheet}, the value of ζ increases and $R^{eff}_{H^+,ca}$ deviates significantly from $R_{sheet}/3$. This behavior reflects the non-linear nature of

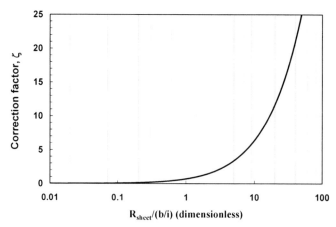

Figure 4. $R^{eff}_{H^+,ca}$ correction factor ζ versus the dimensionless parameter $R_{sheet}/(b/i)$. The plot requires that the ORR follows simple Tafel kinetics, but is valid for all Tafel slopes.[6]

$R^{eff}_{H^+,ca}$, which means that it is not a simple Ohmic (i.e. current independent) resistance.

2.5. WATER TRANSPORT ACROSS THE MEA

As apparent from Figures 1 and 3, both the membrane resistance and the electrode sheet resistance are very strong functions of relative humidity, such that the potential losses in the membrane and the electrodes can only be determined if the *local* RH across membrane and electrodes are known. This requires modeling of the various water transport processes in the MEA: (i) the Fickian diffusion coefficient of water in the membrane, $D_{H2O(memb.)}$; (ii) the Fickian diffusion coefficient of water through the diffusion medium, $D_{H2O(DM)}$; (iii) the electroosmotic drag coefficient in the membrane, n_{H2O}; and, (iv) the thermal gradients normal to the plane of the MEA (from anode to cathode). If these four different water and heat transport properties are known, the relative humidity profile across the MEA can be calculated, and the local RH values can be used to obtain the proton conduction resistances in the membrane and the electrodes. The principal nature of such a profile is shown in Figure 5, indicating that the local RH will be highest in the cathode electrode, where water is being generated.

While it would go beyond the scope of this manuscript to detail all the experimental values and measurement approaches for these four transport properties, we will describe very briefly the magnitude of these parameters.

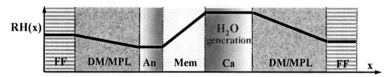

Figure 5. Sketch of the general nature of the RH-profile across the MEA. Note that the thickness of the various components is not drawn to scale.

The water diffusion coefficient in the membrane, $D_{H2O(memb.)}$, generally depends on the water content of the polymer, and measurement techniques have been described in the literature. Its approximate average value is on the order of $3 \cdot 10^{-6} cm^2/s$ at 80°C.[16] The electroosmotic drag coefficient for H_2/air PEMFC operation is ≈ 1,[17] suggesting proton conduction in form of H_3O^+ (note that n_{H2O} would be between 1 and 3 in the case of electrolyzers and direct-methanol fuel cells). The effective diffusion coefficient of O_2 through the cathode diffusion medium can be obtained via limiting current measurements as described, e.g. in Reference 18. Finally, the thermal resistance of the diffusion media and, most importantly, the thermal contact resistance between flow-field/DM and electrode/DM must be measured to assess the temperature gradients across the MEA which in turn will lead to RH-gradients.

2.6. DECONVOLUTION OF VOLTAGE LOSSES IN A H_2/AIR PEMFC

The last remaining terms in Eq. (1) are the gas transport resistances through the diffusion media and the electrodes. When the anode is operated with pure H_2, gas transport losses on the anode electrode are negligible owing to the use of an undiluted gas and the fast diffusivity of H_2. This is not the case for the cathode electrode, where the reactant is diluted with N_2. Thus, transport losses in a H_2-anode (using pure H_2 reactant) are usually negligible compared to those of an air cathode. As mentioned in the introduction, the O_2 transport losses can be separated conceptually into transport losses which would occur if the DM and the electrode were entirely free of liquid water, $\eta_{tx,O2(dry)}$, and another term which is the additional O_2 transport loss in the presence of liquid water in the DM and the electrode, $\eta_{tx,O2(wet)}$. The former term can be calculated precisely using the transport parameters described in 2.5, while the latter part is difficult to quantify as it involves O_2 transport through a partially water-filled porous medium. Therefore, this last term on the right-hand-side of Eq. (1) can only be obtained as a difference term between the measured cell voltage and the other more easily quantifiable voltage loss terms.

A deconvolution of the various voltage loss terms in H_2/air PEMFCs based on short-stack measurements (ca. 30 cells with full-size active area) using state-of-the-art MEA components is shown in Figure 6. At a cell voltage of >0.6 V, current densities as high as 1.5 A/cm^2 can be achieved, which translates into a power density of ≈0.9 W/cm^2 (the maximum power density would be much larger, but voltages of <<0.6 V are not viable with respect to fuel efficiency and heat production).

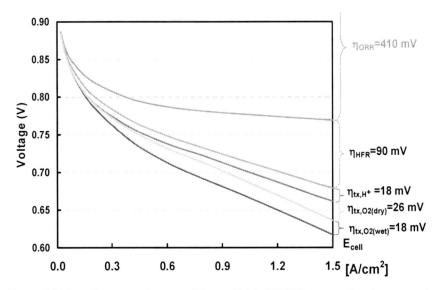

Figure 6. Voltage loss terms in state-of-the-art H_2/air PEMFCs operated under automotive conditions of 80°C, ≈50% inlet-RH, and 150 kPa$_{abs}$. MEAs: membrane thickness of ≈20 μm, and Pt/C catalysts using anode/cathode loadings of ≈0.05/0.4 mg$_{Pt}$/cm^2.

Quite clearly, the major voltage losses are due to the very slow ORR kinetics. Therefore, the Pt requirement in state-of-the-art H_2/air fuel cell stacks is still ≈0.5 g$_{Pt}$/kW (0.45 mg$_{Pt}$/cm^2 divided by 0.9 W/cm^2), and in order to reach the automotive target of <0.2 g$_{Pt}$/kW, the cathode loading would need to be decreased by ≈3x without loss of activity. Alternatively, one might ask whether other transport losses could be minimized such that lower Pt loadings could be used without loss of performance. Examining Figure 6, we see that the Ohmic losses from proton (membrane) and electron conduction (mainly the flow-field/DM contact resistance) amount to ≈90 mV at 1.5 A/cm^2. However, only 30 out of the 90 mV are due to membrane conduction losses, and any substantial reduction of the 90 mV Ohmic losses would require significantly lower electronic contact resistances via the development of electronically highly conductive interfaces. The other transport losses are rather small, so that further optimization of MEAs with Pt/C based catalysts is unlikely to yield very large improvements.

In summary, the above analysis suggests that significant improvements in cathode catalyst activity (≈ 3 to 4x) are necessary in order to meet the automotive cost target of <0.2 g_{Pt}/kW. Some of the promising approaches towards this goal were mentioned above.[11, 12]

3. Direct Methanol Fuel Cell (DMFC) Catalysis and Performance

In contrast to the high power densities obtained with H_2/air PEMFCs, the power densities for state-of-the-art DMFCs operating with air are signifycantly lower, even though the platinum metals loadings are higher:

TABLE 3. DMFC performance data with air feed using Nafion117. Power densities (W/cm^2) and platinum metals (PM) needs (g_{PM}/kW) are given as figures-of-merit at 0.5 and 0.4 V.[23]

T_{cell}	$c_{meth.}$	P_{air}	S_{air}	anode catalyst	cath. catalyst	loading$_{anode}$	loading$_{cath.}$	0.5V performance		0.4V performance		refs.
°C	mol/l	kPa$_{abs}$	--	--	--	mg$_{PM}$/cm^2	mg$_{PM}$/cm^2	W/cm^2	g$_{PM}$/kW	W/cm^2	g$_{PM}$/kW	
90	0.75	300	5	60%wt Pt$_1$Ru$_1$/C	Pt-black	1.0	4.0	0.11	45	0.18	28	[19]
90	0.75	300	2	60%wt Pt$_1$Ru$_1$/C	Pt-black	1.0	4.0	0.17	29	0.18	28	[19]
80	0.5	300	?[1]	Pt$_1$Ru$_1$-black	Pt-black	$\Sigma_{anode/cath.}$=2.6		0.06	43	0.11	24	[20]
100	0.5	300	?[1]	Pt$_1$Ru$_1$-black	Pt-black	$\Sigma_{anode/cath.}$=2.6		0.10	26	0.15	17	[20]
110	1.0	300	?[2]	85%wt Pt$_1$Ru$_1$/C	85%wt Pt/C	$\Sigma_{anode/cath.}$=2.0		0.04	50	0.09	22	[21]
90	0.5	150	>5	PtRu[3]	Pt-black	$\approx$0.7	4.0	0.05	94	0.09	52	[22]

[1] the air stoichiometry was only referred to as "high" and no specific value was given
[2] air stoichiometry was not specified
[3] the used PtRu catalyst was unspecified wrt. composition (assumed 1:1 atomic ratio in above calculation) and support (black or C-supported)

As can be seen from Table 3, DMFC power densities at only 0.4 V are still a factor of 5 to 10 lower than those for a typical H_2/air PEMFC. Also, the Pt-metals (PM) requirement is about a factor of 50 to 100 times higher than for a PEMFC, so that the use of current DMFCs in high power applications is prohibitive. For portable applications in the 20 W power range, the required PM loadings of 20–50 mg$_{PM}$/W (see Table 3), corresponding to 0.4–1 g$_{PM}$ are still too high for large-scale commercialization.

In general, three effects lead to the lower performance of DMFCs vs. H_2/air PEMFCs: (i) a lower, mixed potential at the air cathode due to methanol permeation through the membrane (crossover)[20]; (ii) higher O_2 transport losses on the cathode due the enhanced water permeation when using a water-diluted fuel; and (iii) very slow kinetics of the anodic oxidation of methanol compared to H_2 oxidation. To reduce the effect of methanol crossover, DMFCs are therefore usually operated near the methanol diffusion limited current density, so that the performance at 0.4 V (see Table 3) is not very strongly influenced by crossover. In addition, the effect of water flooding of the cathode is commonly alleviated by the use of high air stoichiometries (see Table 3). Thus, the major cause of the poor performance of current DMFCs are the very slow anode kinetics, which require anode overpotentials of $\eta_{CH3OH} \geq 300$ mV vs. the reversible H_2 electrode potential in order to produce any significant currents:[23]

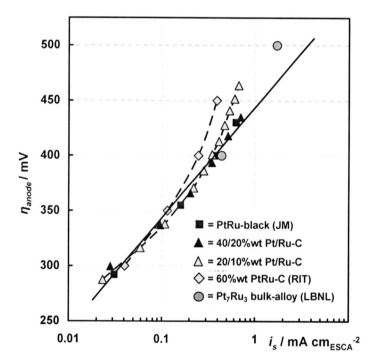

Figure 7. DMFC anode overpotential versus specific current density, i_s (normalized to electrochemical surface area (ECSA)) at or extrapolated to 80°C assuming an activation energy of 60 kJ/mol. *Square and triangular symbols*: 1M H_2SO_4 and 2M CH_3OH at 80°C;[19] *diamonds*: Nafion117 based MEA (CH_3OH anode and H_2 cathode) with 2M CH_3OH at 70°C;[24] *circles*: PtRu bulk-alloy (Ru surface composition of 33% atomic) in 0.5 M H_2SO_4 and 0.5 M CH_3OH at 60°C.[25] Solid line: least-squares regression fit to the PtRu-black (JM) and the 40/20%wt Pt/Ru-C (JM) anode overpotential relation, yielding an apparent Tafel-slope of 100 mV decade^{-1}. (Analysis from Reference 23.)

In order to assess the further development potential of DMFCs, it is instructive to determine the purely kinetically and ohmically limited DMFC performance, which would apply for a perfectly engineered MEA with zero transport losses and zero methanol crossover:

$$E_{cell} = E_{rev} - iR_\Omega - \eta_{CH3OH} - |\eta_{ORR}| \qquad (6)$$

where the reversible potential, E_{rev}, is ≈ 1.165 V,[26] R_Ω is about 0.05 S/cm, η_{CH3OH} is determined from Figure 7, and η_{ORR} from Table 1 using Eqs. (2) and (3). The resulting mass transport and methanol crossover free DMFC performance, $DMFC_{(ideal)}$, is shown in Figure 8, depicting the kinetically (anode and cathode) and ohmically (membrane and contact resistances) limited cell voltage vs. current density performance.[23]

A close comparison between Figure 8 and the performance data in Table 3 shows that the kinetically and Ohmically limited DMFC performance is actually quite close to the state-of-the-art methanol/air performance (i.e. 0.25 A/cm² at 0.4 V, corresponding to 0.1 W/cm²). This suggests that further

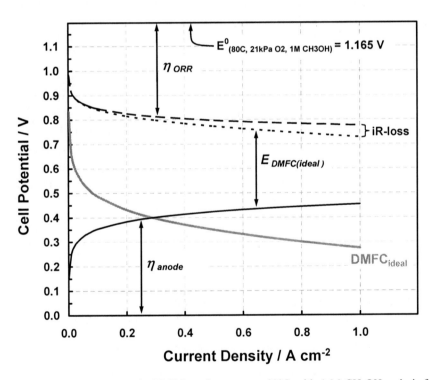

Figure 8. Best-case scenario DMFC performance at 80°C with 1 M CH₃OH and air feed, using state-of-the-art PtRu anode catalyst at 1 mg$_{PtRu}$/cm² (assumed ECSA of 80 m²/g¹$_{PtRu}$) and Pt/C cathode catalyst at 0.4 mg$_{Pt}$/cm² (assumed ECSA of 80 m²/g¹$_{Pt}$). Assumptions: (i) zero methanol cross-over, (ii) zero mass-transport losses on both cathode and anode, and (iii) use of a 25 μm Nafion111 membrane. Details on the analysis are given in Reference 23.

improvements of DMFC performance, at least at high current density where mass-transport losses are negligible, requires significant breakthroughs in either anode or cathode catalyst development, rather than simply MEA design optimization.

4. Conclusions

In this manuscript, we have tried to quantify the various voltage loss terms in state-of-the-art H_2/air PEMFCs and air-fed DMFCs, in order to determine whether new materials development or improved MEA design are required to meet the targets required for large-scale commercialization. From this analysis, it is quite clear that the major R&D challenges are in the areas of catalyst development, in order to reduce the platinum metals requirements to commercially viable levels. While it was not discussed here, another aspect of necessary materials R&D is in the development of novel proton conducting membranes which have sufficient conductivity at low relative humidity, in order to simply the systems design for automotive H_2/air applications.[27]

Acknowledgments

We would like to thank K.C. Neyerlin who developed the ORR and HOR kinetics discussed in this article, as well as other colleagues at General Motors Fuel Cell Activities program whose research over the past five to ten years elucidated many of the concepts discussed in this review. Special thanks also go to Gessie Brisard, for the fact that without her this manuscript would have never been written and for Taixiang Jiang and her for their help in reviewing the manuscript.

References

1. H.A. Gasteiger, S.S. Kocha, B. Sompalli, and F.T. Wagner, Activity Benchmarks and Requirements for Pt, Pt-Alloy, and Non-Pt Oxygen Reduction Catalysts for Pemfcs, *Appl. Catal. B* 56, 9 (2005).
2. P. Piela and P. Zelenay, Researchers Redefine The DMFC Roadmap, *The Fuel Cell Review Aug./Sept.*, 17 (2004).
3. H.A. Gasteiger, W. Gu, R. Makharia, M.F. Mathias, and B. Sompalli, 'Beginning-of-Life MEA performance-efficiency loss contributions', in: *Handbook of Fuel Cells: Fundamentals, Technology, and Applications*, edited by W. Vielstich, A. Lamm, and H.A. Gasteiger (Wiley, Chichester, 2003), vol. 3, pp. 593–610.

4. K.C. Neyerlin, W. Gu, J. Jorne, A. Clark (jr.), and H.A. Gasteiger, Cathode Catalyst Utilization for the ORR in a PEMFC Analytical Model and Experimental Validation, *J. Electrochem. Soc. 154*, B279 (2007).
5. Y. Liu, M. Murphy, D. Baker, W. Gu, C. Ji, J. Jorne, and H.A. Gasteiger, Determination of Electrode Sheet Resistance in Cathode Catalyst Layer by AC Impedance, *ECS Trans. 11(1)*, 473 (2007).
6. K.C. Neyerlin, W. Gu, J. Jorne, and H.A. Gasteiger, Study of the Exchange Current Density for the Hydrogen Oxidation and Evolution Reactions, *J. Electrochem. Soc. 154*, B631 (2007).
7. K.C. Neyerlin, W. Gu, J. Jorne, and H.A. Gasteiger, Determination of Catalyst Unique Parameters for the Oxygen Reduction Reaction in a PEMFC, *J. Electrochem. Soc. 153*, A1955 (2006).
8. J.S. Newman, *Electrochemical Systems* (Prentice Hall, Englewood Cliffs, NJ, 1991).
9. F.T. Wagner, H.A. Gasteiger, R. Makharia, K.C. Neyerlin, E.L. Thompson, and S.G. Yan, Catalyst Development Needs and Pathways for Automotive PEM Fuel Cells, *ECS Trans. 3(1)*, 19 (2006).
10. D. Thompsett, 'Pt Alloys as Oxygen Reduction Catalysts', in: *Handbook of Fuel Cell*, edited by W. Vielstich, A. Lamm, and H.A. Gasteiger (Wiley, NY, 2003), vol. 3, pp. 467–480.
11. V.R. Stamenković, B. Fowler, B.S. Mun, G. Wang, P.N. Ross, C.A. Lucas, and N.M. Marković, Improved Oxygen Reduction Activity on $Pt_3Ni(111)$ via Increased Surface Site Availability, *Science 315*, 493 (2007).
12. R. Srivastava, P. Mani, N. Hahn, and P. Strasser, Efficient Oxygen Reduction Fuel Cell Electrocatalysis on Voltammetrically Dealloyed Pt–Cu–Co Nanoparticles, *Angew. Chem. Int. Ed. 46*, 1 (2007).
13. C.C. Boyer, R.G. Anthony, and A. J. Appleby, Design equations for optimized PEM fuel cell electrodes, *J. App. Electrochem. 30*, 777 (2000).
14. G. Li and P.G. Pickup, Ionic Conductivity of PEMFC Electrodes Effect of Nafion Loading, *J. Electrochem. Soc. 150*, C745 (2003).
15. E.L. Thompson, W. Gu, J. Jorne, and H.A. Gasteiger, Oxygen Reduction Reaction Kinetics in Subfreezing PEM Fuel Cells, *J. Electrochem. Soc. 154(8)*, B783–B792 (2007).
16. T.E. Springer, T.A. Zawodzinski, and S. Gottesfeld, Polymer Electrolyte Fuel Cell Model, *J. Electrochem. Soc. 138*, 2334 (1991).
17. X. Ye and C.Y. Wang, Measurement of Water Transport Properties Through Membrane-Electrode Assemblies, I. Membranes, *J. Electrochem. Soc. 154*, B676 (2007).
18. D. Baker, C. Wieser, K.C. Neyerlin, and M.W. Murphy, The Use of Limiting Current to Determine Transport Resistance in PEM Fuel Cells, *ECS Trans. 3(1)*, 989 (2006).
19. M.P. Hogarth and T.R. Ralph, Catalysis for Low Temperature Fuel Cells, Part III: Challenges for the Direct Methanol Fuel Cell, *Platinum Metals Rev. 46*, 146 (2002).
20. S.C. Thomas, X. Ren, S. Gottesfeld, and P. Zelenay, Direct Methanol Fuel Cells: Progress in Cell Performance and Cathode Research, *Electrochim. Acta 47*, 3741 (2002).
21. R. Dillon, S. Srinivasan, A.S. Aricò, and V. Antonucci, International Activities in DMFC R&D: Status of Technologies and Potential Applications, *J. Power Sources 127*, 112 (2004).
22. M. Baldauf and W. Preidel, Status of the Development of a Direct Methanol Fuel Cell, *J. Power Sources 84*, 161 (1999).

23. H.A. Gasteiger and J. Garche, 'Fuel Cells', in: *Handbook of Heterogeneous Catalysis (2nd edn)*, edited by G. Ertl, H. Knözinger, F. Schüth, and J. Weitkamp (Wiley-VCH, Weinheim, Germany, 2008), in press.
24. J. Nordlund and G. Lindbergh, Temperature-Dependent Kinetics of the Anode in the DMFC, *J. Electrochem. Soc. 151*, A1357 (2004).
25. H.A. Gasteiger, N.M. Marković, P.N. Ross (Jr.), and E.J. Cairns, Temperature-Dependent Methanol Electro-Oxidation on Well-Characterized Pt-Ru Alloys, *J. Electrochem. Soc. 141*, 1795 (1994).
26. S.S. Sandhu, R.O. Crowther, S.C. Krishnan, and J.P. Fellner, Direct Methanol Polymer Electrolyte Fuel Cell Modeling: Reversible Open-Circuit Voltage and Species Flux Equations, *Electrochim. Acta 48*, 2295 (2003).
27. M.F. Mathias, R. Makharia, H.A. Gasteiger, J.J. Conley, T.J. Fuller, C.J. Gittleman, S.S. Kocha, D.P. Miller, C.K. Mittelsteadt, T. Xie, S.G. Yan, and P.T. Yu, Two Fuel Cell Cars in Every Garage? *Interface* (The Electrochemical Society, 2005), *14*, pp. 24–35.

CATALYST DEGRADATION MECHANISMS IN PEM AND DIRECT METHANOL FUEL CELLS

H.A. GASTEIGER*, W. GU, B. LITTEER, R. MAKHARIA,
B. BRADY, M. BUDINSKI, E. THOMPSON,
F.T. WAGNER, S.G. YAN, AND P.T. YU
Acta S.p.A., Via di Lavoria 56/G, 56040 Crespina, Italy
General Motors Corporation – Fuel Cell Activities
Honeoye Falls, 14472 NY, USA

1. Introduction

While much attention has been given to optimizing initial fuel cell performance, only recent research has focused on the various materials degradation mechanisms observed over the life-time of fuel cells under real-life operating conditions. This presentation will focus on fuel cell durability constraints produced by platinum sintering/dissolution, carbon-support oxidation, and membrane chemical and mechanical degradations.

Over the past 10 years, extensive R&D efforts were directed towards optimizing catalysts, membranes, and gas diffusion layers (GDL) as well as combining them into improved membrane electrode assemblies (MEAs), leading to significant improvements in initial performance of H_2/air-fed proton exchange membrane fuel cells (PEMFCs)[1] and methanol/air-fed direct methanol fuel cells (DMFCs).[2,3] While the required performance targets have not yet been met, current PEMFC and DMFC performance are close to meeting entry-level applications and many prototypes have been developed for field testing. This partially shifted the R&D focus from performance optimization to more closely examining materials degradation phenomena which limit fuel cell durability under real-life testing conditions.

The predominant degradation mechanisms are sintering/dissolution of platinum-based cathode catalysts under highly dynamic operating conditions,[4] dissolution of ruthenium from DMFC anode catalysts,[5] the oxidation of carbon-supports of the cathode catalyst during fuel cell startup and shutdown,[6] and the formation of pinholes in proton exchange membranes if

---

*E-mail: hubert.gasteiger@gmail.com

subjected to extensive local relative humidity cycling.[7] These various mechanisms and their impact of fuel cell durability will be discussed in the following.

2. Platinum Dissolution/Sintering

During long-term fuel cell operation, one generally observes a significant loss of active platinum surface area of the cathode catalyst, while the anode catalyst surface area undergoes little changes,[8] consistent with a Pt sintering/dissolution mechanisms based on the increased Pt solubility at high electrode potentials. The loss of active Pt surface area at the cathode electrode leads to cell voltage losses due to reduced kinetics of the oxygen reduction reaction (ORR). Unfortunately, the loss of active Pt surface area is accelerated if the electrode potential is continuously cycled between oxidizing and reducing conditions, as is the case under real-life operating conditions, where dynamic changes of fuel cell current (i.e. fuel cell power) lead to dynamic variations in electrode potential. In a H_2/air PEMFC, dynamic variation of cell current results in dynamic voltage cycling of the cathode electrode potential owing to the slow ORR kinetics, while the anode electrode potential remains nearly constant due the fast H_2 oxidation kinetics (i.e. the small effective overpotential of the anode electrode).[9] In a DMFC, where the anode kinetics are slow (i.e. the anodic methanol oxidation), dynamic variation of cell current leads also to significant voltage cycling of the anode potential.

Thus, for both fuel cell types, major Pt area losses of the cathode catalyst occur under dynamic operating conditions, leading to significant cell performance losses over time. This phenomenon was first realized in the 1970s during the development of phosphoric acid fuel cells,[10] and has become a major issue in the highly dynamic automotive PEMFC applications.[4,7] The voltage losses produced by Pt dissolution/sintering can be quantified by monitoring the Pt mass activity, defined as the Pt cathode loading normalized current at a cell voltage of 0.9 V ($i_m^{(0.9V)}$),[1] versus the number of voltage cycles induced by cycling the current density between 0.02 V (idle, i.e. not net power drawn from the fuel cell system) and 1 A/cm^2 (full power). For fully dynamic, non-hybridized (i.e. no propulsion battery) automotive conditions, approximately 300,000 voltage cycles will occur during vehicle life, which would result in large performance losses of ≈60 mV if a carbon-supported Pt cathode catalyst is used (open blue squares in Figure 1). Fortunately, much lower losses can be realized with Pt-alloy catalysts,[11,12] amounting to only ≈25 mV for the same number of cycles

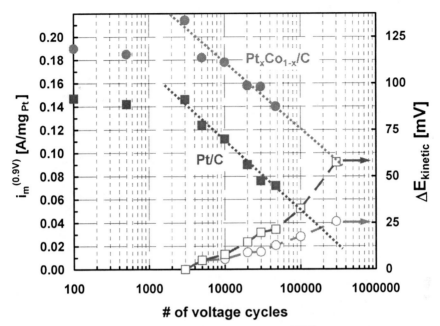

Figure 1. The solid symbols show the Pt mass activity ($i_m^{(0.9\ V)}$) in a H$_2$/air PEMFC vs. number of voltage cycles induced by cycling the cell current density between 0.02 and 1 A/cm^2 at 80°C, 100% RH, and 150 kPa$_{abs}$. The open symbols show the potential loss due to the reduced ORR kinetics vs. number of voltage cycles. Note that the Pt mass activity is defined as current at 100 kPa$_{abs}$ H$_2$/O$_2$ at 100% RH and 80°C.[1] For details see Wagner et al.[12]

(open red circles in Figure 1). Therefore, significant development efforts are made to incorporate Pt-alloy cathode catalysts into PEMFCs. While the dynamic load cycling of DMFCs may be less demanding (shorter required life-time and operation under more stationary conditions), a smaller but analogous benefit would be expected if Pt-alloy cathode catalysts were used.

The underlying mechanism for Pt area loss of the cathode catalyst is based on Ostwald ripening of Pt on the carbon-support as well as on Pt precipitation in the ionomer phase (Figure 2).

This dual mechanism was also observed in phosphoric acid fuel cells (PAFCs),[13, 14] and its occurrence in PEMFCs was demonstrated by a detailed transmission electron microscope (TEM) study.[4] Since the Pt dissolution process known to be accelerated when the electrode is cycled between the oxide-formation/oxide-desorption potential range,[10, 15] any operating conditions which influence the adsorption/desorption of Pt oxides would be expected to lead to differences in the Pt area loss. One example is the observed positive potential shift of the oxide adsorption process on Pt with decreasing relative humidity (RH), shown in Figure 3.[12]

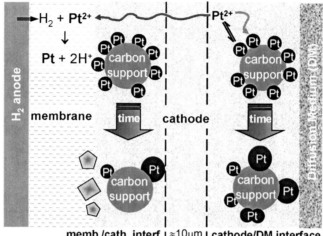

Figure 2. Sketch of the two different Pt-area loss mechanisms occurring in the cathode electrode of a PEMFC: (i) Pt dissolution and Ostwald ripening on the carbon-support (right-hand-side), and (ii) diffusion of dissolved Pt species into the ionomer phase near the cathode/membrane interface and precipitation via reduction by crossover H_2.[4]

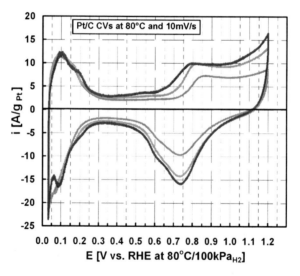

Figure 3. Cyclic voltammetry of a 50% Pt/C catalyst at 80°C as a function of RH.[12]

Based on the data in Figure 3, it is clear that the extent of surface oxide formation during voltage cycling between 0.7 and 0.9 V (idle ↔ full power), decreases with decreasing RH. Therefore, the observed reduced Pt area loss at reduced RH, shown in Figure 4 is not unexpected.[7]

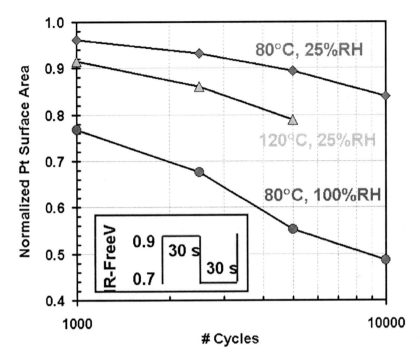

Figure 4. Pt surface area loss during potential cycling between 0.7 and 0.9 V (corrected by the Ohmic resistance loss) of 50% Pt/C catalyst as a function of RH.[7]

While not all of the details of the enhancement of the Pt dissolution mechanism by voltage cycling is well understood,[15,16] it is apparent from the data in Figure 4 that PEMFC operation at 120°C could indeed by viable as far as Pt dissolution is concerned, as long as sufficiently high proton conduction can be obtained at 25% RH or below. Currently available membranes, however, require >50% RH in order to have sufficient proton conductivity for fuel cell applications at high power density.[7]

As mentioned above, no significant Pt area loss is observed for the anode catalyst in a H_2/air PEMFC,[4] but degradation of the anode catalyst does occur in DMFCs, where ruthenium dissolves from the PtRu anode catalysts, diffuses through the membrane, and adsorbs on the Pt cathode catalyst. This leads to performance losses both due to reduced anode kinetics owing to the lower activity of Pt vs. PtRu and due to reduced cathode activity caused by the adsorption of Ru on the Pt cathode catalyst.[5]

3. Carbon-Support Oxidation

The oxidation of the typically used carbon-supports at high electrode potentials was realized as critical degradation mechanism during the development of phosphoric acid fuel cells operating at 200°C. This was mitigated by both keeping the allowable cell voltage below 0.85 V and by using fully graphitized carbon-supports. At the lower PEMFC and DMFC operating temperatures (typically <100°C), carbon-support oxidation is less significant during normal operation, but was discovered to be very critical under start-up and shutdown conditions,[6] where a H_2-front displacing air in the anode compartment after long shutdown periods (or an air-front displacing H_2 during shutdown) temporarily increases the local cathode potential to >1.4 V, leading to extensive carbon-support oxidation. To understand the effect of carbon-support corrosion on cell performance, the carbon-support oxidation kinetics were determined,[17] and the cell performance was measured as a function of carbon weight loss.

Figure 5 shows that a corrosion of only 5–10%wt. of the carbon-support leads to dramatic performance losses, irrespective whether a conventional carbon-support or a graphitized carbon-support is being used, which initially was a very surprising result. Later it became apparent that the reason for this high sensitivity is the fact that the initially highly porous electrode

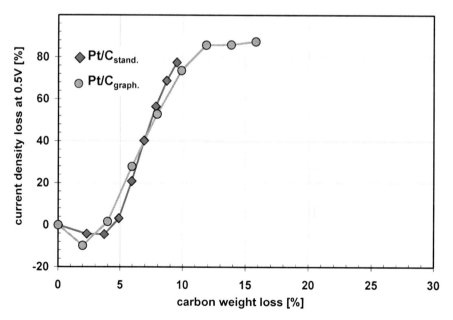

Figure 5. Cell performance loss as a function of carbon-support corrosion. Conditions: H_2/air performance at 80°C, 100% RH, 150 kPa_{abs}.

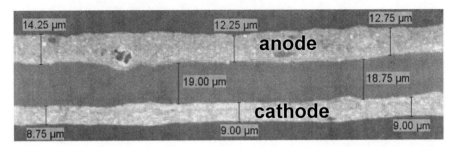

Figure 6. SEM cross-section of an MEA of which 8%wt. of the cathode carbon-support had been corroded. The initial cathode electrode thickness was identical to the anode electrode thickness shown in the picture.

structure (≈60–65% porosity[18]) collapses once <10%wt. of the carbon-support is corroded. This effect, often referred to as *cathode thinning* (induced, e.g. by the high transient cathode potential during start/stop cycles[6]), can be seen in Figure 6, which is a scanning electrode micrograph (SEM) of an MEA after corrosion of 8%wt. of the carbon-support.

By this mechanism, voltage losses in excess of 100 mV can occur over as little as 100 startup/shutdown cycles.[13] Besides the implementation of more corrosion resistant carbon-supports or the development of more stable alternative supports,[13] other systems-based mitigation strategies have been discussed in the literature.[19]

4. Membrane Degradation

The chemical degradation of perfluoro sulfonicacid (PFSA) ionomers, accelerated at low relative humidity and under open-circuit conditions, leads to pinhole formation in the membrane and thus to premature failure of PEMFCs.[20] Over the past two years, however, PFSA membrane additives have been developed, which reduce the chemical degradation rate of PFSAs by a factor of 100–1,000.[21] Therefore, chemical degradation induced membrane failure is not a PEMFC life-limiting mechanism if state-of-the-art MEAs are used.

Nevertheless, membrane pinhole formation can still occur under the highly transient automotive conditions, where a large number of relative-humidity cycles occur. Owing to the high volumetric swelling of ionomeric membranes at high relative humidity and in the presence of liquid water, fluctuations in RH lead induce significant mechanical stresses in the membrane which, even in the absence of chemical degradation, can lead to membrane pinholes.[7]

5. Conclusions

As was outlined above, many of the materials degradation mechanisms in PEMFCs and DMFCs are reasonably well understood at this time. In many cases, these degradation phenomena can be mitigated by careful system design, but for robust and durable fuel cell systems, improved materials are still required. This includes catalysts with improved stability towards voltage cycles, catalyst support materials with higher corrosion resistance, and membranes with better stability during RH cycles.

Acknowledgments

We would like to thank colleagues at General Motors Fuel Cell Activities program whose research over the past years elucidated many of the concepts discussed in this review. Special thanks also go to Gessie Brisard, for the fact that without her this manuscript would have never been written and for Taixiang Jiang and her for their help in reviewing the manuscript.

References

1. H.A. Gasteiger, S.S. Kocha, B. Sompalli, and F.T. Wagner, Activity Benchmarks and Requirements for Pt, Pt-Alloy, and Non-Pt Oxygen Reduction Catalysts for PEMFCs, *Appl. Catal. B 56*, 9 (2005).
2. P. Piela and P. Zelenay, Researchers Redefine The DMFC Roadmap, *The Fuel Cell Review Aug./Sept.*, 17 (2004).
3. J. Müller, G. Frank, K. Colbow, and D. Wilkinson, "Transport/Kinetic Limitations and Efficiency Losses", in: *Handbook of Fuel Cells: Fundamentals, Technology, and Applications*, edited by W. Vielstich, A. Lamm, and H.A. Gasteiger (Wiley, Chichester, UK, 2003), vol. 3, pp. 847–855.
4. P.J. Ferreira, G.J. la O', Y. Shao-Horn, D. Morgan, R. Makharia, S.S. Kocha, and H.A. Gasteiger, Instability of Pt/C Electrocatalysts in Proton Exchange Membrane Fuel Cells – A Mechanistic Investigation, *J. Electrochem. Soc. 152*, A2256 (2005).
5. P. Piela, C. Eickes, E. Brosha, F. Garzon, and P. Zelenay, Ruthenium Crossover in Direct Methanol Fuel Cell with Pt-Ru Black Anode, *J. Electrochem. Soc. 151*, A2053 (2004).
6. C.A. Reiser, L. Bregoli, T.W. Patterson, J.S. Yi, J.D. Yang, M.L. Perry, and T.D. Jarvi, A Reverse-Current Decay Mechanism for Fuel Cells, *Electrochem. Solid State Lett. 8*, A273 (2005).
7. M.F. Mathias, R. Makharia, H.A. Gasteiger, J.J. Conley, T.J. Fuller, C.J. Gittleman, S.S. Kocha, D.P. Miller, C.K. Mittelsteadt, T. Xie, S.G. Yan, and P.T. Yu, Two Fuel Cell Cars In Every Garage? *Interface* (The Electrochemical Society, 2005), *14*, pp. 24–35.

8. T. Tada, "High Dispersion Catalysts Including Novel Carbon-Supports", in: *Handbook of Fuel Cell*, edited by W. Vielstich, A. Lamm, and H.A. Gasteiger (Wiley, Chichester, UK, 2003), vol. 3, pp. 481–488.
9. H.A. Gasteiger, Y. Liu, D. Baker, and W. Gu, "Kinetics and Kinetically Limited Performance in PEMFCs and DEMFCs with State-of-the-Art Catalysts", in this book.
10. K. Kinoshita, J.T. Lundquist, and P. Stonehart, Potential Cycling Effects on Platinum Electrocatalyst Surfaces, *J. Electroanal. Chem. 48*, 157 (1973).
11. P. Yu, M. Pemberton, and P. Plasse, Ptco/C Cathode Catalyst for Improved Durability In PEMFCs, *J. Power Sources 144*, 11 (2005).
12. F.T. Wagner, H.A. Gasteiger, R. Makharia, K.C. Neyerlin, E.L. Thompson, and S.G. Yan, Catalyst Development Needs and Pathways for Automotive PEM Fuel Cells, *ECS Trans. 3(1)*, 19 (2006).
13. J. Aragane, T. Murahashi, and T. Odaka, Change of Pt Distribution in the Active Components of Phosphoric Acid Fuel Cell, *J. Electrochem. Soc. 135*, 844 (1988).
14. J. Aragane, H. Urushibata, and T. Murahashi, Effect of Operational Potential On Performance Decay Rate in a Phosphoric Acid Fuel Cell, *J. Appl. Electrochem. 26*, 147 (1996).
15. M. Uchimura and S.S. Kocha, The Impact of Cycle Profile on PEMFC Durability, *ECS Trans. 11(1)*, 1215 (2007).
16. S. Mitsushima, Y. Koizumi, S. Uzuka, and K. Ota, Dissolution Mechanism of Platinum in Acidic Media, *ECS Trans. 11(1)*, 1195 (2007).
17. P. Yu, W. Gu, R. Makharia, F.T. Wagner, and H.A. Gasteiger, The Impact of Carbon Stability on PEM Fuel Cell Startup and Shutdown Voltage Degradation, *ECS Trans. 3(1)*, 797 (2006).
18. H.A. Gasteiger, W. Gu, R. Makharia, M.F. Mathias, and B. Sompalli, "Beginning-of-Life MEA Performance – Efficiency Loss Contributions", in: *Handbook of Fuel Cells*, edited by W. Vielstich, A. Lamm, and H.A. Gasteiger (Wiley, Chichester, UK, 2003)., vol. 3, pp. 593–610.
19. M.L. Perry, T.W. Patterson, and C. Reiser, Systems Strategies to Mitigate Carbon Corrosion in Fuel Cells, *ECS Trans. 3(1)*, 783 (2006).
20. W. Liu and S. Cleghorn, Effect of Relative Humidity on Membrane Durability in PEM Fuel Cells, *ECS Trans. 1(8)*, 263 (2006).
21. E. Endoh, Highly Durable MEA for PEMFC Under High Temperature and Low Humidity Conditions, *ECS Trans. 3(1)*, 9 (2006).

PRINCIPLES OF DIRECT METHANOL FUEL CELLS FOR PORTABLE AND MICRO POWER

CHAO-YANG WANG
Electrochemical Engine Center (ECEC), and Departments of Mechanical Engineering and Materials Science and Engineering, The Pennsylvania State University, University Park, PA 16802, USA

1. Introduction

A direct methanol fuel cell (DMFC) is an electrochemical cell that generates electricity based on the oxidation of methanol and reduction of oxygen. An aqueous methanol solution of low molarity acts as the reducing agent that traverses the anode flow field. Once inside the flow channel, the aqueous solution diffuses through the backing layer, comprised of carbon cloth or carbon paper. The backing layer collects the current generated by the oxidation of aqueous methanol and transports it laterally to ribs in the current collector plate. The global oxidation reaction occurring at the platinum-ruthenium catalyst of the anode is given by:

$$CH_3OH + H_2O \rightarrow CO_2 + 6H^+ + 6e^- \qquad (1)$$

The carbon dioxide generated from the oxidation reaction emerges from the anode backing layer as bubbles and is removed via the flowing aqueous methanol solution.

Air is fed to the flow field on the cathode side. The oxygen in the air combines with the electrons and protons at the platinum catalyst sites to form water. The reduction reaction taking place on the cathode is given by:

$$3/2 O_2 + 6H^+ + 6e^- \rightarrow 3H_2O \qquad (2)$$

These two electrochemical reactions are combined to form an overall cell reaction as:

$$CH_3OH + 3/2 O_2 \rightarrow CO_2 + 2H_2O \qquad (3)$$

As expected, a DMFC exhibits lower power densities than that of a H_2/air PsEFC. However, the DMFC has the advantages of easier fuel storage, no need for humidification, and simpler design. Thus, DMFC is presently considered a leading contender for portable power application. To compete with lithium-ion batteries, the first and foremost property of a portable DMFC system must be higher energy density in Wh/L. This requirement entails overcoming four key technical challenges: (1) low rate of methanol oxidation kinetics on the anode, (2) methanol crossover through the polymer membrane, (3) water management, and (4) heat management.

2. Fundamentals of a DMFC

The heart of a DMFC is a membrane-electrode assembly (MEA) formed by sandwiching a perfluorosulfonic acid (PFSA) membrane between an anode and a cathode. Upon hydration, the polymer electrolyte exhibits good proton conductivity. On either side of this membrane are anode and cathode, also called catalyst layers, typically containing Pt-Ru on the anode side and Pt supported on carbon on the cathode side. Here the half-cell reactions described in Eqs. (1) and (2) are catalyzed. On the outside of the MEA, backing layers made of non-woven carbon paper or woven carbon cloth, shown in Figure 1, are placed to fulfill several functions. The primary purpose of a backing layer is to provide lateral current collection from the catalyst layer to the ribs as well as optimized gas distribution to the catalyst layer through diffusion. It must also facilitate the transport of water out of the catalyst layer. This latter function is usually accomplished by adding a coating of hydrophobic polymer, polytetrafluoroethylene (PTFE), to the backing layer. The hydrophobic character of the polymer allows the excess water in the cathode catalyst layer to be expelled from the cell by the gas flowing inside the channels, thereby alleviating flooding.

The microstructure of the catalyst layer is of paramount importance for the kinetics of an electrochemical reaction and species diffusion. Figure 2 shows scanning electron microscopy (SEM) images of such microstructures of the DMFC anode and cathode, respectively, where high surface areas for electrochemical reactions are clearly visible.

A cross-sectional SEM of a MEA segment consisting of a backing layer, a microporous layer (MPL) and a catalyst layer, is displayed in Figure 3 [1]. The MPL, with an average thickness of 30 μm, overlays a carbon paper backing layer. The anode catalyst layer of about 20 μm in thickness covers the MPL. In the anode, this MPL provides much resistance to methanol transport from the feed to the catalyst sites, thus reducing the amount of methanol crossover. In the cathode, the MPL helps alleviate cathode flooding by liquid water.

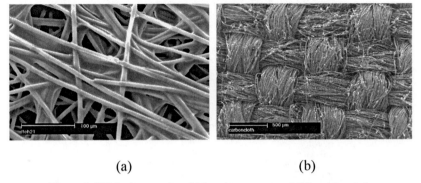

Figure 1. SEM micrographs of (a) carbon paper, and (b) carbon cloth.

(a) anode   (b) cathode

Figure 2. SEM images of electrodes.

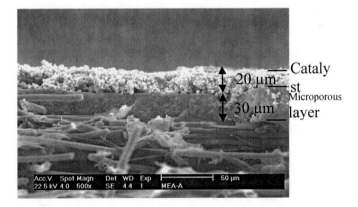

Figure 3. Cross-sectional SEM micrograph of backing layer, microporous layer, and catalyst layer.

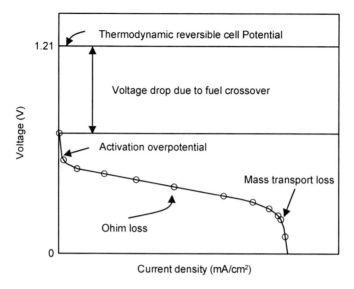

Figure 4. Schematic of DMFC polarization curve.

Figure 4 displays a voltage vs. current density polarization curve of a typical DMFC. The thermodynamic equilibrium cell potential for a DMFC is approximately equal to 1.21 V. However, the actual open circuit voltage in DMFCs is much lower than this thermodynamic value, largely due to fuel crossover. Methanol crossover is an important topic in DMFCs and thus will be fully elaborated in this lecture [2]. On closed circuit, the polarization curve can be categorized into three distinctive regions: kinetic control, ohmic control, and mass transport control. The kinetic control region of DMFC is dictated by slow methanol oxidation kinetics at the anode as well as oxygen reduction kinetics at the cathode. In this region a DMFC suffers the voltage loss second only to the low open circuit voltage caused by methanol crossover. More detailed discussion of this aspect is provided in this lecture. The area where cell voltage decreases nearly linearly in the polarization curve is recognized as the ohmic control region. For a DMFC where the polymer electrolyte is usually well hydrated, the voltage loss in this section is minimal. The last portion is referred to as the mass transport control region, whereby either methanol transport on the anode side results in a mass transport limiting current, or the oxygen supply due to depletion and/or cathode flooding becomes a limiting step. The cell voltage drops drastically in the mass transport control region.

3. Water and Methanol Transport

It is now understood that the fuel concentration allowed in the anode of a DMFC is limited by not only methanol crossover but also water crossover through the polymer electrolyte. A simple analysis of methanol-to-water molecular ratio needed in the fuel supply, first presented by Liu et al. [3, 4], illustrates this new finding. In this analysis, an ideal membrane is considered that features zero methanol crossover but water crossover characterized by α. Here, the net water transport coefficient, α, is defined as the net water flux through the membrane from the anode to cathode normalized by the protonic flux. The highest concentration of methanol solution in the anode must require that the H_2O to CH_3OH molecular ratio be greater than $(1 + 6\alpha)$. Table 1 gives the corresponding MeOH molarity for various α-values. It is clear that for $\alpha \approx 3$ as in typical DMFCs based on Nafion 117, the maximum operational MeOH concentration is only about 3 M. And this maximum molarity is limited by excessive water crossover, not by methanol crossover. Likewise, for hydrocarbon membranes with water crossover $\alpha \sim 1$, the highest MeOH concentration to be used is around 6 M, beyond which the anode will lose proportionally more water than methanol thus leading to the enrichment of methanol concentration in the anode catalyst layer and hence, to excessive methanol crossover, in spite of lower methanol permeability of hydrocarbon membranes. In order to enable direct use of 17 M methanol fuel, α must be reduced to zero. Further, when $\alpha = -1/6$, there is no need to add water in the anode feed or pure methanol operation becomes theoretically possible, in which situation the water molecule needed to oxidize one methanol molecule will come from the product water of oxygen reduction reaction on the cathode.

To further illustrate the decisive role played by water crossover, Figure 5 schematically displays the methanol concentration profile in the anode as a function of the water crossover rate through the membrane. It is clearly indicated that the MeOH concentration in the anode catalyst layer, which directly influences MeOH crossover through the membrane, is a strong function of water crossover. Thus, the methanol crossover current density also strongly depends upon water crossover through the membrane. By carefully reducing the latter, it is possible to mitigate methanol crossover even through thin Nafion membranes and allow for direct use of highly concentrated methanol solution.

Minimizing water crossover through a DMFC membrane therefore emerges as an extremely critical requirement for portable DMFCs. This finding calls for new R&D directions for DMFCs using high concentration methanol. It is apparent that novel membrane development should not focus solely on the methanol permeability reduction. Rather, water transport properties of new membranes must receive close attention. It may also be inappropriate to define the membrane selectivity based on the ratio of proton conductivity to methanol permeability in studies of better membranes for DMFCs. In addition, note that the net rate of water crossover through the membrane depends not only upon the properties of the membrane, but also upon such components as catalyst layers and backing layers.

TABLE 1. Dependece of maximum allowable anode methanol molarity on α at 20°C.

Molarity (M)	H_2O/MeOH molar ratio	α
1	53.31	8.72
2	25.53	4.09
4	11.64	1.77
6	7.01	1.00
8	4.70	0.62
10	3.31	0.39
17	1.02	0.00
25	0.0	-0.17

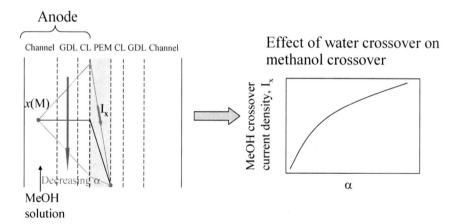

Figure 5. Schematic diagrams of the effects of water crossover on methanol concentration profiles in the anode and methanol crossover current density.

A systems approach is therefore required to design high performance high concentration methanol fuel cells (HC-MFC). Such an approach entails computational modeling of methanol, water and heat transport processes occurring in a DMFC, as discussed extensively in Wang [5], Lu and Wang [2], and references therein. Over the past six years, a suite of simulators for portable DMFCs have been developed at ECEC that consider:

- Two-phase flow in both anode and cathode channels. Both homogeneous and drift flux models are used to simulate CO_2 bubble dynamics in the anode channel, whereas the mist flow or annular film model is employed for liquid water removal through the cathode gas channel.
- Two-phase flow and transport in diffusion layers using the M^2 model pioneered at ECEC. This model considers capillary-driven two-phase flow, both liquid- and gas-phase transport of methanol, and the interfacial coverage of cathode GDL surface by liquid water.
- Electrochemical kinetics of methanol oxidation at the anode and oxygen reduction as well as methanol oxidation reactions at the cathode (leading to a mixed potential).
- Current distribution, methanol crossover rate distribution, and the distribution of the net water transport coefficient, α.
- Electron transport through the GDL and bipolar plate materials.
- Heat transport with heat generation and heat loss via water evaporation and cell heat dissipation.

These fully two-phase models are either one-dimensional (in the through-plane direction) for quick design turnaround, or multi-dimensional for gaining a detailed understanding of the internal physico-chemical processes and for optimizing both anode and cathode flowfields. The DMFC simulators account particularly for intimate coupling between methanol, water and heat transport processes, and are able to predict the effects of material properties, cell geometry (e.g. the channel and land dimensions), and operating conditions on polarization curves, methanol crossover rate, and the net water transport rate through the membrane.

Based on the above described fundamental insight on methanol and water transport, we demonstrated stable operation of a HC-MFC where highly concentrated methanol (i.e. 15 M to neat methanol) is fed directly in the anode without loss in power density [6].

4. Conclusions

This paper discusses the basic principles of direct methanol fuel cells (DMFC) for portable and micro power. Major technological challenges facing DMFCs for portable power are described, and it is shown that the fundamental transport processes of methanol, water and heat, along with methanol oxidation kinetics, hold the key to successfully address these challenges. We then explained how minimizing water crossover through the membrane gives rise to the possibility of using highly concentrated methanol directly and hence leads to much higher energy density for next-generation portable DMFCs.

References

1. C. Lim and C.Y. Wang, Development of high-power electrodes for a liquid-feed direct methanol fuel cell, *J. Power Sources*, Vol. 113, pp. 145–150, 2003.
2. G. Lu and C.Y. Wang, Two-phase microfluidics, heat and mass transport in direct methanol fuel cells, Chapter 9 in Transport Phenomena in Fuel Cells, B. Sunder and M. Faghri (eds.), WIT Press, pp. 317–358 (2005).
3. F.Q. Liu, G.Q. Lu, and C.Y. Wang, Low crossover of methanol and water through thin membranes in direct methanol fuel cells, *J. Electrochem. Soc.*, Vol. 153, pp. A543–553, 2006.
4. G.Q. Lu, F.Q. Liu, and C.Y. Wang, Water transport through Nafion 112 membrane in direct methanol fuel cells, *Electrochem. & Solid-State Lett.*, Vol. 8, pp. A1–A4, 2005.
5. C.Y. Wang, Fundamental models for fuel cell engineering, *Chem. Rev.*, Vol. 104, pp. 4727–4766, 2004.
6. C.Y. Wang and F.Q. Liu, A paradigm shift in direct methanol fuel cell design for portable power, Chapter 10 in *Proc of 8th Small Fuel Cell Symposium*, Washington, DC, 2006.

COMPUTATIONAL MODELING OF TWO-PHASE TRANSPORT IN PORTABLE AND MICRO FUEL CELLS

CHAO-YANG WANG
Electrochemical Engine Center (ECEC), and Department of Mechanical Engineering, The Pennsylvania State University, University Park, PA 16802, USA

1. Introduction

The direct methanol fuel cell (DMFC) is considered a leading contender for next-generation portable and micro power sources, offering a combination of simplicity, robustness and high energy density due to the use of liquid methanol. The basic principles of a DMFC can be found in the literature[1] and thus are not repeated here. In order to compete with lithium-ion batteries, a portable DMFC system must overcome several key technical challenges: (1) low rate of methanol oxidation kinetics, (2) methanol crossover through the polymer membrane, (3) water crossover from the anode to cathode,[2,3] and (4) thermal management. While new materials are being pursued to solve these problems, innovative designs can also be developed with the materials presently available. As a result, there is an urgent need for understanding, prediction, and optimization of various interactive transport and electrochemical processes that occur in portable DMFCs.

Much DMFC research in the past has focused upon the first two issues, methanol oxidation kinetics and methanol crossover, by studying electrocatalysis and electrolyte membrane materials.[4-14] The more recent interest in small-scale DMFC systems for application to portable and micro power[15,16] entails a unique design regime under lower temperatures and ambient pressure as well as a better understanding of methanol, water and heat transport. For this purpose, visualization of two-phase flow in the DMFC anode was carried out by Argyropoulos et al.[17] Lu and Wang.[18]

In tandem with experimental efforts, mathematical modeling of DMFCs has received much attention with the goal of having a design tool to design and optimize cell structures under a myriad of operating conditions and form factors. Focusing on either one or two dimensions, early DMFC modeling works were developed to study the mass transport phenomena, electro-chemical processes, and their interactions.[19-24] However, the two-phase effects, recently found to be of paramount importance to understand DMFC behaviors, were not considered in these earlier models.

Wang et al.[25] were the first to successfully apply the multi-phase mixture (M^2) modeling framework of Wang and Cheng[26] to simulate two-phase flow and transport in the air cathode of a polymer electrolyte fuel cell. They suggested that capillary action is the dominant mechanism for liquid water transport through the gas diffusion layer (GDL). Later, this model was extended for a DMFC by Wang and Wang[27] and simulation results showed the importance of the gas-phase transport of methanol in the two-phase anode GDL. A more thorough review of DMFC modeling is recently given by Wang.[1]

Water transport, in addition to methanol and oxygen transport, has emerged as an important modeling issue for portable DMFCs where water budget must be considered in order to attain high energy density. Water transport has recently been considered by Liu and Wang.[28]

2. Computational Model

The present 3D model is extended from that of Wang and Wang[27] and based on the M^2 formulation of Wang and Cheng, which is particularly suitable and popular for two-phase fuel cell modeling. The specific assumptions made in this model include: (i) incompressible gas mixture, (ii) laminar flow due to a Reynolds number of the order of several hundreds, (iii) isotropic and homogeneous porous GDL, characterized by an effective porosity and permeability, and (iv) negligible potential drop due to ohmic resistance in the electronically conductive solid matrix of GDL and catalyst layers, as well as bipolar plates. Furthermore, a homogeneous flow is assumed for the two-phase flow through cathode channels, and either homogeneous or drift flux model is employed for two-phase flow in anode channels.

Mass Conservation – A generic mass conservation equation, valid for all components including channels, backing layers, catalyst layers of both anode and cathode, and the membrane, can be written as

$$\frac{\partial(\varepsilon\rho)}{\partial t} + \nabla \cdot (\rho \mathbf{u}) = \dot{m} \tag{1}$$

where

$$\dot{m} = \begin{cases} M^{MeOH}S^{MeOH} + M^{H_2O}S^{H_2O} + M^{CO_2}\dfrac{j}{6F} & \text{anode catalyst layer} \\ M^{H_2O}S^{H_2O} + M^{O_2}S^{O_2} + M^{CO_2}\dfrac{j_{xover}}{6F} & \text{cathode catalyst layer} \end{cases} \tag{2}$$

Detailed expressions for various species sources or sinks, S^k, are provided in Table 1. Due to the species consumption and production inside a DMFC, we have different mass source/sink terms, $\dot{m}$, applied in the anode and cathode catalyst layers, respectively. In the anode catalyst layer, the mass source is caused by methanol consumption and crossover through the mem-brane, water consumption and crossover through the membrane, and carbon dioxide generation by the anodic reaction. In the cathode catalyst layer, on the other hand, the mass source term includes water generation and flux from the anode, oxygen consumption by the cathodic reaction, and carbon dioxide generation due to the parasitic oxidation reaction of crossover methanol.

Momentum Equation – The momentum equation can be given by

$$\frac{1}{\varepsilon}\left[\frac{\partial(\rho \mathbf{u})}{\partial t} + \frac{1}{\varepsilon}\nabla \cdot (\rho \mathbf{u}\mathbf{u})\right] = -\nabla p + \nabla \cdot \tau + S_\mathbf{u} \quad (3)$$

where

$$S_\mathbf{u} = \begin{cases} 0 & \text{channels} \\ -\dfrac{\mu}{K}\mathbf{u} & \text{backing and catalyst layers} \end{cases} \quad (4)$$

and

$$\mathbf{u} = 0 \quad \text{membrane} \quad (5)$$

Here the fluid velocity in the backing and catalyst layers is described by Darcy's law and applied for single- and two-phase flow, while in the membrane it is assumed to be zero due to the negligible convective velocity through nanopores of the membrane.

Species Transport Equation – The general conservation equation for a species can be written, in the form of mass fraction, as,

$$\frac{\partial}{\partial t}(\rho Y^k) + \nabla \cdot (\gamma \rho \mathbf{u} Y^k) = \nabla \cdot \left[\rho_l D_{l,\text{eff}}^k \nabla Y_l^k + \rho_g D_{g,\text{eff}}^k \nabla Y_g^k\right] \\ - \nabla \cdot \left[(Y_l^k - Y_g^k)\mathbf{j}_l\right] + \dot{m}^k \quad (6)$$

where Y^k stands for the mixture mass fraction of methanol, water and oxygen in the two-phase mixture. Note that the advection correction factor, γ, is equal to unity in the channel regions due to the homogeneous flow assumption made earlier, but non-unity in backing and catalyst layers since the M^2 model for a porous medium is a two-fluid model. In addition, the effective diffusion coefficients in the liquid and gas phases are given, respectively, by

$$D_{l,\text{eff}}^k = (\varepsilon s)^{1.5} D_l^k \quad \text{and} \quad D_{g,\text{eff}}^k = [\varepsilon(1-s)]^{1.5} D_g^k \tag{7}$$

Based on the relation between species mass fraction and molar concentration,

$$\rho Y^k = c^k M^k \tag{8}$$

and the two-phase property definition

$$\rho Y^k = \rho_l Y_l^k s + \rho_g Y_g^k (1-s) \tag{9}$$

we have the following species equation in terms of molar concentration:

$$\frac{\partial}{\partial t}\left[c_l^k s + c_g^k(1-s)\right] + \nabla \cdot \left\{\gamma \mathbf{u}\left[c_l^k s + c_g^k(1-s)\right]\right\} =$$
$$\nabla \cdot \left[D_{l,\text{eff}}^k \nabla c_l^k + D_{g,\text{eff}}^k \nabla c_g^k\right] - \nabla \cdot \left[\left(\frac{c_l^k}{\rho_l} - \frac{c_g^k}{\rho_g}\right)\mathbf{j}_l\right] + S^k \tag{10}$$

In the above, constant liquid and gas densities, ρ_l and ρ_g, have been assumed.

Defining the mixture molar concentration, c^k, as

$$c^k = c_l^k s + c_g^k (1-s) \tag{11}$$

the species equation (Eq. (6)), can be rewritten as

$$\frac{\partial c^k}{\partial t} + \nabla \cdot \left\{\gamma \mathbf{u} c^k\right\} = \nabla \cdot \left[D_{l,\text{eff}}^k \nabla c_l^k + D_{g,\text{eff}}^k \nabla c_g^k\right] - \nabla \cdot \left[\left(\frac{c_l^k}{\rho_l} - \frac{c_g^k}{\rho_g}\right)\mathbf{j}_l\right] + S^k \tag{12}$$

where c_l^k and c_g^k stand for species molar concentrations in the liquid and gas phases, respectively.

The second term on the RHS of Eq. (12) represents species transfer caused by relative motion of liquid to gas phase under capillary action in the porous backing and catalyst layers. In this term, the capillary-diffusional flux of the liquid phase, $\mathbf{j}_l$, as defined in Eq. (13), is directly proportional to the gradient in capillary pressure, p_c, and thus, is related to the surface wetting characteristics of the porous materials. That is

$$\mathbf{j}_l = \frac{\lambda_l \lambda_g K \rho}{\mu} \nabla p_c \tag{13}$$

where the definition of various two-phase properties is listed in Liu and Wang[28].

Methanol Transport – Assuming the vapor-liquid equilibrium of methanol on the anode side and invoking the Henry's law, the methanol conservation equation in the anode side can be specifically rewritten as

$$\frac{\partial}{\partial t}\left[c_l^{MeOH}(s+\frac{1-s}{k_H})\right]+\nabla\cdot\left\{\gamma\mathbf{u}c_l^{MeOH}(s+\frac{1-s}{k_H})\right\}=$$
$$\nabla\cdot\left[(D_{l,eff}^{MeOH}+\frac{D_{g,eff}^{MeOH}}{k_H})\nabla c_l^{MeOH}\right]-\nabla\cdot\left[\left(\frac{1}{\rho_l}-\frac{1}{k_H\rho_g}\right)c_l^{MeOH}\mathbf{j}_l\right]+S^{MeOH} \quad (14)$$

where

$$S^{MeOH}=-\frac{j}{6F}-\frac{j_{xover}}{6F} \quad \text{anode catalyst layer} \quad (15)$$

Assuming completion consumption of crossover methanol at the cathode catalyst layer and averaging the methanol diffusive flux through the membrane along the catalyst layer thickness, the net methanol crossover flux through the membrane, caused by electro-osmotic drag and diffusion, can be estimated by

$$\frac{j_{xover}}{6F}=\nabla(n_d^{MeOH}\frac{i}{F})+\frac{(D_m^{MeOH}\frac{c_l^{MeOH}|_{int}}{\delta_m})}{\delta_{cata}} \quad (16)$$

where the methanol electroosmotic drag coefficient, n_d^{MeOH}, is proportional to the methanol concentration such that

$$n_d^{MeOH}=n_d^{H_2O}\frac{c_l^{MeOH}|_{int}}{c_l^{H_2O}} \quad (17)$$

and $c_l^{MeOH}|_{int}$ denotes the methanol concentration at the interface between the anode catalyst layer and membrane.

Water Transport – Assuming that the water concentration in the gas phase is always saturated and constant at given temperature and pressure, we have that

$$c_g^{H_2O}=c_{g,sat}^{H_2O}=const \quad (18)$$

Then, the water conservation equation in both anode and cathode sides can be derived from Eq. (12) as

$$\frac{\partial c^{H_2O}}{\partial t} + \nabla \cdot \{\gamma \mathbf{u} c^{H_2O}\} = \nabla \cdot \left[D_{l,\mathit{eff}}^{H_2O} \nabla c_l^{H_2O}\right] - \nabla \cdot \left[\left(\frac{c_l^{H_2O}}{\rho_l} - \frac{c_{g,sat}^{H_2O}}{\rho_g}\right) \mathbf{j}_l\right] + S^{H_2O} \quad (19)$$

where the non-zero source term, S^{H_2O}, only exists in the anode and cathode catalyst layers, given by

$$S^{H_2O} = \begin{cases} -\dfrac{j}{6F}(1+6\alpha) & \text{anode catalyst layer} \\ \dfrac{j}{2F} + \dfrac{j_{xover}}{3F} + \alpha\dfrac{j}{F} & \text{cathode catalyst layer} \end{cases} \quad (20)$$

and the net water flux through the membrane is caused by electro-osmotic drag, diffusion and hydraulic permeation due to different hydraulic pressures between the anode and cathode, such that

$$\begin{aligned}
\alpha \frac{i}{F} &= n_d^{H_2O} \frac{i}{F} + N_{m,\mathit{diff}}^{H_2O} - N_{m,pl}^{H_2O} \\
&= n_d^{H_2O} \frac{i}{F} + \frac{\rho_m}{EW} D_m^{H_2O} \frac{\lambda_a^{H_2O} - \lambda_c^{H_2O}}{\delta_m} \\
&\quad - \frac{\rho_l K_m}{M^{H_2O} \mu_l \delta_m} \left[\frac{2\sigma_a \cos\theta_a}{r_a} J(s_a) - \frac{2\sigma_c \cos\theta_c}{r_c} J(s_c)\right]
\end{aligned} \quad (21)$$

where $J(s)$ is Leverett function, given as

$$J(s) = \begin{cases} 1.417(1-s) - 2.120(1-s)^2 + 1.263(1-s)^3 & \theta < 90° \\ 1.417s - 2.120s^2 + 1.263s^3 & \theta > 90° \end{cases} \quad (22)$$

Note that Eq. (21) indicates that the water hydraulic permeation flux is directly proportional to the hydraulic permeability of the membrane, a fundamental parameter requiring experimental measurements, and inversely proportional to the membrane thickness and the pore sizes of anode and cathode backing layers. In practice, the cathode pore size can be engineered to enhance hydraulic permeation by using a highly hydrophobic microporous layer. Furthermore, the net water transport coefficient through the membrane, α, varies spatially over the membrane.

Oxygen Transport – On the cathode side, oxygen solubility in liquid water is very small and thus, oxygen transport in the liquid phase is neglected in this study. Then, we have that

$$\frac{\partial c^{O_2}}{\partial t} + \nabla \cdot \{\gamma \mathbf{u} c^{O_2}\} = \nabla \cdot \left[D^{O_2}_{g,eff} \nabla c^{O_2}_g\right] - \nabla \cdot \left[\left(\frac{c^{O_2}_g}{\rho_g}\right) \mathbf{j}_l\right] + S^{O_2} \quad (23)$$

where

$$S^{O_2} = -\frac{j_c + j_{xover}}{4F} \quad \text{cathode catalyst layer} \quad (24)$$

Electrochemical Kinetics – The anodic transfer current density can be expressed by Tafel approximation of Butler-Volmer equation, such that

$$j = \frac{aj^{ref}_{o,a} c^{MeOH}_l \big|_{cata} \exp(\frac{\alpha_a F}{RT}\eta_a)}{c^{MeOH}_l \big|_{cata} + K_c \exp(\frac{\alpha_a F}{RT}\eta_a)} \quad (25)$$

where the rate constant K_c conveniently controls the transition from the zero-order kinetics of methanol oxidation under high methanol concentration and low overpotential to the first-order kinetics under low methanol concentration and high overpotential. The anode overpotential is defined as

$$\eta_a = \Phi_s - \Phi_e - U^o_a \quad (26)$$

Similarly, the cathodic transfer current density can be written as

$$j_{xover} + j_c = aj^{ref}_{o,c} \left(\frac{c^{O_2}_g\big|_{cata}}{c^{O_2,ref}_g}\right)(1-s)\exp(-\frac{\alpha_c F}{RT}\eta_c) \quad (27)$$

where the term $(1 - s)$ accounts for the fraction of catalytic surfaces rendered inactive by the presence of liquid water in the cathode catalyst layer, and cathode overpotential is

$$\eta_c = \Phi_s - \Phi_e - U^o_c \quad (28)$$

Under the assumption of a perfectly conductive electronic phase of anode and cathode catalyst layers, the electronic phase potential, Φ_s, becomes zero for the anode and is equal to the cell voltage for the cathode.

The electrolyte phase potential, Φ_e, is given by the proton transport equation

$$0 = \nabla \cdot (\kappa_{eff} \nabla \Phi_e) + S_\Phi \quad (29)$$

where

$$S_\Phi = \begin{cases} j & \text{anode catalyst layer} \\ j_c + j_{xover} & \text{cathode catalyst layer} \end{cases} \quad (30)$$

Eq. (30) indicates that protons are generated in the anode catalyst layer and consumed in the cathode catalyst layer. The proton conductivity of the membrane, κ_{eff}, is assumed constant in this work since the membrane is well hydrated in a liquid-feed DMFC.

Phase Saturations – Liquid saturation is a key parameter in the two-phase flow model. Here, we can obtain the liquid saturation from the mixture water molar concentration via

$$s = \frac{c^{H_2O} - c_{g,sat}^{H_2O}}{c_l^{H_2O} - c_{g,sat}^{H_2O}} \quad (31)$$

where $c_l^{H_2O}$ is simply calculated by

$$c_l^{H_2O} = \frac{\rho_l}{M^{H_2O}} \quad (32)$$

Boundary Conditions – All governing equations of the 3D model are summarized in Table 1, with eight unknowns: $\mathbf{u}$ (three components), p, c_l^{MeOH}, c^{O_2}, c^{H_2O} and Φ_e. Their corresponding boundary conditions are described as follows:

Flow Inlet Boundaries – The inlet velocity $\mathbf{u}_{in}$ in a flow channel is expressed by the respective stoichiometric flow ratio, i.e. ξ_a or ξ_c, defined at a reference current density, i^{ref}, as

$$\xi_a = \frac{c_l^{MeOH} \mathbf{u}_{in,a} A_{cross,a}}{\frac{i_{ref} A}{6F}} \quad \text{and} \quad \xi_c = \frac{c^{O_2} \mathbf{u}_{in,c} A_{cross,c}}{\frac{i_{ref} A}{4F}} \quad (33)$$

where $A_{cross,a}$ and $A_{cross,c}$ are the flow cross-sectional areas of the anode and cathode flow channels, respectively. The anode inlet methanol concentration is given as an operating parameter and the cathode oxygen molar concentration determined by the cathode inlet pressure, temperature, and relative humidity according to the ideal gas law.

Outlet Boundaries – Fully developed or no-flux conditions are applied

$$\frac{\partial \mathbf{u}}{\partial n} = 0, \ \frac{\partial p}{\partial n} = 0, \ \frac{\partial c^k}{\partial n} = 0 \text{ and } \frac{\partial \Phi_e}{\partial n} = 0 \qquad (34)$$

Walls – No-slip and impermeable velocity condition and no-flux condition are applied

$$\mathbf{u} = 0, \ \frac{\partial p}{\partial n} = 0, \ \frac{\partial c^k}{\partial n} = 0 \text{ and } \frac{\partial \Phi_e}{\partial n} = 0 \qquad (35)$$

3. Results and Discussion

Among the most desirable outputs from the 3D DMFC model are the current density distribution, the crossover current density distribution, and the inter-relationship between them. These are shown in Figure 1 for the cell voltage of 0.4 V. The current density is relatively uniform in the channel region of the membrane, but varies greatly in the land areas of the membrane, due primarily to the severe methanol transport limitation there. Moreover, the highest local current density does not occur at the inlet although there is the highest methanol concentration there. This can be explained by methanol crossover. Near the inlet, methanol crossover is most severe due to both methanol diffusion driven by the highest methanol concentration and electro-osmotic drag under the relatively high current density, as can be seen from Figure 1b. The large methanol crossover current leads to a severe mixed potential caused by the parasitic methanol oxidation reaction at the cathode catalyst layer, thus reducing the operating current density of the cell. With decreasing methanol concentration along the flow direction, methanol crossover decreases quickly, from 0.30 A/cm^2 at the inlet to about 0.15 A/cm^2 in the middle of the flow direction. Therefore, the highest local current density occurs in the middle section of the flow due to a combination of weak methanol crossover, still relatively high methanol concentration, and zero-order kinetics of the anode reaction.

In the land areas of the membrane, the current density distribution is controlled by the local methanol concentration because, due to insufficient methanol transport, the MOR is already a first-order reaction under small methanol concentrations there. Under the given operating conditions, the lowest current density under the lands of the current collector is only about 60% of that in the channel area of the membrane. Therefore, the geometry of lands and flow field are important for uniform current distribution and high cell performance.

TABLE 1. Summary of governing equations in the 3D model.

Mass conservation equation

$$\frac{\partial(\varepsilon\rho)}{\partial t} + \nabla \cdot (\rho \mathbf{u}) = \dot{m}$$

where

$$\dot{m} = \begin{cases} M^{MeOH} S^{MeOH} + M^{H_2O} S^{H_2O} + M^{CO_2} \dfrac{j}{6F} & \text{anode catalyst layer} \\[6pt] M^{H_2O} S^{H_2O} + M^{O_2} S^{O_2} + M^{CO_2} \dfrac{j_{xover}}{6F} & \text{cathode catalyst layer} \end{cases}$$

Momentum conservation equation

$$\frac{1}{\varepsilon}\left[\frac{\partial(\rho\mathbf{u})}{\partial t} + \frac{1}{\varepsilon}\nabla\cdot(\rho\mathbf{u}\mathbf{u})\right] = -\nabla p + \nabla\cdot\tau + S_\mathbf{u}$$

where

$$S_\mathbf{u} = \begin{cases} 0 & \text{channels} \\ -\dfrac{\mu}{K}\mathbf{u} & \text{backing and catalyst layers} \end{cases} \quad \text{and } \mathbf{u} = 0 \text{ membrane}$$

MeOH transport equation

$$\frac{\partial}{\partial t}\left[c_l^{MeOH}(s + \frac{1-s}{k_H})\right] + \nabla \cdot \left\{\gamma \mathbf{u} c_l^{MeOH}(s + \frac{1-s}{k_H})\right\} =$$

$$\nabla \cdot \left[(D_{l,eff}^{MeOH} + \frac{D_{g,eff}^{MeOH}}{k_H})\nabla c_l^{MeOH}\right] - \nabla \cdot \left[\left(\frac{1}{\rho_l} - \frac{1}{k_H \rho_g}\right) c_l^{MeOH} \mathbf{j}_l\right] + S^{MeOH}$$

where

$$S^{MeOH} = -\frac{j}{6F} - \frac{j_{xover}}{6F} \quad \text{in anode catalyst layer}$$

Water transport equation

$$\frac{\partial c^{H_2O}}{\partial t} + \nabla \cdot \left\{\gamma \mathbf{u} c^{H_2O}\right\} = \nabla \cdot \left[D_{l,eff}^{H_2O} \nabla c_l^{H_2O}\right] - \nabla \cdot \left[\left(\frac{c_l^{H_2O}}{\rho_l} - \frac{c_{g,sat}^{H_2O}}{\rho_g}\right)\mathbf{j}_l\right] + S^{H_2O}$$

where

$$S^{H_2O} = \begin{cases} -\dfrac{j}{6F}(1+6\alpha) & \text{anode catalyst layer} \\[6pt] \dfrac{j_c}{2F} + \dfrac{j_{xover}}{3F} + \alpha\dfrac{j}{F} & \text{cathode catalyst layer} \end{cases}$$

(Continued)

Oxygen transport equation

$$\frac{\partial c^{O_2}}{\partial t} + \nabla \cdot \{\gamma \mathbf{u} c^{O_2}\} = \nabla \cdot [D_{g,\mathit{eff}}^{O_2} \nabla c_g^{O_2}] - \nabla \cdot \left[\left(\frac{c_g^{O_2}}{\rho_g}\right)\mathbf{j}_l\right] + S^{O_2}$$

where

$S^{O_2} = -\dfrac{j_c + j_{xover}}{4F}$ cathode catalyst layer

Proton transport equation

$$0 = \nabla \cdot (\kappa_{\mathit{eff}} \nabla \Phi_e) + S_\Phi$$

where

$S_\Phi = \begin{cases} j & \text{anode catalyst layer} \\ j_c + j_{xover} & \text{cathode catalyst layer} \end{cases}$

Electrochemical kinetics

$$j = \frac{aj_{o,a}^{\mathit{ref}} c_l^{MeOH}|_{cata} \exp(\dfrac{\alpha_a F}{RT}\eta_a)}{c_l^{MeOH}|_{cata} + K_c \exp(\dfrac{\alpha_a F}{RT}\eta_a)}$$

where

$\eta_a = \Phi_s - \Phi_e - U_a^o$ (with $\Phi_s = 0$ in the anode)

$$j_{xover} + j_c = aj_{o,c}^{\mathit{ref}} (\frac{c_g^{O_2}|_{cata}}{c_g^{O_2,\mathit{ref}}})(1-s)\exp(-\frac{\alpha_c F}{RT}\eta_c)$$

where

$\eta_c = \Phi_s - \Phi_e - U_c^o$ (with $\Phi_s = V_{cell}$ in the cathode)

A representative example of methanol concentration contours predicted by the model at different cross-sections are shown in Figure 2. Along the flow direction, the methanol concentration in the anode channel is dominated by convection while it is mainly determined by diffusion in the anode backing layer. In Figure 2 it is evident that methanol concentration in the backing is higher under the flow channel than under the lands, demonstrating the land effect in limiting methanol transport in the anode. Also, the gradient of methanol concentration in the channel cross-section increases from the inlet to the outlet. Although the average methanol concentration in the liquid exiting from the channel is still high, about 1.2 M (see Figure 2c), the concentration inside the catalyst layer is very low near the exit, about 0.1 M, due to the insufficient methanol transport under the lands of the anode. Much more details of the computational results can be found in Liu and Wang.[28]

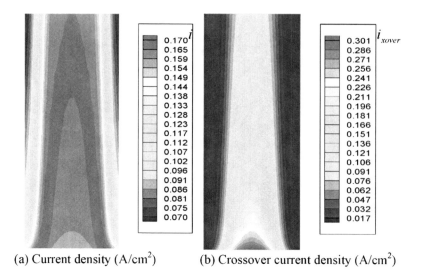

Figure 1. Current density (a) and crossover current density (b) distributions in the middle of the membrane at cell voltage of 0.4 V.

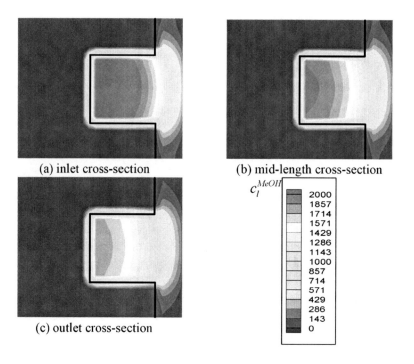

Figure 2. Methanol concentration (mol/m^3) distributions near the inlet (a), middle of the cell length (b) and near the outlet (c), in the anode.

4. Conclusions

It is indicated that performance and design of a liquid feed direct methanol fuel cell (DMFC) is controlled not only by electrochemical kinetics and methanol crossover but also by water transport and by their complex interactions in the design regime for portable electronics applications. In this paper, a three-dimensional (3D), two-phase model is discussed for DMFCs, in particular considering water transport and treating the catalyst layer explicitly as a component rather than an interface without thickness. The DMFC model is based on the multi-phase mixture formulation and encompasses all components in a DMFC using a single computational domain. 3D numerical simulations are described to explore mass transport phenomena occurring in DMFCs for portable applications as well as to reveal an interplay between the local current density and methanol crossover rate.

References

1. C. Y. Wang, Fundamental models for fuel cell engineering, *Chem. Rev.*, 104, 4727–4766 (2004).
2. G. Q. Lu, F. Q. Liu, and C. Y. Wang, Water transport through Nafion 112 membrane in DMFCs, *Electrochem. Solid-State Lett.*, 8, A1–A4 (2005).
3. F. Q. Liu, G. Q. Lu, and C. Y. Wang, Low crossover of methanol and water through thin membranes in direct methanol fuel cells, *J. Electrochem. Soc.*, 153, A543–A553 (2006).
4. G. T. Burstein, C. J. Barnett, A. R. Kucernak, and K. R. Williams, Aspects of the anodic oxidation of methanol, *Catal. Today*, 38, 425–437 (1997).
5. S. Wasmus and A. Kuver, Methanol oxidation and direct methanol fuel cells: a selective review, *J. Electroanal. Chem.*, 461, 14–31 (1999).
6. A. Hamnett, Mechanism and electrocatalysis in the direct methanol fuel cell, *Catal. Today*, 38, 445–457 (1997).
7. H. N. Dinh, X. Ren, F. H. Garzon, P. Zelenay, and S. Gottesfeld, Electrocatalysis in direct methanol fuel cells: in-situ probing of PtRu anode catalyst surfaces, *J. Electroanal. Chem.*, 491, 222–233 (2000).
8. L. Liu, C. Pu, R. Viswanathan, Q. Fan, R. Liu, and E. S. Smotkin, Carbon supported and unsupported Pt–Ru anodes for liquid feed direct methanol fuel cells, *Electrochim Acta*, 43, 3657–3663 (1998).
9. A. S. Arico, P. Creti, E. Modica, G. Monforte, V. Baglio, and V. Antonucci, Investigation of direct methanol fuel cells based on unsupported Pt–Ru anode catalysts with different chemical properties, *Electrochim. Acta*, 45, 4319–4328 (2000).
10. D. Chu and R. Jiang, Novel electrocatalysts for direct methanol fuel cells, *Solid State Ionics*, 148, 591–599 (2002).
11. S. R. Narayanan, H. Frank, B. Jeffries-Nakamura, M. Smart, W. Chun, G. Halpert, J. Kosek, and C. Cropley, in: *Proton Conducting Membrane Fuel Cells I*, edited by S. Gottesfeld, G. Halpert, and A. R. Landgrebe (PV 95-23, 278, The Electrochem. Soc. Proc. Series, Pennington, NJ 1995).

12. X. Ren, T. A. Zawodzinski Jr., F. Uribe, H. Dai, and S. Gottesfeld, in: *Proton Conducting Membrane Fuel Cells I*, edited by S. Gottesfeld, G. Halpert, and A. R. Landgrebe (PV 95-23, 278, The Electrochem. Soc. Proc. Series, Pennington, NJ 1995).
13. J. -T. Wang, S. Wasmus, and R. F. Savinell, Real-time mass spectrometric study of the methanol crossover in a direct methanol fuel cell, *J. Electrochem. Soc.*, 143, 1233–1239 (1996).
14. S. Hikita, K. Yamane, and Y. Nakajima, Measurement of methanol crossover in direct methanol fuel cell, *JSAE Rev*, 22, 151–156 (2000).
15. S. C. Kelly, G. A. Deluga, and W. H. Smyrl, A miniature methanol/air polymer electrolyte fuel cell, *Electrochem. Solid-State Lett.*, 3, 407–409 (2000).
16. G. Q. Lu, C. Y. Wang, T. J. Yen, and X. Zhang, Development and characterization of a silicon-based micro direct methanol fuel cell, *Electrochim. Acta*, 49, 821–828 (2004).
17. P. Argyropoulos, K. Scott, and W. M. Taama, Gas evolution and power performance in direct methanol fuel cells, *J. Appl. Electrochem.*, 29, 663–671 (1999).
18. G. Lu and C. Y. Wang, Electrochemical and flow characterization of a direct methanol fuel cell, *J. Power Sources*, 134, 33–40 (2004).
19. J. Wang and R. F. Savinell, in: *Electrode Materials and Processes for Energy Conversion and Storage*, S. Srinivasan, D. D. Macdonald, and A. C. Khandkar (PV 94-23, 326, The Electrochem. Soc. Proc. Series, Pennington, NJ 1994).
20. S. F. Baxter, V. S. Battaglia, and R. E. White, Methanol fuel cell model: anode, *J. Electrochem. Soc.*, 146, 437–447 (1999).
21. A. A. Kulikovsky, J. Divisek, and A. A. Kornyshev, Two-dimensional simulation of direct methanol fuel cell. A new (embedded) type of current collector, *J. Electrochem. Soc.*, 147, 953–959 (2000).
22. A. A. Kulikovsky, Two-dimensional numerical modelling of a direct methanol fuel cell, *J. Appl. Electrochem.*, 30, 1005–1014 (2000).
23. K. Scott, P. Argyropoulos, and K. Sundmacher, A model for the liquid feed direct methanol fuel cell, *J. Electroanal. Chem.*, 477, 97–110 (1999).
24. P. Argyropoulos, K. Scott, and W. M. Taama, Hydrodynamic modelling of direct methanol liquid feed fuel cell stacks, *J. Appl. Electrochem.*, 30, 899–913 (2000).
25. Z. H. Wang, C. Y. Wang, and K. S. Chen, Two-phase flow and transport in the air cathode of proton exchange membrane fuel cells, *J. Power Sources*, 94, 40–50 (2001).
26. C. Y. Wang and P. Cheng, Multiphase flow and heat transfer in porous media, *Adv. Heat Transfer*, 30, 93–196 (1997).
27. Z. H. Wang and C. Y. Wang, Mathematical modeling of liquid-feed direct methanol fuel cells, *J. Electrochem. Soc.*, 150, A508–A519 (2003).
28. W. Liu and C. Y. Wang, Three-dimensional simulations of liquid feed direct methanol fuel cells, *J Electrochem. Soc.*, 154, B352–B361 (2007).
29. J. Yuan and B. Sunden, Analysis of intermediate temperature solid oxide fuel cell transport processes and performance, *J. Heat Transfer* 127, 1380–1390 (2005).
30. V. V. Kharton, F. M. B. Marques, and A. Atkinson, Transport properties of solid oxide electrolyte ceramics: a brief review, *Solid State Ionics* 174, 135–149 (2004).

OPTIMIZATION OF DIRECT METHANOL FUEL CELL SYSTEMS AND THEIR MODE OF OPERATION

SHIMSHON GOTTESFELD[1]* AND COSTAS MINAS [2]
[1]*STA, MTI Microfuel Cells, Albany, NY, USA & President, Fuel Cell Consulting , LLC, Niskayuna, NY, USA*
[2]*MTI Microfuel Cells, Albany, NY, USA*

1. Key Technical Challenges in the Development of Micro-DMFC Technology and Some Solutions Provided

The merits of the DMFC as the technology of choice for micro-fuel cell power systems in hand held consumer electronics devices have been described in the Introductory Technical Remarks for this School. The DMFC scheme in Figure 1 reveals it is important key feature: direct and complete conversion of methanol fuel to CO_2 in a methanol/air cell operating at a temperature well under 100°C. With neat liquid methanol having a heating value of near 5 Wh/CC – i.e. theoretical energy content in one CC of liquid methanol equal to that of a relatively large cell phone battery – and with the simple nature of a "direct" fuel cell that can use this energy-rich liquid fuel directly, one might expect easy reduction to practice of a DMFC power pack (fuel + fuel cell + auxiliaries) of energy density higher than that of a Li-ion battery. In reality, however, three main barriers to the reduction to practice of DMFC systems of superior energy density, have been: (i) high methanol permeability of mainstream ionomeric membranes used in polymer electrolyte fuel cells, (ii) the water management challenge posed by water product generated on the cathode side of the DMFC while being required for the cell process on the anode side, and (iii) the moderate power density of DMFC stacks (compared with hydrogen fueled stacks), defined by the modest rate of the anodic oxidation of methanol. This chapter describes innovative approaches developed during the last decade for the resolution of the first two barriers and, in the context of the third challenge, describes briefly the state-of-the-art of electrocatalyst research and development for polymer electrolyte DMFCs.

---

*E-mail: shimshon.gottesfeld@gmail.com

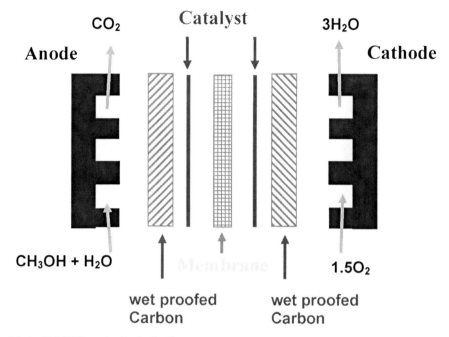

Main DMFC technical challenges
1. **Significant "methanol crossover"**
   Caused by: high methanol permeability of the membrane
2. **"Water management" challenge**
   Caused by: need in the anode of water generated in the cathode
3. **Moderate stack power density**
   Caused by: modest rate of the methanol electro-oxidation process

Figure 1. Scheme of the DMFC and some key technical challenges on route to reduction to practice.

1.1. CAN HIGH FUEL UTILIZATION BE OBTAINED WITH "LEAKY" MEMBRANES?

Figure 1 provides a scheme of the DMFC as background, together with the three critical technical issues associated with it is reduction to practice. The significance of high rate of methanol permeation ("cross-over") across the membrane electrolyte should be obvious. When the current density of methanol oxidation at the anode catalyst is J_{cell} and the rate of methanol cross-over is $J_{c.o}$, only a fraction of the fuel feed is converted to current, this fraction typically referred to as "fuel utilization", given by:

$$\eta_{fuel} = \frac{J_{cell}}{[J_{cell} + J_{c.o}(J_{cell})]} \qquad (1)$$

For poly {PFSA} membranes like Nafion, the rate of methanol permeation measured at C_{Meoh} = 1 M next to one surface of the membrane (and C_{Meoh} = 0 on the other side), is equivalent to 100–200 mA/cm^2. Since the current density of a DMFC operating at acceptable conversion efficiencies is also in the range 100–200 mA/cm^2, cursory examination of Eq. (1) would suggest that fuel utilization is expected to be as low as 50% when such membranes are used for DMFCs. It is important to recognize, however, that, as stated in Eq. (1), the rate of methanol cross-over, Jc.o, is not an invariable value in an operating DMFC, but, rather, a sensitive function of cell current. Figure 2 demonstrates the origin of this dependence of $J_{c.o}$ on J_{cell}. The figure shows the type of methanol concentration profiles expected across the cell in a DMFC under conditions of zero cell current (solid lines) and under a current close to the anode limiting current (dashed lines). It can be readily seen that, when the cell limiting current, J^*_{cell}, is set by the limited permeability of methanol through the anode backing layer (the anode "GDL"), the concentration of methanol next to the anode side of the membrane can be lowered substantially when operating near this pre-designed anode limiting current. It can be shown that, under such operation conditions, corresponding to the the dashed lines in Figure 2, Eq. (1) can be written using an explicit dependence of $J_{c.o}$ on J_{cell}, as:

$$\eta_{fuel} = \frac{J_{cell}}{\left[J_{cell} + J_{crossover@J_{cell}=0}\left(1 - J_{cell}/J^*_{anback}\right)\right]} \qquad (2)$$

Using some specific numbers as example, it can be seen from Eq. (2) that, with a cross-over rate at zero cell current, $J_{crossover}@J_{cell}$l = 0, as high as 50% of the design J_{cell}, operation of the cell at $J_{cell}/J^*_{lim,an}$ = 0.85 will enable to reach, with such " leaky" membrane, fuel utilization as high as 93% .

Implementation of such approach of lowering the concentration of the reactant methanol next to the anode side of the membrane so as to lower the rate of cross-over, is facilitated by the concentration independence of the catalytic oxidation of methanol: "zero order" vs. methanol of this catalytic process has been documented for methanol concentration > 0.01 M. The benefit of this feature of the anode process is the insignificant effect of a low local concentration of methanol on the rate of the electro-catalytic methanol oxidation process taking place at the anode catalyst surface.

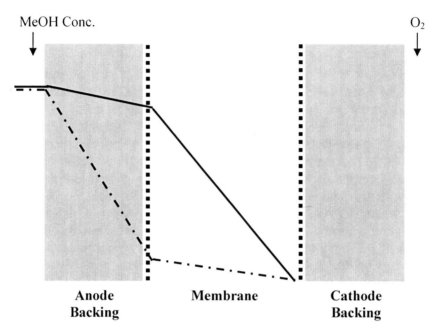

Figure 2. Methanol concentration profiles at zero cell current (solid) and under cell current close to the anode limiting current (dashed); "Enhanced methanol utilization in direct methanol fuel cells", Ren and Gottesfeld, US Patent # 6,296,964, 2001.

In a DMFC where the conversion efficiency is enhanced at higher cell current through the lowering of the rate of methanol cross-over (Eq. 2), both cell power and conversion efficiency increase with the product $V_{cell} \times J_{cell}$ and, consequently, the *peak power and peak efficiency points are one and the same*. In other words, operating the DMFC at the highest cell current short of the limiting current "cliff" defined by anode transport features (Figure 2), achieves simultaneously maximum cell power and maximum conversion efficiency.

The significance to DMFC technology, of the capability to achieve high fuel utilization in a cell employing a "methanol leaking membrane", has been quite substantial. Such capability provided an answer to a key question, raised repeatedly earlier on, about the criticality of an ionomeric membrane of low methanol permeability for the development of a viable, polymer electrolyte DMFC technology. The prevailing opinion has been that such a membrane is indeed a critical prerequisite and, therefore, there is not much point in pursuing polymer electrolyte DMFC technology as long as such membrane is not available. It turns out, however, that reducing substantially methanol cross-over in an ionomeric membrane is always accompanied by significant loss of conductivity of the less permeable ionomer vs. the conductivity of poly {PFSA} ionomers. The resulting drop in both membrane performance and electrode performance, typically outweigh any benefit of a lowered methanol cross-over rate. Consequently, the pursuit of an ionomeric membrane of lower methanol permeability which would enable higher cell performance, resulted in limited success to date. At the same time, the ability to use for DMFC technology commercially available poly {PFSA} membranes, developed and commercialized earlier for hydrogen/air PEFCs, opened the door for much faster development of DMFC technology.

1.2. THE CHALLENGE OF WATER MANAGEMENT IN DMFCS

The DMFC electrode processes described in Figure 1 reveal the nature of the water management challenge in an operating DMFC. Water is generated in the cell cathode, while being required as both reactant and proton conducting facilitator at the anode, as well as for effective hydration through the thickness dimension of the membrane. Consequently, a DMFC system has to provide a solution for the demand of water at the anode, by either carrying the water required for the anode process "on board", i.e. as part of the system, or by system component(s) that will direct the required amount of water from the cathode to the anode so that the overall water balance in the cell can be sustained without carrying any water on board. In principle, the latter approach is possible because the cell as a whole is a net generator of water and the challenge is essentially to properly re-distribute the cathode water. Optimized water distribution demands that, at least, 1 mol of water is transferred from the cathode to the anode per mole of methanol consumed, while the excess net cell production of 2 mol of water per mole of methanol consumed is released out the cell in vapor form.

The mainstream DMFC technology approach to water management has been to develop solutions for utilizing exclusively water generated by the cell to achieve the required cell water content and cell water distribution under operation conditions. The other option mentioned above, of carrying water in the fuel cartridge, results in a significant sacrifice of system energy content as a significant fraction of the fuel in the tank/cartridge is replaced by water. Implementation of a sub-system for collecting water from the cathode and pumping it to the anode, involves, however, further complexity, a number of additional possible failure modes and extra volume and weight penalties. A general scheme of a DMFC system involving collection and pumping of water from cathode to anode around the cell is shown in Figure 3. As the figure shows, such a DMFC system involves a re-circulated anode feed stream, typically of low methanol concentration around 0.1 M, maintained by dosing fuel from the methanol cartridge and adding water pumped from the cathode. It is easy to see why such system design may be especially problematic when the total volume of the micro-fuel cell power pack is

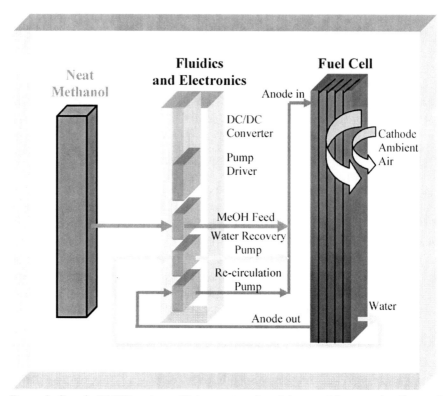

Figure 3. Generic DMFC system utilizing neat methanol *but* requiring complex flow and flow control subsystems.

limited to 25 CC, or smaller, as such external "plumbing" then takes up volume critically required for the micro fuel cell stack and for the fuel.

To answer the demand of miniaturization and simplicity, a much simpler and more compact alternative micro-DMFC system design was developed starting in 2001 at MTI Micro Fuel Cells (NY, USA). This novel approach was built around physical properties of components of the membrane/electrode assembly, that, when properly implemented can induce spontaneous flux of cathode generated water from the cathode to the anode, through the thickness dimension of the cell. The so named Mobion technology platform, removes any need to collect and pump water outside the cell, i.e. water management in this DMFC platform becomes totally passive. Furthermore, this mode of DMFC operation relies on direct feed to the anode of 100% methanol, with the mixing with water required for the anode process occurring within the anode. A scheme of the Mobion DMFC system is presented in Figure 4. The key component for totally passive water balance, schematically described in Figure 4 by a light colored layer on the cathode edge, is an element that can fulfill the combined role of "pushing" enough of the cathode-generated liquid water through the thickness dimension of the cell to the anode and, at the same time, allow the net excess water generated to escape the cell in vapor form. Clearly, the overall gain in system volume and system simplicity and the significantly lower part count achieved by replacing the mainstream DMFC platform (Figure 3) with the Mobion DMFC platform (Figure 4), is substantial. In fact, the latter platform makes the DMFC system resemble a battery, whereas the mainstream platform has characteristics which more strongly resemble a chemical reactor.

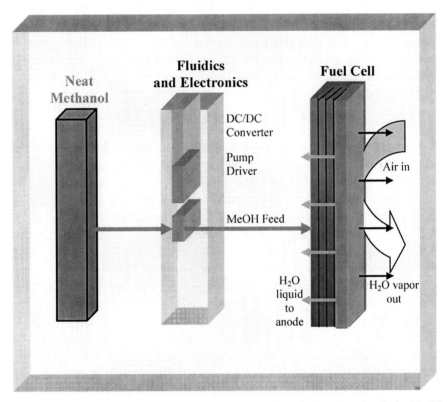

Figure 4. Highly simplified DMFC System (Mobion) with cathode MEA of optimized build enabling effective passive water management.

A question frequently raised with regard to the Mobion platform, is whether such passive redistribution of water across the thickness dimension of the cell (Figure 3) does not result in lower performance of the cell, compared with the performance of a DMFC operating with addition of pumped water into the anode (Figure 4). The question is justified, in principle, because unless the spontaneous, materials property dependent flux of water from the cathode to the anode within the cell (Figure 4) is totally sufficient, both the power and the conversion efficiency of this DMFC platform will be inferior. The reason for such sensitivity to effective water supply to the anode, is that the anode process not only needs water as a reactant (Figure 1), but needs to operate at a high local water/methanol ratio to ensure complete oxidation to CO_2 It turns out that the recorded performance of Mobion type DMFCs has practically matched the performance obtained with re-circulated liquid feed streams, reaching a combination of areal MEA power density of 60 mW/cm^2 and energy to the load of 1.6 Wh per CC of methanol consumed. Areal power densities over 100 mW/cm^2 are readily obtainable with the Mobion DMFC platform, at some sacrifice in energy conversion efficiency. To move from operation point of higher conversion efficiency to operation point of higher power, the (only) "knob" required in the Mobion system, is for controlling the rate of (100%) methanol dosing into the anode (Figure 4).

1.3. ON THE MODEST POWER DENSITY OF THE DMFC

The penalty for the very attractive features of a low temperature, direct oxidation fuel cell operating on a convenient, energy rich fuel like methanol, is the relative low rate of the anode process (methanol electro-oxidation) vs. the anode process in a hydrogen/air cell (hydrogen electro-oxidation). At conversion efficiencies exceeding 30%, the power obtained per cm^2 of MEA (areal power density) in a methanol/air cell (DMFC), is typically 3–5 times lower vs. that obtained in a hydrogen/air cell. Overall, the widely preferred choice of DMFC technology for development of micro-fuel cells, is based on the significantly higher energy content and ease of refilling and distribution of liquid 100% methanol fuel, as compared with any begining form of hydrogen storage for such applications, e.g. metal hydrides. However, once DMFC technology is chosen, one challenge is what can and should be done at present and in the near future to accommodate the relatively lower power density of the DMFC stack. As of today, the tools available are in the domains of stack engineering and optimized system design and operation. Packaging of a given area of MEA into a smaller stack volume is enabled, in principle, by design of stack of minimized thickness per cell, or, in fuel cell parlance, the smallest "pitch" per cell. This should be implemented while ensuring that the resulting very small dimensions of the flow channels will not create higher flow barriers, e.g. because of increase in the channel surface/volume ratio. A pitch per cell of 2 mm is probably a challenging but rather realistic target in this regard, and it could lead to volume power density of a DMFC stack as high as 300 W/l (assuming 100 mW/cm^2 of MEA and an active fraction of the overall stack volume at 60%). While this is still around factor 3 lower vs. a high quality hydrogen/air stack, a DMFC stack of 300 W/l will allow sufficient volume for the energy rich methanol fuel to enable an overall DMFC system for a variety of applications with system energy density x2–x3 that of a Li-ion battery (see discussion in the Introductory Remarks).

Another very helpful approach to best design for the actual characteristics of the DMFC stack, including the power output, the energy conversion efficiency and their inter-dependence, is to apply overall system optimization based on actual, measured stack and peripheral component characteristics. A DMFC system optimization program developed at MTI Micro fuel Cells has targeted evaluation of the minimum volume for a DMFC "straw man" system, including stack, fuel cartridge and insulation around the stack. The input parameters included key thermal properties: the insulator heat conduction and the rate of heat removal from the outer surface by natural convection, key electrochemical characteristic expressed by polarization curves recorded for the particular cell through a range of inner cell temperature considered (50–80°C), and key stack build parameters: cell "pitch" of 4.5 mm and active volume fraction of 60%. The two variables in the optimization were the cell voltage and the inner cell temperature. Basically, an optimized cell voltage reflects a trade-off between stack power output and stack energy conversion efficiency, whereas the optimized cell temperature reflects a trade-off between enhanced stack power density at higher cell temperature (leading to lower stack volume per given power demand) and

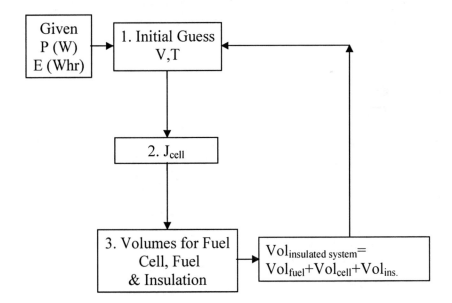

Figure 5. Optimization algorithm for a DMFC "straw man" system, targeting minimum system volume through an optimized pair of V_{cell} and T_{cell}, the latter within the temperature range of 50–80°C. Input parameters include thermal conductivity of the insulating envelope around the stack, rate of heat removal from outer surface by natural convection and DMFC polarization characteristics recorded in the range of cell temperatures considered.

the increase in the volume of insulation required to sustain a higher steady state stack temperature. The optimization algorithm is described in Figure 5.

Results for optimized pairs of cell voltage and inner cell temperature enabling minimum system volume for a {1 W, 30 Wh} DMFC system of specific physical and electrochemical characteristics, are described by the contour plots in Figure 6. The figure shows that there is no deep minimum associated with some pair of V_{cell}; T_{cell}. The smallest system volume, of 43 CC, is obtained near the low end of the temperature range considered, meaning that the increase in cell performance between 50°C and 80°C results in stack volume savings that are overwhelmed by an increase in volume of insulation with the target cell temperature. This situation is likely to be reversed only when advanced insulating materials considered here are replaced by an evacuated sealed enveloped around the stack – a pretty demanding technology by itself.

With the 1 W; 30 Wh system packaged into 43 CC, the overall system energy density achieved at the optimized V_{cell}; T_{cell}, is near 0.7 Wh/CC, about twice that of a Li-ion battery. However, this number is higher than expected in reality, as several components of an actual complete DMFC system have not been included in this straw-man system, including the methanol dosing element, control electronics and the outer envelope of the overall system. The performance of the DMFC stack can certainly use some further improvement in both power density and conversion efficiency, to enable tighter stack packaging and a smaller volume of fuel per some required total energy. Stack engineering is one possible tool, as explained above in this section. The height per cell assumed in the optimization described, was 4.5 mm and, if reduced by 50% the volume of the stack (at 60% active volume) will drop by 30%. Furthermore, the ohmic resistance of the cell considered in the above system optimization exercise, was rather high and this caused excessive losses in cell voltage, i.e. excessive loss in conversion efficiency per some gain in cell power. Consequently, there is certainly remaining room for further DMFC performance improvement even with present day DMFC electrode technology.

A further boost of DMFC performance could be achieved, at least in principle, with novel, more active catalysts, particularly for the process in the methanol anode. The catalyst of choice to date for methanol electrooxidation, has been the PtRu metal alloy of composition near 1:1 atomic ratio of Pt:Ru. It is understood that the special characteristics of this catalyst are derived from a "bi-functional" surface activity, that enables fast cleavage of C-H bonds on Pt surface sites and relatively effective removal of the CO moiety remaining adsorbed on Pt site(s) by surface OH groups formed from water on Ru surface sites. Accordingly, the mechanism of the anodic process in a DMFC at a PtRu catalyst is written as:

(I) CH_3OH (+ Pt surface sites) = $4H^+ + 4e^- + Pt\text{-}CO$
(II) $Ru + H_2O = Ru\text{-}OH + H^+ + e^-$
(III) $Pt\text{-}CO + Ru\text{-}OH = CO_2 + H^+ + e^-$

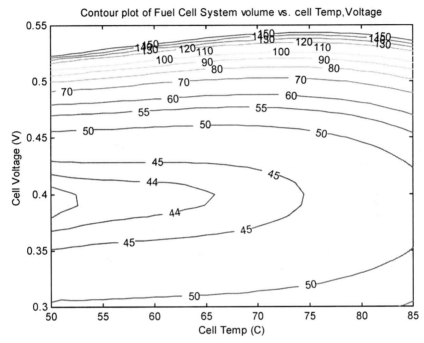

Figure 6. Contour plots for the volume of a {1 W, 30 Wh} DMFC system calculated for a range of V_{cell}; T_{cell} pairs by the optimization algorithm described in Figure 5.

There are very good indications pointing to step III in the sequence being the bottle neck in the series of steps (I)–(III) and, consequently, the search for a more active anode catalysts has centered on a catalyst surface that would further facilitate oxidative removal of CO adsorbed essentially at the Pt site. In principle, this would be achieved with a catalyst that lowers the affinity of CO to the surface and/or enhances the rate of the bimolecular surface process (III) by introduction of a CO electro-oxidation enhancing function.

Such a search for advanced electrocatalysts has relied most recently to larger degree on DFT calculations, as predictors of the effects of, for example, a third element added to Pt and Ru, where the resulting ternary alloy will exhibit enhanced CO removal characteristics vs. PtRu. A result of such a DFT calculation done by the group of Prof. Matt Neurock at the University of Virginia, resulted in the prediction that addition of gold atoms to the PtRu alloy surface is expected to raise the CO electro-oxidation activity. This was projected based on both a weakening of the bond of CO_{ads} to the catalyst surface and a contribution of the gold atoms to enhancement of the rate of CO electro-oxidation. Such very recent developments in DFT calculations of electronic properties of electro-catalyst/electrolyte interfaces, have the potential to improve the effectiveness of experimental searches for better electro-catalysts.

One very important caveat to remember here, however, is that, in the "real world" of an operating fuel cell, it is rare to have a surface of an alloy catalyst maintain a composition and, particularly, a structure which has been pre-tailored to achieve superior catalytic properties. The likelihood that the structure, or even the composition of the ternary alloy catalyst surface can be maintained in an operating methanol anode, is rather small. The relative stability of the PtRu binary alloy is, in fact, a very important feature of this catalyst of choice to date. It is therefore going to be a quite demanding task to improve on it with some ternary alloy.

References

1. US Patents: 6,296,964 Enhanced methanol utilization in direct methanol fuel cell, Ren and Gottesfeld, 2001.
2. US Patents: 6,981,877 Simplified direct oxidation fuel cell system, Ren et al., 2006.
3. US Patents: 7,282,293 Passive water management techniques in direct methanol fuel cells, Ren et al., 2007.

DEVELOPMENT OF A 5-W DIRECT METHANOL FUEL CELL STACK FOR DMB PHONE

YOUNG-CHUL PARK, DONG-HYUN PECK,
SANG-KYUNG KIM, AND DOO-HWAN JUNG*
*Advanced Fuel Cell Research Center, Korea Institute of
Energy Research (KIER), 71-2, Jang-dong, Yuseong, Daejeon
305–343, Korea*

1. Introduction

As high-tech multifunctional and miniature devices such as laptop computers, digital video recorders, digital multimedia phones, PDAs, and electronic game players become more widely used, new power sources need to be developed that have much longer run time and stronger power than those that current power sources, namely lithium batteries, provide. A fuel cell, which is a device that generates electricity by a chemical reaction, is considered the promising candidate for replacing them. Among various fuel cells, direct methanol fuel cell (DMFC) is the most suitable power source because it does not require any fuel processing equipment and can be operated at low temperatures. Also DMFC has advantages of easy transportation and storage of the fuel, and reduced system weight and size.[1,2] However, in order to develop a micro fuel cell needs much examination due to a limited weight and size and a high performance, durability. To meet these requirements, the MEA has to perform well and the stack design has to be very compact.

We have developed a 5-W DMFC stack for DMB phones. This paper describes the design of the micro fuel cell system and stack and flow field. We also investigated the feasibility of metal bipolar plates to decrease the volume and cost of the stack. It was fabricated with five cells having an electrode area of 131 cm^2 and has the internal manifolds for supply of air and fuel. The electrochemical performance of the single cell and the stack was evaluated.

---

*Tel.: +82-42-860-3180; e-mail: doohwan@kier.re.kr

2. What is a DMFC?

2.1. CONCEPT OF DMFC

DMFC is similar to the PEMFC in that the electrolyte is a polymer membrane and the charge carrier is the hydrogen ion. As shown in Figure 1, the methanol in DMFC is oxidized in the presence of water at the anode generating CO_2, hydrogen ions and the electrons that travel through the external circuit as the electric output of the fuel cell.

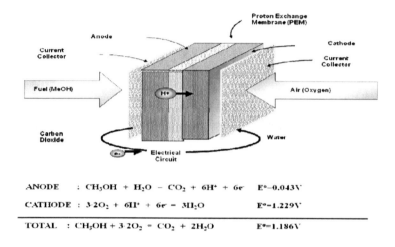

Figure 1. An overall reaction of DMFC.

The hydrogen ions travel through the electrolyte and react with oxygen from the air and the electrons from the external circuit to form water at the anode completing the circuit.[3,4] An overall reaction is represented as below. The net energy density of methanol is higher than any of the other fuels. It is the main advantage of the direct methanol system. Other important advantages include that the system is simpler to use and very quick to refill.[4] However, a slow oxidation reaction of methanol requires a more active catalyst at low-temperatures, which typically demands a larger quantity of expensive platinum catalyst.

2.2. COMPONENTS AND CONSTRUCTION OF DMFC

DMFC generally consists of membrane-electrode assembly (MEA), gas diffusion layer (GDL), bipolar plates. MEA is comprised by a polymer electrolyte membrane and electrode catalyst layers and gas diffusion layers. Polymer electrolyte membrane has to have two properties: conducting H+ ions from the anode to the cathode and providing electrical insulation between the anode and the cathode to force the electrons to move from the anode to the cathode all the way through an external circuit. A sulfonated tetrafluorethylene co-polymer such as Nafion developed by Dupont is usually used as a polymer electrolyte membrane but recently hydrocarbon series which improved a methanol crossover begin to be applied.[5] The best catalyst for both the anode and the cathode is platinum. The platinum is prepared into very small particles on the surface of carbon powders to increase the reaction area and rate of the electrodes.[6]

A carbon backing material such as carbon cloth or paper is usually called the gas diffusion layer. It provides the basic mechanical structure for the electrode and carries away reaction products from the catalyst and reactants towards the catalyst. Also it provides an electrical connection between the catalyst and the bipolar plate. Bipolar plates separate individual cells in a fuel cell stack. Each bipolar plate distributes reactants over the cell surface through the system of channels and collects current produced by individual cell and transports this current from one cell to another.[7] Machined graphite is traditionally used as the bipolar plates due to a good conductivity but metals and composite materials are under development to substitute for it because of its high cost and larger volume. Gasket to prevent liquid and gas leakage is PTFE and end plates are a stainless steel or a copper coated by gold.

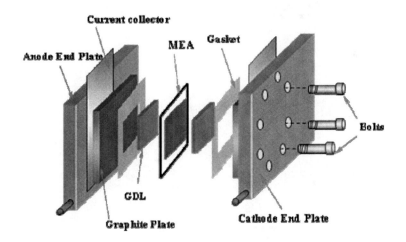

Figure 2. Construction and key components of DMFC.

Figure 2 shows construction of direct methanol fuel cell, having MEA and the modular components of anode and cathode. MEA is mounted within bipolar plates consisting of two electrically conducting field plates in which channels are fabricated, as shown in Figure 2. PTFE gaskets are fitted adjacent to flow channels to act as efficient gas seals around the edges of the Nation membrane.[8, 9]

3. Micro Fuel Cell Stack for DMB Phone

3.1. DEVELOPMENTS OF MICRO FUEL CELL FOR CELLULAR PHONES

Many micro fuel cell cellular phones were demonstrated by many Japanese and Korean electronic companies such as Toshiba, Hitachi, Fujitsu and Samsung in 2005 and 2006.

TABLE 1. Prototype images and characteristics of the micro fuel cell cellular phones.

Company	Prototype	Type	Size (cc)	Power (W)	Demonstrated year
KDDI-Toshiba		DMFC	200	0.3	2005
KDDI-Hitachi		DMFC	122	0.3	2005
NTT DoCoMo-Fujitsu		DMFC	160	9	2005
Samsung		DMFC	100	2	2006

Their fuel cell is based on DMFC technology. They are compact between 100 and 200 cc and produce power between 0.3 and 9 W. The fuel cell of the cellular phone co-developed by Toshiba and KDDI is a hybrid type, which is combined at the back of the handset with lithium ion battery. It can increase the battery run time 2.5 times longer with a single refill. The fuel cell of the cellular phone co-developed by Hitachi and KDDI is also a hybrid type but it is more compact. One feature is that whenever its methanol fuel becomes low, it can be easily refilled from a compact cartridge.[10] The fuel cell co-developed by Fujitsu and NTT DoCoMo enables eight hours of continuous talk and improved the capacity by increasing the methanol concentration from 30% to over 99% and developing a method of recycling the generated water.[11] SAIT and Samsung SDI demonstrated the world's smallest fuel cell mobile devices charger. It produces 2 W and is 5 mm thick and weighs 5.3 ounces. It includes user-replaceable methanol cartridges. It is designed to recharge the battery system in your PDA, cell phone, digital camera or PMP.[12] The images and characteristics of the micro fuel cell cellular phone prototypes are summarized in Table 1

3.2. DESIGN OF MICRO FUEL CELL STACK FOR DMB PHONE

The stack was designed with five cells that have an electrode area of 131 cm^2. It has the internal manifolds for supply of air and fuel. Its dimensions are 48 × 75 × 14.5 mm. Each cell of this DMFC stack has an active area of 17 cm^2. Graphite was used as bipolar plates with two flow paths and a cell pitch is 2 mm. Figures 3 and 4 show a micro fuel cell stack and bipolar plates prepared for DMB phone.

3.3. DESIGN OF FLOW FIELD FOR MICRO FUEL CELL STACK[13]

Three-dimensional computational simulation is used in order to illustrate the causes of performance difference according to the flow-filed design. Parallel, serpentine, parallel serpentine and zigzag shapes were simulated and estimated for the flow field of the anode by comparing the concentration distribution and the current density distribution. Table 2 summarizes specifications of the various flow fields such as electrode area and channel area, rib area, etc.

Figure 3. Micro fuel cell stack for DMB phone.

Figure 4. Bipolar plates for DMB phone.

TABLE 2. Specifications of the flow-fields.

	Electrode area (cm^2)	Width × length (cm^2)	Channel area (cm^2)	Rib area (cm^2)	Rib area ratio
Parallel	2.25	1.5 × 1.5	1.41	0.84	0.373
Serpentine	2.21	1.7 × 1.3	1.25	0.96	0.434
Parallel serpentine	2.255	2.05 × 1.1	1.32	0.935	0.415
Zigzag	2.2575	1.5 × 1.55	1.3875	0.87	0.385

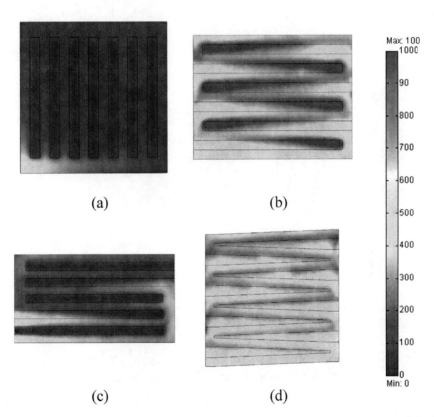

Figure 5. Concentration distributions at the electrode surface when the overpotential of 0.45 V is imposed. Average concentrations on the catalyst layer: (a) 84 mol·m^{-3} for parallel, (b) 303 mol·m^{-3} for serpentine, (c) 166 mol·m^{-3} for parallel serpentine, and (d) 363 mol·m^{-3} for zigzag.

When the overpotential of 0.45 V is applied to the electrode, the calculated methanol concentration distribution in the overall geometry of the serpentine type is that the concentration difference between the flow field and the catalyst layer is big while the difference between fuel input and output in the flow fields is not big. Particularly, the concentration at rib areas of the catalyst layer which contact with a bipolar plate is very low because the fuel can not contact with the backing layer directly.

As the overpotential of 0.45 V is applied, most sections of the parallel and the parallel serpentine reach to the threshold. The serpentine type starts to fall into the concentration polarization although zigzag type is still keeping the theoretical current density without concentration polarization as shown in Figures 5 and 6.

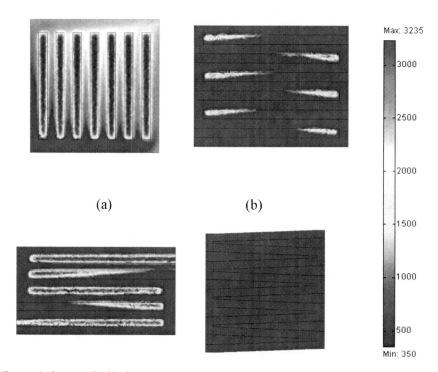

Figure 6. Current distributions at the electrode surface when the overpotential of 0.45 V is applied. Average current densities on the catalyst layer: (a) 2087 A·m^{-2} for parallel, (b) 3,152 A·m^{-2} for serpentine, (c) 2,723 A·m^{-2} for parallel serpentine, (d) 3,235 A·m^{-2} for zigzag.

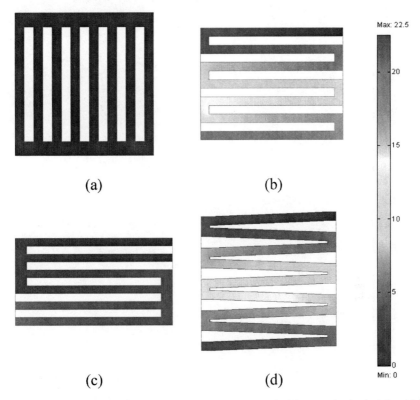

Figure 7. Pressure drop in the flow-field channel when the fuel input velocity is 1.5×10^{-8} $m^3 \cdot s^{-1}$: (a) 0.18 Pa for parallel, (b) 19.70 Pa for serpentine, (c) 4.95 Pa for parallel serpentine, (d) 22.50 Pa for zigzag.

Figure 7 shows pressure drop in the flow field channel when the fuel input velocity is 1.5×10^{-8} $m^3 \cdot s^{-1}$. The parallel type does not need high pressure drop in the channel however the serpentine and the zigzag type require the high pressure drop. The pressure difference between the parallel and the serpentine types reaches about 102 times.

Figure 8 presents an anode overpotential curve according to the flow-field design. Each point is originated from the average current densities calculated from the methanol concentration at the catalyst layer when the corresponding overpotential is imposed. Dotted line is a theoretical activation polarization curve in which the concentration polarization is not considered. According to the simulated results, all types of the flow fields agree with the theoretical values well until the overpotential of 0.35 V whereas they can not overcome the concentration polarization and finally bring out the sudden increment of overpotential. The overpotential curves of the parallel, the parallel serpentine, the serpentine and the zigzag types deviate

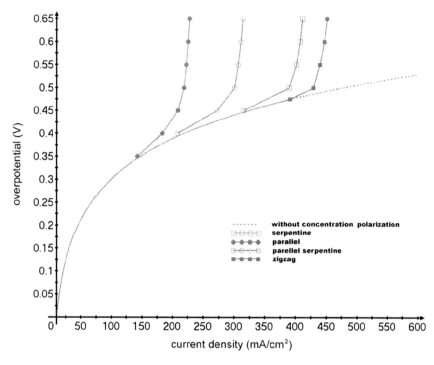

Figure 8. Predicted anode overpotential curve according to the flow-field design.

from the theoretical overpotential curve from 142 mA·cm^{-2} at 0.35 V, 208 mA·cm^{-2} at 0.4 V, 315 mA·cm^{-2} at 0.45 V and 391 mA·cm^{-2} at 0.475 V, respectively. The zigzag type is concluded to be the best flow-field shape in this work.

4. Fabrication of Micro Fuel Cell Stack

4.1. MEA FABRICATION

Membrane-Electrode Assembly (MEA) is one of key components in direct methanol fuel cell, which is comprised of a polymer electrolyte membrane and electrode catalyst layers and gas diffusion layers. The performance of MEA is mainly affected by these components. A preparation condition of electrode catalyst layers and MEA are very important because they influence the high proton conductivity of Nafion binder in the electrode and the electrochemical active surface area for catalyst reaction. Also a microstructure and porosity of MEA, which influence a mass transfer process, can be changed during preparation process of MEA. Therefore a preparation condition of electrode catalyst layers and MEA was optimized to attain the best performance of DMFC.[14, 15]

4.1.1. Preparation of Catalyst Ink

The preparation of catalyst ink slurry is very important because the performance of MEA greatly depends on this process. If the catalysts become agglomerated during the preparation of the ink, the catalyst utilization will be reduced for both methanol oxidation and oxygen reduction reactions and it results in poor performance.[16]

Figure 9 shows a scheme of preparing the catalyst ink. PtRu black (Johnson Matthey, HISPEC 6000) was used as the anode catalyst and Pt black (Johnson Matthey, HISPEC 1000) the cathode catalyst. The catalyst, Nafion solution (Dupont, 5 wt%), distilled water, 1-propanol, 2-propanol were mixed to make an electrode catalyst slurry and the mixture was homogenized with an ultrasonic processor (Hielscher, UP 100H). Then it was stirred with a magnetic stirring system to prepare appropriate catalyst ink.

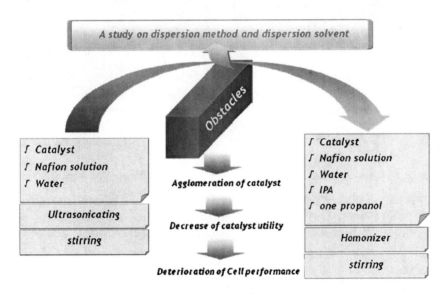

Figure 9. Scheme of preparing catalyst slurry.

4.1.2. Coating of Electrode Catalyst Layer

The catalyst layer could be formed on the porous gas diffusion media (e.g. carbon paper or cloth) by means of various methods such as brush painting, decal transfer, spraying and screen printing.[17, 18] But we formed the electrode catalyst layers on the carbon paper by a bar-coating method. We did not use other methods to minimize the catalyst loss during the coating and to achieve more uniform coatings. Figure 10 shows the semi-automatic bar coating machine we used and a stripe of the catalyst layer formed by it on the gas diffusion layer. The cathode catalyst ink was applied to the hydrophobic diffusion layers (SGL 25BC). Then the catalyst-coated cathode gas diffusion layer was dried in an oven at 70°C. The Pt metal loading was 5 mg Pt/cm^2 for the cathode. The anode catalyst ink was deposited on the carbon paper (Toray 060) with 5 wt% PTFE. The PtRu metal loading was 4 mgPt/cm^2 for the anode. The catalyst loading could be controlled by bar-thickness and coating speeds. The thickness of the catalyst layer by the bar-coating is approximately 50 μm and it is very uniform throughout the surface of the electrode.

(a) Semi-automatic bar coating machine   (b) Electrode catalyst layer

Figure 10. Semi-automatic bar coating machine and electrode catalyst layer by bar coating machine.

4.1.3. *Fabrication Conditions of Membrane and Electrode Assembly*

Hot-pressing is a simple way to assemble electrodes and a membrane, gas diffusion layers and it assures a good contact between the electrodes and membrane. Preparation conditions of MEA and the properties of the prepared MEAs are summarized in Table 3. It has been shown that physical and electrical properties of the hot-pressed MEAs depend on the lamination conditions.[19, 20] The thickness of these MEAs was reduced with an increase of hot-pressing pressure from 690 to 420 μm, while hot-pressing temperature did not influence a change of the thickness of MEA. Cell resistivity at open circuit voltage was slightly affected by hot-pressing pressure but it is not related with hot-pressing temperature and time. It was learned that a contact resistance between a membrane and an electrode was not greatly affected by the change of hot-pressing conditions. The performance of the prepared MEAs depended on preparation conditions of MEA.

TABLE 3. Summary of performance and the properties of the MEAs with the different preparation conditions.

No	Hot-pressing conditions			Cell resistance of OCV at 60°C (mΩ)	Thickness of MEAs (μm)	Power density of at 60°C (mW/cm^2)
	Pressure (kg/cm^2)	Temp (°C)	Time (second)			
P1	0	0	0	24	690	132
P2	7	135	90	22	650	141
P3	27	135	90	20	580	147
P4	44	135	90	20	510	142
P5	64	135	90	19	480	141
P6	74	135	90	19	480	138
P7	83	135	90	20	470	140
P8	94	135	90	19	430	144
P9	104	135	90	19	430	139
P10	115	135	90	19	420	131
C1	115	115	90	20	430	103
C2	115	125	90	19	430	120
C3	115	135	90	19	420	132
C4	115	145	90	19	420	138
C5	115	150	90	19	420	144
C6	115	155	90	19	420	138
T1	44	135	30	20	–	135
T2	44	135	90	19	–	150
T3	44	135	180	20	–	147
T4	44	135	300	20	–	144
T5	44	135	450	19	–	135
T6	44	135	600	20	–	121

The performance of MEAs hot-pressed below 27 kg/cm^2 and above 100 kg/cm^2 was found to be inferior and the hot-pressed MEA at 27 kg/cm^2 showed the highest power density of 147 mW/cm^2 when tested at 60°C.

When pressed at 150°C, the highest power density of 144 mW/cm^2 was obtained but those of the pressed MEAs at temperatures below 135°C were much lower. In terms of hot-pressing time, the highest power density was obtained at 90s, while at 600s the power density was substantially lower. These results indicate that the optimum hot-pressing pressure, temperature, and time are 27 kg/cm^2, 150°C, and 90 seconds, respectively.

4.2. FABRICATION OF MICRO FUEL CELL STACK

The stack was fabricated with five cells having an electrode area of 131 cm^2, and it has the internal manifolds for supply of air and fuel. Its dimension is 48 × 75 × 14.5 mm. Each cell of this DMFC stack has an active area of 17 cm^2/cell. The micro fuel cell stack was assembled with MEAs, carbon cloths, gaskets and machined graphite bipolar plates. The MEA was sandwiched between a bipolar plates with two current collectors positioned at each end of the stack. Teflon gaskets were placed between bipolar plates to prevent liquid and gas leakage to the exterior and cross leakage between the fluids in stack. Serpentine type with two flow paths was used for fuel distribution on both anode and cathode sides. The end plates were made from a stainless steel. A thickness of the bipolar plate was 2 mm.[21]

5. Operating Test of Micro Fuel Cell Stack

The performance of a single cell with an electrode area of 9 cm^2 and five-cell stack with an electrode area of 17 cm^2/cell was evaluated at various temperatures with the methanol concentration of 1 M. A 1 M methanol solution with pure water was pumped into the anode channel of the cell, and air or oxygen was supplied into the cathode channel.

Figure 11 shows the photographs (a) and current-voltage curve (b) of the single cell for stack, fed with 1 M CH$_3$OH, and air at various temperatures. The electrode area of this single cell is 9 cm^2. The maximum power of the single cell was 1.18 W (0.32 V @ 3.7 A) at 60°C. As shown in the figure, the power densities of the single cell were 131 and 116 mW/cm^2 at constant cell voltages of 0.32 and 0.40 V, respectively. The performance of this single cell was more than double higher than that of cells in the five-cell stack. Therefore to increase the performance of the stack, we improved the sealing method, and optimized the tightening pressure of the stack. Additional work is conducted on the optimization of the diffusion layers and electrodes, and manufacturing process of the stack.

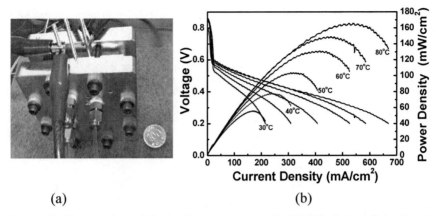

(a)   (b)

Figure 11. Photographs and the performances curve of a DMFC single cell in air and at various temperature (electrode area: 9 cm^2, anode: 1 M CH_3OH, 4 cc/minute, cathode: 300 cc/minute air).

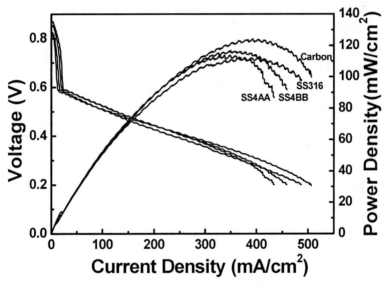

Figure 12. Polarization curves of each single cells with graphite or metal bipolar plates at 60°C.

Figure 12 shows the current-voltage curves of single cell with graphite or metal bipolar plates (stainless steel 316 and 400 series) which design to prepare the micro fuel cell stack. The single cell with graphite performed best. The performance of the single cell with stainless steel 316 plates shows a lower power density curve than that of stainless steel 400 series plates. Even though the cells with the metal plates did not perform as well as that with graphite plates, we plan to continue to investigate the use of metal plates due to the advantages they offer such as lower cost and reduced system volume. We will try also the metal bipolar plates coated with corrosion-resistant materials.

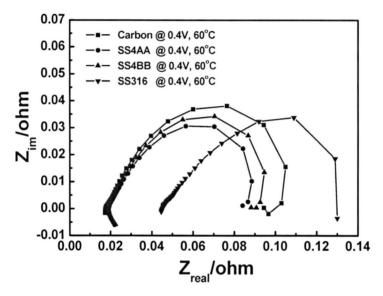

Figure 13. Nyquist diagram of each single cell with graphite or metal bipolar plates using electrochemical impedance spectroscopy at 0.4 V, 60°C.

Figure 13 shows the results of the impedance measurements of single cells with graphite or metal bipolar plates (stainless steel 316 and 400 series). The single cell with graphite plates (17 mΩ) shows the lower resistance than that with metal bipolar plates of stainless steel 316 (45 mΩ) and 400 series (19 mΩ). It reveals that performance difference of graphite and metal bipolar plates arises from the difference in contact resistance between MEA and bipolar plates. Therefore our upcoming research for metal bipolar plates would be focused on overcoming this problem by coating the plates with corrosion-resistant materials.

Figure 14 shows a photograph of the 5 W class DMFC stack with internal manifolds. This DMFC stack consists of five cells with an active area of 17 cm^2/cell. It has graphite bipolar plates of 2 mm and its dimensions are is 48 × 75 × 14.5 mm. The stack has the internal manifolds for supply of air and fuel.

Figure 14. Photographs of the 5 W class DMFC stacks with internal manifolds.

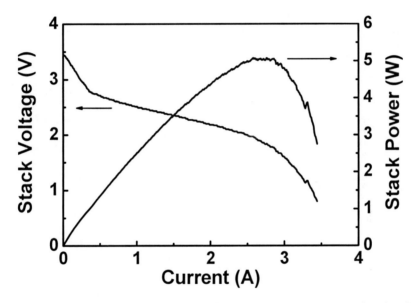

Figure 15. Polarization curves and power of five-cell stack using 1 M methanol and air at 52°C.

Figure 15 shows a current-voltage curve of the five-cell stack. It was fed with 1 M methanol and air (= 2λ) as fuels.

The stack performance was measured by an electrochemical test system (Wona Tech.). The temperature of this stack rose from 22 to 52°C during the stack operation. The maximum power of this stack was 5.07 W (1.95 V @ 2.6 A) and the power density was 60 mW/cm^2. The power output of the stack was 5.0 W (2 V @ 2.5 A) at 2 V of the stack (0.4 V/cell). The maximum power density of this stack was more than two times lower than that of a single cell. It means that if the sealing method and the tightening pressure of the stack, the internal resistance could be improved, the performance of the stack will be much higher.

Figure 16 shows the stack voltage of the five-cell stack at constant current of 1.9 A at 52°C. As shown in the figure, the power and the voltage of the stack were 4.1 W and 2.18 V at a constant load of 1.9 A, respectively. In this figure, the average voltage of each cell was approximately 0.4 V in the stack, and it shows a constant cell voltage of each cell in the stack.

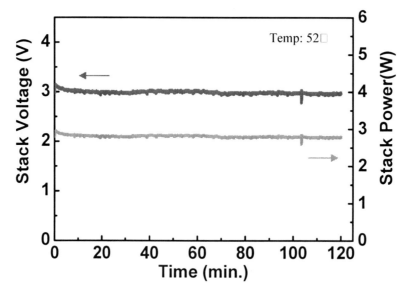

Figure 16. Stack voltage of the five-cell stack at constant current of 1.9 A at 52°C.

Figure 17 shows the cell voltage distribution of each cell in the five-cell stack at OCV and constant current of 1.5 A and 2 A at 48°C. As shown in the Figure, each cell in the five-cell stack has the uniform cell voltage at respective constant current. The performance of this single cell was more than double higher then that of cells in the stack. If we can decrease the internal resistance of the stack, the performance of the stack will be much better.

DEVELOPMENT OF A 5-W DIRECT METHANOL FUEL CELL 287

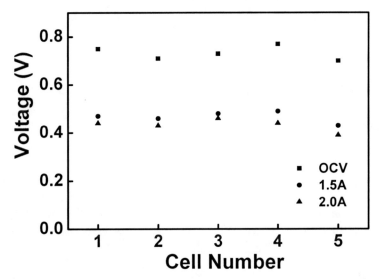

Figure 17. Cell voltage distribution of each cell in the five-cell stack at OCV and constant current of 1.5 A and 2 A at 48°C.

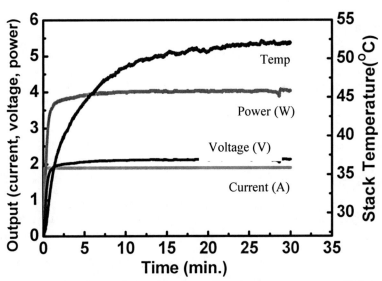

Figure 18. Starting characteristics of the five-cell stack at constant current of 1.9 A.

Figure 18 shows starting characteristics of the five-cell stack at constant current of 1.9 A. The power and the voltage of the stack show a same behavior and also they held constant values which were 4.1 W and 2.18 V at a constant load of 1.9 A, respectively. Temperature of the stack gradually increases with operating time until 12 minutes and then it held a constant temperature of 52°C. From the Figure, it can be seen that the performance of the single cell linearly improved with increasing operating temperature of the cell, while that of the stack was not greatly affected by the temperature of the stack.

6. Conclusions

We developed a 5W-class direct methanol fuel cell (DMFC) stack for the micro power sources of portable electronic devices. Its dimensions are is 48 × 75 × 14.5 mm. Each cell of this DMFC stack has an active area of 17 cm^2/cell. The stack has the internal manifolds for supply of air ($\lambda = 2$) and fuel (1 M CH$_3$OH). This stack was 5.07 W (1.95 V @ 2.6 A) and the power density was 60 mW/cm^2. The power output of the stack was 5.0 W (2 V @ 2.5 A) at 2 V of the stack (0.4 V/cell). Each cell in the five-cell stack has the uniform cell voltage at respective constant current. The power and the voltage of the stack were 4.1 W and 2.18 V at a constant load of 1.9 A, respectively.

References

1. M.S. Wilson and S. Gottesfeld, High Performance Catalyzed Membranes of Ultra-low Pt Loadings for Polymer Electrolyte Fuel Cells, Electrochem. Soc., 139, L28 (1992).
2. D.H. Jung, S.Y. Cho, D.H. Peck, D.R. Shin, and J.S. Kim, Preparation and performance of a Nafion/montmorillonite nanocomposite membrane for direct methanol fuel cell, J. Power Sources, 118, 205–211 (2003).
3. S. Srinivasan, Fuel Cells (Springer, New York, 2006), pp. 442–443
4. J. Larminie and A. Dicks, Fuel Cell System Explained (2nd edn, Wiley, Chichester, England, 2003), pp. 141–143.
5. C. Heitner-Wirguin, Recent advances in perfluorinated ionomer membranes: structure, properties and applications, J. Membr. Sci., 120 1–33 (1996).
6. H. Liu, C. Song, L. Zhang, J. Zhang, H. Wang, and D.P. Wilkinson, A review of anode catalysis in the direct methanol fuel cell, J. Power Sources, 155, 95–110 (2006).
7. A.A. Kulikovsky, Voltage loss in bipolar plates in a fuel cell stack, J. Power Sources, 160, 431–435 (2006).
8. W. Vielstich, A. Lamm, and H.A. Gasteiger, Handbook of Fuel Cell, (Wiley, Chichester, England, 2003), pp. 306–7.

9. M. Hogarth, P. Christensen, A. Hamnett, and A. Shukla, the design and construction of high-performance direct methanol fuel cells. 1. Liquid-feed systems, J. Power Sources, 69, 113–124 (1997).
10. (September 28, 2005), http://www.embeddedstar.com, KDDI, Toshiba, Hitachi Showcase Mobile Phone Fuel Cells at CEATEC Japan.
11. (July 6, 2005) http://www.nttdocomo.com, NTT DoCoMo Enhances Prototype Micro Fuel Cell for FOMA Handsets.
12. (December 6th, 2006) http://www.techshout.com, Smallest Fuel Cell Mobile Charger developed by Samsung and SAIT Team.
13. M. Hyun, S.K. Kim, D. Jung, B. Lee, D. Peck, T. Kim, and Y. Shul, Prediction of anode performances of direct methanol fuel cells with different flow-field design using computational simulation, J. Power Sources, 157, 875–885 (2006).
14. S.Q. Song, Z.X. Liang, W.J. Zhou, G.Q. Sun, Q. Xin, V. Stergiopoulos, and P. Tsiakaras, Direct methanol fuel cells: the effect of electrode fabrication procedure on MEAs structural properties and cell performance, J. Power Sources, 145, 495–501 (2005).
15. T. Frey, K.A. Friedrich, L. Jörissen, and J. Garche, Preparation of direct methanol fuel cells by defined multilayer structures J. Electrochem. Soc., 152, A539–544 (2005).
16. Z. Poltarzewski, P. Staiti, V. Alderucci, W. Wieczorek, and N. Giordano, Nafion distribution in gas diffusion electrodes for solid-polymer-electrolyte-fuel-cell applications, J. Electrochem. Soc., 139, 761–765 (1992).
17. J. Zhang, G. Yin, Z. Wang, and Y. Shao, Effects of MEA preparation on the performance of a direct methanol fuel cell, J. Power Sources, 160, 1035–1040 (2006).
18. Q. Mao, G. Sun, S. Wang, H. Sun, G. Wang, Y. Gao, A. Ye, Y. Tian, and Q. Xin, Comparative studies of configurations and preparation methods for direct methanol fuel cell electrodes, Electrochim. Acta, 52, 6763–6770 (2007).
19. J. Zhang, G.P. Yin, Z.B. Wang, Q.Z. Lai, and K.D. Cai, Effects of hot pressing conditions on the performances of MEAs for direct methanol fuel cells, J. Power Sources, 165, 73–81 (2007).
20. A. Kuver, I. Vogel, and W. Vielstich, Distinct performance evaluation of a direct methanol SPE fuel cell: a new method using a dynamic hydrogen reference electrode, J. Power Sources, 52, 77–80 (1994).
21. C.Y. Chen, J.Y. Shiu, and Y.S. Lee, Development of a small DMFC bipolar plate stack for portable applications, J. Power Sources, 159, 1042–1047 (2006).

DYNAMIC CHARACTERISTICS OF DMFC/BATTERY SYSTEM FOR NOTEBOOK PC

YOUNG-RAE CHO, MIN-SOO HYUN, AND DOO-HWAN JUNG[*]
Advanced Fuel Cell Research Center, Korea Institute of Energy Research, 71-2 Jang-dong, Yuseong-gu, Daejeon 305-343, Republic of Korea

1. Introduction

Many research groups have been working on direct methanol fuel cells for the portable applications because of the convenience of using the liquid fuel, methanol. Particularly the direct methanol fuel cells are studied for military applications because of increasing use of power-hungry, state-of-the-art, and wearable electronic digital equipment by today's soldiers. Life time of the potable power supply devices becomes more important as the amount of power consumption of the electronic and communicational equipment increases. At present, direct methanol fuel cells (DMFCs) which uses liquid methanol without a reformer are considered to meet these requirements best.[4] In order to achieve rapid start-up and high reliability required for these applications, we combined a DMFC with a lithium-ion battery. In the present work, the sharing of power consumption at the start-up and the response characteristics of each component of the fuel cell/battery hybrid system to power consumption were investigated. The characteristics of battery charging at the rated operation also were determined when hybrid system was running a notebook PC.

Recent DMFC investigations have identified important system engineering issues such as long term stability, under static and dynamic loads, and response to the dynamic operation.[2] Argyropoulos et al.[3] have shown that the dynamic performance of the DMFC is affected by complex interactions of electrode kinetics and mass transport processes.[3] It was also shown that the dynamic response of the DMFC cell voltage is significantly affected by the methanol solution flow rate, methanol concentration and applied cathode air pressure.[2]

---

[*]E-mail : doohwan@kier.re.kr

Lee et al.[1] of KIER (Korea Institute of Energy Research) have shown the dynamic characteristics of the cell voltage depending on load changes in battery-DMFC hybrid system. They also have investigated the relationship between load share characteristic and battery capacity.[1]

Published information on the dynamic characteristics of load share of and the transient response by the DMFC/battery hybrid system is sparse.[2] Therefore, the dynamic characteristics of load share, transient response by the configuration of DMFC/battery hybrid system were confirmed experimentally. To observe the load response characteristic of system and the dynamic response of fuel cell stack, the configuration of system was composed of two major types: general configuration and fuel cell protection configuration. In a general configuration the power from a fuel cell is higher than total system and battery initially supplies power to BOPs (Balance of Plants) and load. Then the activated fuel cell charges the battery. Therefore, a fuel cell plays an important role in system and a battery partially contributes to the total system power as an auxiliary power source. In fuel cell protection configuration, the power provided by a fuel cell is lower than total system, battery initially supply power for BOPs, load and its prevent voltage drop from fuel cell and the power of battery is partially a complement to system power. The power of battery and fuel cell is alike. Because the power of fuel cell is insufficient, the battery has to be charged separately. The battery prevents voltage drop from the fuel cell and limits the output power of the fuel cell. If the battery does not maintains the fuel cell voltage, the fuel cell voltage decreases while its current increases. In this case, the power provided by the fuel cell is not stable.

In this paper, the fuel cell/battery hybrid system was composed of general configuration and fuel cell protection configuration. The general configuration and fuel cell protection configuration generally used by means of configuration of fuel cell/battery hybrid system. The dynamic characteristics of the hybrid system were determined according to fluctuation of load and status of fuel cell. The fluctuation of load, transient response of fuel cell/battery hybrid system by means of difference in system configuration was confirmed. As the results of this study can be applied to design the interface module of the fuel cell/battery hybrid system and to determine the design requirement in the fuel cell stack for portable applications.[1]

2. Portable DMFC Stack for Notebook PC

2.1. DEVELOPMENTS OF POTABLE FUEL CELL FOR NOTEBOOK PC

The fuel cell notebook prototypes have been shown by Japanese, Korean and Taiwanese electronic companies such as TOSHIBA,[5] NEC,[6] Antig, HITACHI, SANYO,[8] Panasonic,[9] SAMSUNG,[10] and LG.[11] These companies used DMFC technology for their prototypes. They had a compact size between 50 and 250 cc and a power between 10 and 75 W. Toshiba developed the world's first prototype of a small form factor direct methanol fuel cell for portable PCs, a clean energy breakthrough with the potential to end reliance on rechargeable batteries.[5] NEC developed a fuel cell that boasts the world's highest output power density and the refinement of the packaging technology allowed reduction in size of the fuel cell module by 20% while maintaining sufficient power to drive a notebook PC.[6] Sanyo and IBM developed a hybrid system in which the direct methanol fuel cell will work alongside a slim-line lithium-polymer rechargeable battery. That's presumably because of the bulk of the fuel cell. A concept model described by Sanyo and IBM weighs in at 2.2 kg and is larger that the ThinkPad they used to demonstrate it. The 28 × 27 × 1.6–5.4 cm pack incorporates a 130 cc fuel tank which holds enough methanol and water to provide 16 V and 12–72 W of power for up to eight hours' operation.[8] Panasonic developed DMFC (Direct Methanol Fuel Cell) powered laptop in 2006. This laptop is last 20 hours out of a full tank of fuel (about 6 3/4 ounces). Although its DMFC power hardware is still a bit bigger than a typical battery pack, it is indeed half the size of comparable devices which are available.[9] Samsung developed a high-density fuel cell storage system for its Q35 laptop. The power system is contained in a large dock that the laptop sits on the top. Samsung claims that the fuel cell offers an energy density of 650 Wh/l, which is about four time as much as competing offerings. The total energy storage is an impressive 12,000 Wh, which depending on the laptop's power settings and usage could theoretically power the laptop continuously for a whole month without the need for recharging.[10] The fuel cell developed by LG comprises of a reaction vessel and a removable 200 cc methanol fuel reservoir. Together, they can generate 25 W of power for more than 10 hours. The cell itself lasts for more than 4,000 hours. The fuel-cell unit weighs less than 1 kg.[11] The images and characteristics of these fuel cell notebook prototypes are summarized in Table 1.

TABLE 1. Prototype images and characteristics of the fuel cell notebook.

Company	Prototype	Power (W)	Type	Fuel tank (cc)	Release
TOSHIBA		13–20	Active	100	2003
NEC		14–24	Passive	250 (30% MeOH)	2003
Antig		16	Active	70	2005
SANYO		12–75	Active	130	2005
Panasonic		13–20	Active	200	2006
SAMSUNG		20	Active	200	2006
LG		25–75	Active	200	2005

2.2. POTABLE FUEL CELL STACK FOR NOTEBOOK PC

The stack was fabricated with 20 cells having an active area of 30 cm^2/cell, and it has the internal manifolds for supply of air and fuel. Its dimension is

$55 \times 70 \times 80$ mm (308 cm^2). Figure 1 shows a 30 W-class DMFC stack for note book PCs. To use the DMFC as a portable power source, its voltage and current characteristics were measured.[4] Figure 2 shows polarization curve of a DMFC for notebook PC.

Figure 1. Direct methanol fuel cell stack for notebook PC.

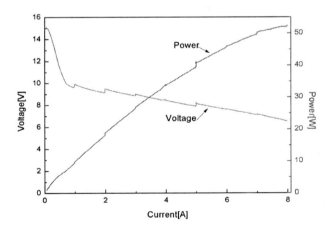

Figure 2. Polarization curve for DMFC stack.

3. Configuration of Potable Fuel Cell System

Because the fuel cell system is composed of stack, methanol pump, air pump, concentration sensor, and processor, for operation it needs an auxiliary power source. The battery takes charge of both an auxiliary power source for start-up of fuel cell system and load until fuel cell is activated. The fuel cell system shows different dynamic characteristics depending on the configuration.

3.1. GENERAL CONFIGURATION

The general configuration is composed of a fuel cell stack, a battery, converters, and a notebook PC. The fuel cell powers BOPs and the notebook PC, charges the battery, and makes up the conversion loss. Before the fuel cell is fully activated to power the notebook PC, the battery is used as the auxiliary power source. When the fuel cell is fully activated, the battery is charged. Because the fuel cell supplies the power more than the battery, the capacity and cost of the battery are lower. Figure 3 shows the general configuration of the fuel cell/battery hybrid system.

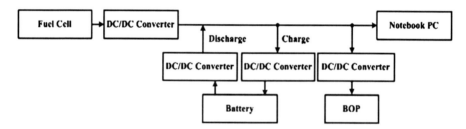

Figure 3. General configuration of the fuel cell/battery hybrid system.

3.2. FUEL CELL PROTECTION CONFIGURATION

The fuel cell protection configuration is shown in Figure 4. In the fuel cell protection configuration, the fuel cell voltage is maintained by the battery. Hence, the fuel cell system operates stably. The battery supplies the power to the notebook PC and BOPs until the fuel cell is activated. But the battery used in this mode is larger than the general configuration because the fuel cell's power is lower than the total system.

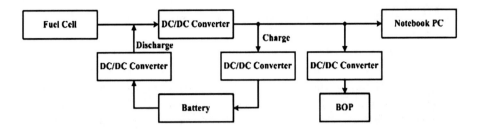

Figure 4. Fuel cell protection configuration of fuel cell/battery hybrid system.

4. Operational Test of Portable Fuel Cell System for Notebook PC

The dynamic characteristics of a 30 W-class DMFC system for a notebook PC were experimentally determined. Its fuel cell stack was manufactured for this experiment. The battery used as an auxiliary power is a lithium-ion battery of 11.1 V/5700 mAh. The notebook PC was operated from the booting to the power-off. The fuel cell and battery connect the switch, and it's operated by the state of fuel cell.

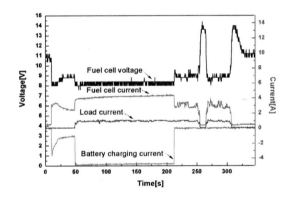

Figure 5. Dynamic characteristics of the fuel cell protection configuration (battery charge/discharge).

Figure 5 shows dynamic characteristics of the fuel cell protection configuration (battery charge/discharge). The battery voltage maintains 8.4 V using a DC to DC converter and the fuel cell voltage changes with the amount of current. The battery starts to discharge when the fuel cell voltage reached 8.4 V. When the fuel cell voltage is higher than the battery voltage, the battery was charged. The fuel cell is protected by the battery with preventing high current output of the fuel cell. Figure 6 shows the characteristics of voltage-current in the fuel cell protection configuration (not activated). The not-activated fuel cell originates significant voltage drop due to the current increase. The fuel cell current increases with the fuel cell activation, but the battery current decreases slowly. Figure 7 shows power of the fuel cell protection configuration (not activated). The change of fuel cell current is stable. On the other hand, the battery current is not stable relatively because the battery should supply the insufficient current immediately. In addition, the response characteristics of the fuel cell are slower than battery due to the slow reaction and long activation time.

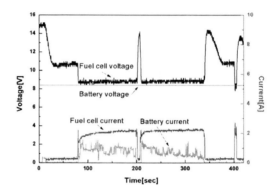

Figure 6. Characteristics of voltage-current in the fuel cell protection configuration (not activated).

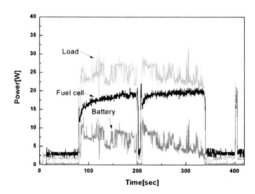

Figure 7. Characteristics of power in the fuel cell protection configuration (not activated).

Figure 8 shows the characteristics of voltage-current in the fuel cell protection configuration (activated). When the fuel cell voltage is higher than the battery voltage, the battery scarcely supplies power. Because the response characteristics of battery are faster than the fuel cell, the battery

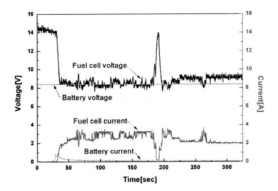

Figure 8. Characteristics of voltage-current in the fuel cell protection configuration (activated).

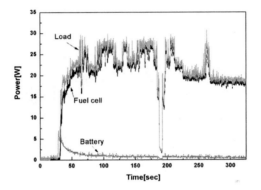

Figure 9. Characteristics of power in the fuel cell protection configuration (activated).

initially supplies current. Figure 9 shows the characteristics of the power in the fuel cell protection configuration (activated). The fuel cell having the rated power of 50 W stably operates the system. Therefore, the battery power decreases and the fuel cell power gradually increases.

Figure 10 shows the characteristics of voltage-current in the general configuration (not activated). The general configuration is a parallel connection between the fuel cell and the battery. The cell voltage of the fuel cell is significantly influenced by the amount of current. When the power of the fuel cell is larger than that of the total system, the fuel cell is stably operated. Figure 11 shows the characteristics of power in the general configuration (not activated). In contrast to the fuel cell protection configuration, the battery supplies the power constantly. The total system current is not appropriately distributed to the fuel cell and the battery because the current

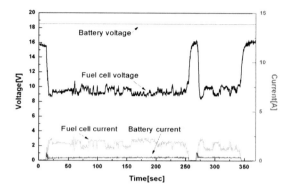

Figure 10. Characteristics of voltage-current in the general configuration (not activated).

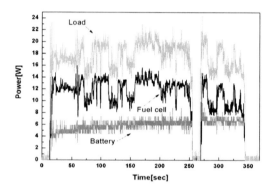

Figure 11. Characteristics of power in the general configuration (not activated).

consumption is affected by the internal resistance of the each device. So the fuel cell is not activated initially. However, the fuel cell stably operates the system because the fuel cell power is higher than the demanded power by the total system.

Figure 12 shows the characteristics of voltage-current in the general configuration (not activated and overload). When the load increases about two times, the fuel cell voltage decreases to 2.5 V and its current increases more than 10 A. When the fuel cell voltage decreases to 2.5 V, the fuel cell will be damaged. Figure 13 shows the characteristics of power in the general configuration (not activated and overload). Generally voltage of fuel cells decreases with the current increase. Because the total power is higher than the fuel cell power, the fuel cell power is supplied into the system prior to the battery power. When the total power of the system is higher than the rated power of fuel cell, the fuel cell performance decreases and the system remains unstable.

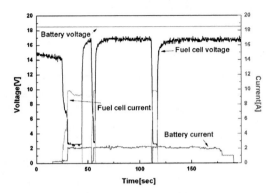

Figure 12. Characteristics of voltage-current in the general configuration (not activated and overload).

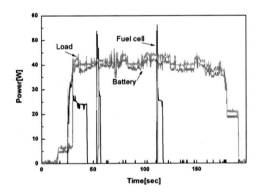

Figure 13. Characteristics of power in the general configuration (not activated and overload).

5. Conclusions

The fuel cell/battery hybrid system that consisted of a stack, a battery, a notebook PC, converters and BOPs were assembled in two different configurations; general configuration and fuel cell protection configuration. The dynamic characteristics depending on the configuration of system, load fluctuation, and the state of fuel cell were examined experimentally.

In the general configuration, the total system current was not appropriately distributed to the fuel cell and the battery because the current consumption was affected by the internal resistance of the each device. So the fuel cell was not activated initially. However, the fuel cell stably operated the system because the fuel cell power was higher than the demanded power by the total system. Consequently, in the general configuration, the proper fuel cell and battery should be selected for the stabilization of the system.

In the fuel cell protection configuration the fuel cell was activated by gradually supplying power to the system. Generally voltage of fuel cells decreases with the current increase. On the other hand, the fuel cell current was restricted within a rated range by maintaining a constant voltage in the fuel cell protection configuration. Before the full activation, the fuel cell operated below the rated power. The decision of charge or discharge in the battery was determined by the value of the fuel cell voltage.

The configuration of the fuel cell/battery system determined according to the fuel cell power, the battery power and the load. The fuel cell has an immediately effect on the stability and reliability of the fuel cell battery system. In the near future, we can be looking forward to commercializing a fuel cell by means of improvement of a fuel cell performance and stability of a system.

References

1. B.D. Lee, D.H. Jung and Y.H. Ko, Analysis of DMFC/battery hybrid power system for portable applications, *Journal of Power Sources*, 131, 207–212 (2004).
2. Jaesung Han and Eun-Sung Park, Direct methanol fuel-cell combined with a small back-up battery, *Journal of Power Sources*, 112, 477–483 (2002).
3. P. Argyropoulos, K. Scott and W.M. Taama, The effect of operating conditions on the dynamic response of the direct methanol fuel cell, *Electrochimica Acta*, 45, 1983–1998 (2000).
4. P. Argyropoulos, K. Scott and W.M. Taama, Dynamic response of the direct methanol fuel cell under variable load conditions, *Journal of Power Sources*, 87, 153–161 (2000).
5. Toshiba (March 5, 2003); http://www.toshiba.co.jp
6. NEC (September 17, 2003); http://www.nec.co.jp
7. Gizmodo (March 2, 2006); http://gizmodo.com
8. Tony Smith (April 11, 2005); http://www.theregister.co.uk
9. Red (January 7, 2006); http://www.portablegadgets.net
10. Jeremy Reimer (December 27, 2006); http://arstechnica.com
11. Tony Smith (September 29, 2005); http://www.theregister.co.uk

A REVIEW ON MINIATURIZATION OF SOLID OXIDE FUEL CELL POWER SOURCES-I: STATE-OF-THE-ART SYSTEMS

X. Y. ZHOU[1]*, A. PRAMUANJAROENKIJ[1]
AND S. KAKAÇ[2]
[1]*Department of Mechanical and Aerospace Engineering
University of Miami, Coral Gables, FL 33124*
[2]*TOBB University of Economics and Technology
Sogutozu, Ankara, Turkey*

1. Introduction

Solid oxide fuel cells (SOFCs) can play an indispensable role in the futurist's hydrogen economy. SOFC operating at 800–1,000°C is a very efficient power source because the heat generated from SOFCs can be used for co-generation of power with a turbine and the rejected heat can still be used for heating.[1–4] Thus, it is envisioned that large-scale stationary SOFC systems can achieve a high energy efficiency (>60%) and high fuel efficiency (>80%). Due to their high operating temperatures, SOFCs can directly utilize biofuels, natural gas, syngas (H_2 + CO), and hydrogen for power generation or operate on reformate from a relatively simple fuel processor. Thus, deployment of SOFC technology does not depend on availability of hydrogen transport and storage technologies. The major obstacles for deployment of SOFCs are the high manufacture costs, voluminous structure, and limited reliability and durability. Should these obstacles be removed, SOFCs can be universal power sources for diver-sified and decentralized large-scale energy sources that are available during the transition period to and in the hydrogen economy.

The solid oxide fuel cell (SOFC) technology is rooted in the early discovery by E. Bauer and H. Preis in the late 1930s.[4–5] The current status of the technology is largely contributed to the continuous effort by Westinghouse Electric Corporation to make tubular SOFCs since late 1950s.[4–5] In the late 1990s and at the turn of the century, Westinghouse developed a few successful commercial prototypes of 100–300 kW stationary cogeneration systems with an ultimate goal to be deployment of large-scale (>10 MW) stationary

---

*E-mail: xzhou@gmail.com

power plants. These traditional solid-oxide fuel cells operate at high temperatures, generally in the range 800–1,000°C.[5] At these temperatures the fuel cells can effectively catalyze hydrocarbon fuels and operate at high thermodynamic efficiency through the secondary use of otherwise waste heat. Because of their high operating temperatures and the primary ceramic components that are vulnerable to thermal and mechanical shock related damages, research and development of the SOFC technology were initially focused on large-scale stationary power generation systems. In 2000, Delphi Automotive Systems developed a 5 kW automotive auxiliary power unit (APU) using anode-supported planar SOFCs, to power an electric air conditioning system without the need for operating the vehicle engine.[6, 7] Today, we witness a bifurcation of development direction for SOFC technology. The first is to scale up the stationary SOFC co-generation stations along the path set forth by Westinghouse. The second is to develop medium-scale and small-scale SOFC power sources via discovering a variety of new design concepts, materials, and manufacture methods. According to a recent statistics,[4] among at least 15 SOFC manufacturers in the world, only two of them focus on large-scale statio-nary power stations and others are involved in development of medium-scale or small-scale power sources. This second direction is clearly aimed to occupy the presumed application domains for polymer electrolyte fuel cells (PEFCs).

Development of new power sources including PEFCs, SOFCs, and high performance batteries heavily relies on market and technological niches. According to Agnolucci and McDowall,[8] *"In the **technological niches**, actors promote technologies thought to offer potential **future** benefits while in the **niche markets** consumers with particular and specialist needs value technologies for their **present** performance characteristics"*. They stressed that the survival of a technical niche depended on positive expectations about future benefits of the core technology. The survival of a present market niche depends on the following factors.[8]

1. Internal economies of scale
2. External economies of scale and industry development
3. Expectations
4. Learning around user and institutional context
5. Network effects and infrastructure
6. Revenue.

This paper will review current status of minimization of SOFCs in the context of technological and market niches and of competition from other power sources. The goal is to identify the major technical barriers or problems and propose solutions to the problems that permit SOFC technology to succeed in this competition.

2. Technological Niche in Hydrogen Economy and in the Transition Period

The quest for sustainable energy technologies is a dual-purpose mission that seeks energy security for a sustainable economic growth in future and at the same time curb greenhouse gas emissions from production and consumption of the energy products,[9] Current primary energy supplies of 11.7 Giga tonnes of oil equivalent (Gtoe)] include oil (32.3%), coal (23.9), natural gas (20.6%), biomass (11.7%), nuclear power (5.4%), hydroelectric power (5.4%), and combination of geothermal, wind, and solar energies (0.7%).[8] Today's economy is supported by three major energy networks of production, distribution, and utilization: (1) electricity network (38.1%), (2) heat and industry use network (44.3%), and (3) oil network (18.1%) mainly for transportation use.[9] The overall energy efficiency for the electricity network is 30% whereas the energy efficiency for transportation uses (automobiles, ships, trains, and airplanes) is only 15–22%.

It is predicted that in 2050, the total primary energy consumption will be doubled up to about 25 Gtoe.[9] Each of the oil, natural gas (NG), and coal resources by its known reserve can only meet energy demand for 35, 50, and 75 years respectively. Nuclear energy is initially thought an abundant source but in fact only marginally sufficient for powering the world economy. All of these energy sources in combination will probably sustain for a few hundreds of years. However, these energy sources are considered as non-renewable and not environment friendly. If current ways of energy consumption is maintained, increase of greenhouse gases will result in global warming and consequent environmental disasters.[10] Nuclear energy source does not produce greenhouse gases but it causes long-term radioactive pollution.

A popular vision of the ultimate solution for the global energy shortage and greenhouse problems is realization of "hydrogen economy" in which the energy sources are renewable and the energy distribution networks are electricity network and hydrogen network, which run in a cooperative way.[9–11] Both electricity and hydrogen are clean secondary energy sources. The electricity network will convert renewable energy into electrical power and supply electrical power for industrial, commercial, and residential grid users and to electrolysis systems for hydrogen production. The hydrogen network will convert renewable energy into hydrogen and supply hydrogen for stationary electrical power generation and heat generation for off-grid users. The hydrogen network will also replace the current oil network for fueling the transportation vehicles. Both stationary power generation and heating and fueling transportation systems will count for about 40% and 20% of energy consumption respectively. The hydrogen economy relies on the power sources that convert hydrogen into electricity and on electrolyzers that convert electricity into hydrogen. A common vision is that polymer electrolyte fuel cells and solid oxide fuel cells are the best choices for power sources.

Current capabilities of the renewable energies are very low. Biomass meets 11.3% of energy consumption. Hydroelectric, geothermal, solar, tide, and wind power sources in combination only meet 6.1% or energy consumption. If the increase rate of energy consumption is maintained, it is a question whether the renewable energy sources can satisfy the energy demand of the world beyond 2050.[9, 10] Biomass will not be sufficient for the overall energy demand before the issue of the negative impact of massive implantation of biomass crops on food supply is addressed. Wind, tide, hydroelectric, and geothermal energies in combination will meet only a small portion of the energy demand and will not be the dominant energy resources in the world economy, although each of these may be the dominant energy resources locally. The expectation for the future capability of solar energy resources is high. On the one hand there is abundant solar energy. It is estimated that the energy demand in 2050 could be met by solar hydrogen produced on just 0.5 percent of the world's land area.[12] On the other hand, the efficiency of the photovoltaic (PV) devices is very low (<5%) and the cost is very high so that a massive deployment of PV devices will be economically not feasible at least in near future. In a very long period, the world's economy will still rely on non-renewable energy sources although the percentage in the total energy supplies will hopefully decrease. PEFCs operate on pure hydrogen fuel in the temperature range between 80 and 100°C and enable a high efficiency (40–50%). Advantages of hydrogen-based PEFCs over other types of fuel cells include zero-emission, high efficiency, high mass specific power density, simplicity in design, easy packaging features, and minimized corrosion. Hydrogen-based PEFCs are best choices for both stationary and mobile power generations if hydrogen transport and storage technologies are in place. The rejected heat from the PEFC systems is ideal for heating houses because the small temperature difference between the PEFC systems and the standard heating radiators (70°C) ensures a high thermal efficiency.[10] However, one of the disadvantages of the PEFCs is the degradation problem due to catalyst poisoning. The deployment of PEFC systems largely depend on the availability of poison-free hydrogen production, distribution, and storage infrastructures. The realization of hydrogen-based fuel cell automobiles heavily relies on the hydrogen production and storage technologies.

In comparison with other forms of secondary energy sources (electrical in batteries, methanol, ethanol, gasoline, etc.) under consideration, hydrogen has the greatest mass specific energy but lowest volume specific energy density.[9, 14, 15] Compressed hydrogen (300 bar) can only carry 60 Wh l^{-1}

whereas conventional NiCd rechargeable batteries can carry 90 Wh l^{-1}. Ultrahigh compressed hydrogen technology (350–700 bar) is under research whose goal is to permit the storage of hydrogen up to 11 wt%. Compression storage also requires 10% of the low heating value (LHV) of hydrogen, which reduces the overall or "well-to-wheels" energy efficiency, and a high cost (\$3,000 kg^{-1}). Liquid hydrogen or cryogenic storage at 20 K is less voluminous but the only contains 5 wt% of hydrogen. The energy required for the liquefaction is 30% LHV, which is unacceptable economically. Another problem with cryogenic storage is the "boil-off" problem or loss of the hydrogen due to constant boiling. Hydrogen storage with rechargeable hydrides (AlH_3, NH_4BH_4, $NaAlH_4$, etc.) is a neat and safe method but the storage rate is only up to 3 wt%. Porous carbon was at one time considered as an interesting material for hydrogen storage. However, the expectations of carbon nanotubes fail, because the initial results of 30–60 wt% of stored hydrogen are now considered to have been an experimental error.[16]

Taking into consideration of the current status of the renewable energy technology and advancement patterns of the key renewable energy technologies including PEFCs and portable hydrogen storage in the past,[4] it is very likely that there will be a very long transition period from a non-renewable energy dominant economy to renewable energy dominant one. In this transition period, there will be a steady reduction of consumption of fossil fuels and gradual expansion of the capability of the renewable energies. In this very long period, the non-renewable energy sources will be the major energy sources before a world economy on the networks of renewable energy production, distribution, and consumption are established. In addition, conventional energy technologies such as internal combustion engine (ICE) will probably dominate a part of this transition period. A new set of technologies that not only can utilize a diversity of energy sources during the transition process but also meet the performance and functional requirements or standards in hydrogen economy are desired. Development of these technologies will minimize the total cost of the world and help enable a transition as smoothly as possible.

SOFC operating at 800–1,000°C is another promising choice for stationary power generation because the heat generated from SOFCs can be used for co-generation of power with a turbine and the rejected heat can still be used for heating houses. SOFCs can play an important role in the futurist's hydrogen economy. Generally speaking, the SOFC systems can achieve a higher efficiency (50–60%). Due to their high operating temperatures, SOFCs can directly utilize biofuels, natural gas, syngas (H_2 + CO), and hydrogen for power generation or operate on reformate from a relatively simple fuel processor. Thus, deployment of SOFC technology does not depend on availability of hydrogen transport and storage technologies.

The major obstacles for deployment of SOFCs are the high manufacture costs, voluminous structure, and limited reliability and durability. Should these obstacles be removed, SOFCs can be universal power sources for diversified and decentralized energy sources that are available during the transition period and in the hydrogen economy. Thus, the technological niches for the SOFCs are both the high efficiency power generation and heating in "hydrogen economy" and highly efficient utilization of non-renewable and renewable fuels during the transition period.

3. Market Niches for New Power Sources

PEFCs including DMFCs, SOFC, and high performance batteries are competitors in the market of power sources. The market niche for the new power sources consists of three major application areas: (1) large-scale stationary power plants (>10 MW); (2) medium-size power sources or plants (1 kW–1 MW); and (3) small-scale or portable power sources for particularly portable electronics. In the case of SOFCs, the medium-size and portable systems are considered as miniaturization of SOFCs with respect to the large-scale SOFC systems. Miniaturization of SOFC or micro-SOFC is not a rigorous concept. A micro-SOFC simply means that the SOFC's size or power range is much smaller than that for the large-scale stationary SOFCs.

Implement of the new technologies in the large-scale power generation should have a great enhancement of the overall efficiency of the energy network and reduction of emissions.[5] However, this is an area requires a massive investment and a very long R&D period. Market diffusion for the new technologies at the present time is difficult. High efficiency, high reliability and durability, and low cost are the emphases of the technology whereas high power density and energy density are not. This is an arena for a couple of international industrial giants. High temperature fuel cells [SOFCs and molten carbonate fuel cells (MCFCs)] are considered to be more advantageous than PEFCs mainly because the cogeneration scheme for the high temperature fuel cells can enable a high energy efficiency (~60%) and the high temperature fuel cells have a good fuel flexibility and resistance to poisoning which allow utilization of various types of fuels (natural gas, biofuels, and syngas).

The area of portable power sources for electronic devices and other portable devices probably requires the lowest investment for commercialization. Consequently, there is an intensive competition between all kinds of power sources including photovoltaic cells, fuel cells and batteries. The major winners of the competition will be the system with very high power density, high energy density, and low cost. A high energy efficiency and low emission are not particularly important because the total energy output and emission are considered as being negligible in comparison to those of the medium-scale and large-scale power generation systems. Technologes under development include photovoltaic cells, micro-SOFC, PEFC with hydrogen storage, direct methanol fuel cell (DMFC), direct formic acid fuel cell, primary air/metal batteries, and secondary Li-ion rechargeable batteries. According to an analysis by Flipsen,[18] the prospective for PEFCs and DMFCs to replace Li-ion batteries is not promising because the specific powers of both PEFC and DMFC are 10 times lower than that for Li-ion batteries (3 vs. 300 W kg^{-1}) although their specific energies are 3–4 times higher (500 vs. 150 Wh kg^{-1}) than that for Li-ion batteries. In addition, the requirements for the operation conditions for DMFCs are more stringent and complex than those for Li-ion batteries. The rationale for the R&Ds for small PEFCs and DMFCs is primarily based on a general vision that PEFC will be one of the major power source types in the hydrogen economy as being expressed in Dunn's review paper.[10] Diffusion of the PEFC technology into this area will facilitate the expansion of the technology into other areas. In other words, PEFCs and DMFCs have future technological niche but barely have present market niche. R&D of micro-SOFC targeted in this area is making first step marked by the report of a butane-fuelled handhold SOFC developed by Lilliputrian Inc.[19] However, in this report, important technical parameters such as power density, energy density, efficiency, and durability were not specified.

The fuel cell power sources for automobiles can be considered as the medium-scale fuel cells. The so-called well-to-wheel efficiency is evaluated in order to compare the propulsion power sources for automobiles in terms of energy saving and environmental impacts.[25-27] Evaluation of well-to-wheel efficiency counts the work required for production, transportation, and power generation. According to the data present in Table 1, replacing the IC engines in transportation vehicles with *ideal* PEFC, SOFC, and battery based power systems will result in a significant energy saving. If the PEFC and SOFC are non-ideal with a low efficiency, the benefits of switching from IC engines to fuel cells will be not obvious and market niche for the fuel cells in the applications of power sources in automobiles will disappear.

TABLE 1. Well-to-wheel efficiency for transport power systems.

Processes	IC engine (%)	PEFC-compression H_2 (%)	PEFC-on-board reforming (%)	SOFC-on-board reforming (%)	Battery (%)
Production	85	70	85	85	50
Compression		90			
Distribution	99	90	99	99	90
Transfer	~100	97	~100	~100	
AC-DC conversion					92
Charging					80
Fuel processing			70	85	
FC stack or engine	18	50	50	50	
BOP		90	90	90	
Electric power train		90	90	90	90
Regeneration					110
Total	15	22	24	29	33

The applications in the second area also include power sources for off-board and on-board auxiliary power units (APUs), backup power generators or uninterruptible power supply (UPS), combined heat and power (CHP) generation for residential building and other uses, and re-chargers. The hurdle for commercialization of the SOFC, PEFC, and high performance battery systems for these applications are relatively low. High efficiency, power density, and energy density are emphasized for these applications.

Fuel cell APU and UPS are often considered as very promising early applications of fuel cells.[8, 21, 28, 29] In this market, fuel cells are used as a complement for the main internal combustion engine (ICE) that is today's main source of propulsion. Instead, they provide power and heat for on-board services, such as entertainment, heating, air conditioning, and so on, for which ICEs are not particularly efficient. Thus, APUs can improve generation efficiency, reduce emissions, extend the engine life and eliminate noise.

CHP is another very promising market for fuel cells.[30, 31] Solid oxide fuel cell (SOFC) based micro combined heat and power (micro-CHP) systems exhibit fundamentally advantages over other common micro-CHP technologies.[30] One of them is that SOFCs have a low heat-to-power ratio and may benefit from avoidance of thermal cycling. Ideal SOFC-based micro-CHP may be best suited to slow space heating demands, where the heating system is on constantly during virtually all of the winter period.

The APU, UPS, and CHP are areas that attract many companies working on commercialization of PEFC or SOFC based APUs, UPSs, CHPs, back-up systems, and mobile or portable battery re-chargers. Some researchers[10, 17] claim that PEFCs are the frontrunners while many others[8, 20–23] suggest that SOFCs are better choices.

4. State-of-the-art SOFC Systems Versus Other Systems

Commercialization of fuel cell systems depends on how much the fuel cell systems will outperform the existing systems in the technological and market *niches*. The design of a fuel cell power source unit should consider at least six parameters: (1) net electrical efficiency that is expressed as electric power produced, minus parasitic power loss, divided by the lower heating value (LHV) of the feed and fuels in the systems; (2) fuel efficiency that is expressed as the electric power plus heat that is re-used for heating, divided by the LHV of the feed and fuels; (3) dynamic behavior that include the system's response to load change and typical start-up time; (4) lifetime or durability; (5) size and weight that are described by power density or specific power and energy (weight or mass specific); and (6) cost that is expressed in \$/kW.[24] Table 2 lists the performances of existing power sources or prototypes including IC engine, ICE based backup generator, micro-gas turbine (MGT), natural gas (NG) based PEFC generator, PEFC based direct methanol fuel cell (DMFC) re-charger, Li-ion batteries, and SOFC power sources.

TABLE 2. State-of-the-art power sources.

Power sources (Manu-facturer)	Power (kW)	Power density (W l^{-1})	Specific Power (W kg^{-1})	Specific energy (Wh kg^{-1})	Fuel type	η_f (%)	η_e (%)	
Auto ICE	80	250	800	~1,000	Gasoline, diesel	–	~18	[12]
Autohybrid ICE	100	250	950	~1,200	Gasoline, diesel	–	~27	[25]
Micro-turbine (S. Cal. Edison)	30	30–71	110	~1,000	NG	–	24–27	[26]
ICE-CHP (IntelliGEN Power)	15	20	60	300–700	Gasoline, diesel, NG, LPG	–	~30	[32]

(Continued)

System					Fuel			Ref
PEMFC-CHP	1.5	2	6	400–900	NG	~80	~30	[33, 34]
DMFC (Motorola)	0.002	2.2	3	490	Methanol		~20	[35]
Li-ion battery	1	320–500	200–300	150	NA			[18, 36]
Stationary SOFC (Siemens-Westinghouse)	125–500	1		450–1,050	NG	>80	44–47	[37]
Stationary SOFC (MHI)	75	1			NG	>80	>50	[38]
SOFC-CHP (Acumentrics)	5	3	13	450–1,050	NG, LPG, ethanol, methanol	>80	30–50	[39]
SOFC-CHP (CFCL)	1–5	2.2	6.7	450–600	NG	~80	~35	[40]
SOFC-CHP (SOFCo)	10	20.5	–	450–600	LPG	–	~25	[41]
SOFC-CHP (Sulzer Hexis)	1	4	–	450–600	Biogas	~80	25–30	[42]
SOFC-APU (Delphi-Battelle)	1.8	30	27	208–634	Gasoline, diesel, NG, LPG	~75	25–38	[43]
SOFC-portable (Mesoscopic)	0.25	20.6	57	450	JP–8		25	[44]
SOFC-portable (Adaptive Materials, Inc)	0.020	7.69	16	701	Gasoline		20	[45]

The specific energy is evaluated using:

$$Energy\ density = \frac{(Power\ output \times time)}{(Weight\ of\ power\ source\ device + Weight\ of\ fuel)} \quad (1)$$

For the stationary and combined heat and power (CHP) systems including European FC's PEMFC system, CFCL's SOFC system, and Siemens-Westinghouse SOFC system, it is assumed that the fuel supply is continuous or may sustain for a long period of time. Thus, the energy density for the CHP and stationary systems is the greater limit for Eq. (1). For the APU systems and automobile propulsion system, it is assumed that the weight of the fuel in the fuel tank is the same as that of the power generation device.

The currently existing IC power systems are based on mature technologies that have been optimized since they occupied the markets 100 years ago. These IC systems have very high power densities and energy density but a low efficiency. Large-scale gas turbine is a mature technology. Microturbine technology is relative new but has already achieved an energy efficiency of 24–27%. This is a strong competitor for the CHP market. Li-ion battery is also a commercialized technology with a very high power density but a relative low energy density. Combination of the IC engine and Li-ion battery enables hybrid power systems that have high power density, high energy density, and an efficiency of ~30%.

It is clear that both hydrocarbon based SOFC and PEFC systems are running far behind of the IC systems and hybrid IC/battery systems in terms of power density and cost for the market niches while the advantage of a high efficiency for the fuel cells is also not obvious at present time.

The energy (electrical) efficiencies of the SOFC based CHP systems are generally greater than that of the IC engine based CHP (e.g. the IntelliGEN system). The trends of energy efficiency vs. power density and energy density are exhibited respectively in Figures 1 and 2. As can be seen in Figures 1 and 2, the power density and specific power of the SOFC-APU systems is close to that for IC engine power systems where SOFC-CHP systems are running far behind of the IC engines in terms of power density and specific power. The high price for the SOFC systems ($6,800/kW)[9] vs. the price of the IC systems ($1,300/kW) is another major obstacle for the SOFC systems to compete with the IC systems in the CHP market. Moreover, further increasing the power density of the SOFC power sources correlates with reduction of the energy (electrical) efficiency (SOFCo and Delphi's systems vs. other systems). It is worth noting that even the 25–38% efficiency for the Delphi's system was obtained using hydrogen fuel and the efficiency for the system on hydrocarbon fuels should be even lower.[43]

The state-of-the-art SOFC systems seemly can only enable an energy efficiency of 25% with a power density similar to that of an IC engine according to the trend shown in Figure 1. In terms of saving energy and reducing emissions or well-to-wheel efficiency, there will be no obvious benefit for using the state-of-the-art SOFC systems to replace medium or small-scale IC engine based power sources. The trend of efficiency reduction to a low level with increasing power density is a major technical obstacle for miniaturization of SOFCs.

The PEFC systems based on hydrocarbon fuels can't compete with the hybrid IC/battery, IC CHP system, and SOFC systems in terms of power density and electrical efficiency. The requirement for CO and S free hydrogen gas for the PEMFC stack substantially increases the complexity, size, energy loss, and cost of the PEFC system.

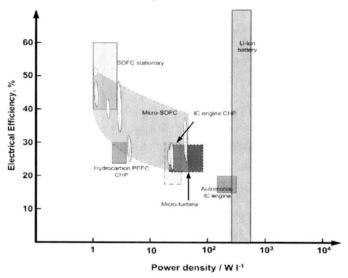

Figure 1. Energy efficiency of power systems vs. power density.

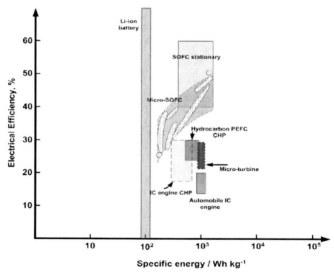

Figure 2. Energy efficiency of power systems vs. specific energy.

5. Conclusions

PEFC in coupling with hydrogen network is considered as the scenario that is best fit the desire for a renewable and efficient hydrogen economy. The technological niches for SOFCs are that the co-generation scheme can enable a high efficiency (60%) and that SOFCs can operate on a variety of fuels that will exist in a probably very long transition period. The current investment and R&D efforts are mainly sustained by the vision of these technological niches.

In general, R&D of industrial SOFC power generation systems is still in an immature stage with little market niche. The large-scale stationary systems show some benefits in terms of fuel and electrical efficiencies over IC engine and gas turbine based systems. Further development is now focused on enhancing the electrical efficiency via co-generation with a turbine, reliability, durability, and cost effectiveness.[5] Market diffusion is heavily resisted by the mature IC and gas turbine technologies and by the high cost for the R&D programs.

Miniaturization of SOFC is a relative concept that may cover the R&D efforts for medium-scale and small-scale SOFC power sources. Although miniaturized or micro SOFC systems (in comparison with the stationary systems) have a good chance to occupy the APU, UPS, and CHP markets, they are facing strong competitions from the improved IC engine, sterling power generation systems, micro-turbines, batteries, and hybrid systems. The following barriers need to be removed before commercialization of miniaturized SOFCs. First, the new development of the SOFC technologies should permit the reduction of size or increase of power density while maintain the electrical efficiency at around 40% to ensure a marginal benefit edge in efficiency and emissions. The trend that the efficiency decreases with increasing power density has to be broken with technological breakthroughs. Second, the reliability and the durability should be enhanced to be comparable with the IC engines. Finally, the manufacture cost should be reduced 3–7 folds to a competitive level ($1,000–2,000/kW).

Specifically for the CHP application, the power density or specific power of SOFC based systems (e.g. the SOFCo's system) can be close to that of IC engine based systems but the efficiency of the SOFC systems is similar to that of IC engines of gas turbines. Further technology intensification on the power density, efficiency, reliability, durability, compactness, and costs will determine the survival of the market niche for the SOFC based systems.

For the APU and UPS applications, SOFCs show an advantage in terms of energy density but much lower power densities than Li-ion battery. Further reduction of the size and mass and increase of the energy efficiency are desired.

For the application of SOFCs in automobile/transportation systems, it seems to be a long way to go when we look at the status of the current industrial SOFC systems in comparison to the commercialized hybrid IC engine/battery systems. The power density of the SOFC power sources has to be increased up to 5–10 folds while the electrical efficiency of 40–50% has to be maintained. The fuel cell stacks may have to be combined with battery stacks in order to minimize the energy loss and enable a total efficiency significantly greater than that of a hybrid IC/battery power system. Technological breakthroughs are desired in the terms of materials, fuel cell structures, fuel processing, and system control schemes. The second part of this review emphasizes the technical problems and new technical developments on system, structure, and material that may significantly enhance the both power density and efficiency.

References

1. K. Kendall, N.Q. Minh, and S.C. Singhal, in: *High-Temperature Solid Oxide Fuel Cells-Fundamentals, Design and Applications*, edited by S. Singhal and K. Kendall (Elsevier Science, Oxford, England, 2003), pp. 197–229.
2. M.A. Khaleel and J.R. Selman, in: *High-Temperature Solid Oxide Fuel Cells-Fundamentals, Design and Applications*, edited by S. Singhal and K. Kendall (Elsevier Science, Oxford, England, 2003), pp. 293–332.
3. C. Song, Fuel processing for low-temperature and high-temperature fuel cells Challenges, and opportunities for sustainable development in the 21st century. Catalysis Today 77, 17–49 (2002).
4. EG&G Technical Services, Inc., Fuel Cell Handbook, 7th edn US Department of Energy, Office of Fossil Energy, National Energy Technology Laboratory, Morgantown, West Virginia, 2004.
5. M.C. Williams, J.P. Strakey, and S.C. Singhal, US distributed generation fuel cell program. Journal of Power Sources 131, 79–85 (2004).
6. J. Zizelman, J. Botti, J. Tachtler, and W. Strobl, Automotive Engineering International 14, September, 2000.
7. M.C. Williams, J.P. Strakey, and W.A. Surdoval. The US Department of Energy, Office of Fossil Energy Stationary Fuel Cell Program. Journal of Power Sources 143, 191–196 (2005).
8. Paolo Agnolucci and William McDowall, Technological change in niches: auxiliary power units and the hydrogen economy. Technological Forecasting & Social Change 74, 1394–1410 (2007).
9. Gregorio Marbán and Teresa Valdés-Solís, Towards the hydrogen economy? International Journal of Hydrogen Energy 32, 1625–1637 (2007).

10. Seth Dunn, Hydrogen futures: toward a sustainable energy system. International Journal of Hydrogen Energy 27, 235–264 ((2002).
11. Carl-Jochen Winter, Electricity, hydrogen—competitors, partners? International Journal of Hydrogen Energy 30, 1371–1374 (2005).
12. P. Kruger, Electric power requirements in the United States for large-scale production of hydrogen fuel. International Journal of Hydrogen Energy 25, 1023–33 (2000).
13. William McDowalla and Malcolm Eamesb, Towards a sustainable hydrogen economy: a multi-criteria sustainability appraisal of competing hydrogen futures. International Journal of Hydrogen Energy, proof for publication, 2007.
14. S. Flipsen, Power sources compared: the ultimate truth? Journal of Power Sources 162, 927–934 (2006).
15. R.K. Dixon, The US Hydrogen program; Department of Energy., http://www.hydrogen.energy.gov/
16. de la Casa-Lillo MA, Lamari-Darkrim F, Cazorla-Amorós D, and Linares-Solano A. Hydrogen storage in activated carbons and activated carbon fibers. Journal of Physical Chemistry B, 106, 10930–10934 (2002).
17. C.-J. Brodrick, T.E. Lipman, M. Farshchi, N.P. Lutsey, H.A. Dwyer, D. Sperling, I. Gouse, S. William, D.B. Harris, and F.G. King, Evaluation of fuel cell auxiliary power units for heavy-duty diesel trucks. Transportation Research, Part D: Transport and Environment 7 (4), 303–316 (2002).
18. S.F.J. Flipsen, Power sources compared: the ultimate truth?, Journal of Power Sources 162, 927–934 (2006).
19. Breakthrough Fuel Cell on a Chip™ Technology Recognized as "Life-Changing Innovation", www.fuelcellworks.com, Dec. 2006.
20. Arthur D. Little, Conceptual design of POX SOFC 5 kw net system final report to the department of energy national energy technology laboratory January 8, 2001 (2001).
21. F. Baratto, U.M. Diwekar, and D. Manca, Impacts assessment and trade-offs of fuel cell-based auxiliary power units Part I system performance and cost modelling, Journal of Power Sources 139, 205–213 (2005).
22. P. Lamp, J. Tachtler, O. Finkenwirth, S. Mukerjee, and S. Shaffer, Development of an auxiliary power unit with solid oxide fuel cells for automotive applications, Fuel Cells 3 (3), 146–152 (2003).
23. J. Zizelman, S. Shaffer, and S. Mukerjee, Solid oxide fuel cell auxiliary power unit: a development update, Fuel Cell Power for Transportation 2002, Detroit, MI, Society for Automotive Engineers Technical Paper Series, 2002.
24. P.F. van den Oosterkamp, Critical issues in heat transfer for fuel cell systems, Energy Conversion and Management 47 (20), 3552–3561 (2006).
25. Tomaz˘ Katras˘nik, Hybridization of powertrain and downsizing of IC engine – A way to reduce fuel consumption and pollutant emissions – Part 1, Energy Conversion and Management 48, 1411–1423 (2007).
26. Stephanie L. Hamilton, Project Title: Micro Turbine Generator Program, Southern California Edison, Proceedings of the 33rd Hawaii International Conference on System Sciences, 2000.
27. Well-to-Wheel Energy Use and Greenhouse Gas Emissions of Advanced Fuel/Vehicle Systems, General Motor, Argonne National Laboratory, Exxon, Shell, 2001.
28. T. Aicher, B. Lenz, F. Gschnell, U. Groos, F. Federici, and L. Caprile, Fuel processors for fuel cell APU applications. Journal of Power Sources 154, 503–508 (2006).
29. E. Varkarakia, N. Lymberopoulosa, E. Zouliasa, D. Guichardotb, and G. Polic, Hydrogen-based uninterruptible power supply. International Journal of Hydrogen Energy 32, 1589–1596 (2007).

30. A.D. Hawkes and P. Aguiar, B. Croxford, M.A. Leach, C.S. Adjiman, N.P. Brandon, Solid oxide fuel cell micro combined heat and power system operating strategy: options for provision of residential space and water heating. Journal of Power Sources 164, 260–271 (2007).
31. N.M. Sammes and R. Boersma, Small-scale fuel cells for residential applications. Journal of Power Sources 86, 98–110 (2000).
32. http://www.homegeneratorsystems.com/products/intelligen/15kw/index.cfm#feets.
33. T. Susai, A. Kawakami, A. Hamada, Y. Miyake, and Y. Azegami, Development of a 1 kW polymer electrolyte fuel cell power source. Journal of Power Sources 92, 131–138 (2002).
34. http://www.fuelcellmarkets.com/european_fuel_Cell/1,1,2994.html
35. Chenggang Xie, Joseph Bostaph, and Jeanne Pavio, Development of a 2 W direct methanol fuel cell power source. Journal of Power Sources 136, 55–65 (2004).
36. T. Sack and T. Matty, Li-ion battery technology for compact high power sources 96, 47–51 (2001).
37. S. Veyo, Westinghouse SOFC Field Unit Status, Westinghouse Science & Technology Center, 2004.
38. Y. Yoshida, N. Hisatome, and K. Takanobu, Development of SOFC for Products, Mitsubishi Heavy Industries, Ltd., Technical Review Vol. 40 No. 4 (Aug. 2003).
39. ACUMENTRICS 5000 POWER SYSTEM http://www.fuelcellmarkets.com/ article_default_view.fcm?articleid=7195&subsite=425.
40. Julian Dinsdale, Karl Föger, Raj Ratnaraj, Jonathan Love, and Alison Washusen Ceramic Fuel Cells Limited (CFCL), COMMERCIALISATION OF CFCL'S ALL-CERAMIC STACK TECHNOLOGY Paper presented at the Fuel Cell Seminar, Miami, November 2003.
41. Cummins Power Generation 10kWe SOFC Power System Commercialization Program March 22, 2002.
42. Markus Jenne, Demonstration Project-Sulzer Hexis SOFC System for Biogas (Fermentation Gas) Operation, ESF Workshop January 29–30th, 2003.
43. Steven Shaffer, Update on Delphi's Development of a Solid Oxide Fuel Cell Power System, Honolulu, Hawaii, 2006 Fuel Cell Seminar, 2006.
44. Mesoscopic Devices, MesoGen™ 250 W military SOFC battery charger, 2005.
45. Aaron Crumm, Portable Fuel Cell Systems-Solid Oxide Fuel Cells, Adaptive Materials, Inc., 2005.

A REVIEW ON MINIATURIZATION OF SOLID OXIDE FUEL CELL POWER SOURCES-II: FROM SYSTEM TO MATERIAL

X.Y. ZHOU[1]*, A. PRAMUANJAROENKIJ[1], AND S. KAKAÇ [2]
[1]*Department of Mechanical and Aerospace Engineering University of Miami, Coral Gables, FL 33124*
[2]*TOBB University of Economics and Technology Sogutozu, Ankara, Turkey*

1. Introduction

Solid oxide fuel cell (SOFC) is an electrochemical power source that meets the technical expectations for no or low emission and high energy efficiency in future hydrogen economy or renewable energy economy. The high fuel flexibility of SOFC permits the SOFC technology to establish and expand its market niche and to serve as a primary power generation technology in a long transition period. SOFC technology is expected to play an important role in this transition period in circumventing the problems of energy shortage and energy consumption related emissions when the capacity of the renewable energy production is still small. SOFC technology would play an essential role in facilitating the transition from current non-renewable energy technologies to hydrogen energy technologies that harness unlimited renewable energy sources.

Current commercialization of miniaturized or micro-SOFC is targeted the markets for auxiliary power units (APU), uninterrupted power sources (UPS), and combined heat and power (CHP) for small-scale and medium-scale power generation. However, miniaturization of conventional large SOFC systems is facing significant technical barriers. One of them is the trend that enhancement of specific power or power density will result in a reduction of energy efficiency of the systems. For example, state-of-the-art SOFC systems that have comparable power density to that of IC engine based APUs and CHPs are as efficient as these IC engine based power sources. If this trend can't be reversed, it would be difficult for the SOFC technology to maintain and expand its market niches, which is crucial for

---

* E-mail: xzhou@gmail.com

the development of the renewable energy technologies. This paper will review literature and identify the current technical limitations that hinder the advancement of the micro-SOFC technology and reflect recent developments that will lead to solutions. This review will include accounts on system, device, component, and material levels.

2. SOFC Systems

2.1. MAJOR SUBSYSTEMS

As shown in Figure 1, a self-contained SOFC power generation system consists of four parts: (1) fuel container; (2) fuel processor; (3) fuel cell stacks; and (4) balance of plant (BOP) that includes all the peripherals necessary or critical for system operations.[1-5] The fuel container is often excluded from technical discussions because this part is simple and of minor technical interest except for the case of hydrogen storage. The fuel processor for the SOFC may include a preheating/vaporization unit, cracking/pre-reforming unit for mitigating carbon deposition problem in the reforming system, reformer, desulfurization units, and energy and water recuperators.[6] There may be a desulfurization system at the upstream and another one at the downstream of the reforming system.

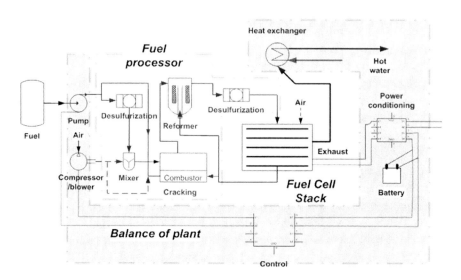

Figure 1. Schematic of a self-contained SOFC power generation system.

The fuel cell stack or stacks is the heart of the power generation system. The fuel cell stack receives hydrogen rich fuel or syngas (H_2 + CO) from the fuel processor into anode and converts the fuel directly into electricity via coupling with an oxidant (O_2 in air) in cathode. The off-gas from the anode of the fuel cell stack is recycled for combustion in a burner to generating heat for fuel cell processing. The heat emission from the fuel cell stacks is also recovered using heat exchangers for fuel processing in order to cut energy loss and enhance the efficiency. The electrical power is partially used for operation of the BOP and the fuel cell system itself.

The BOP may include insulations, liquid pumps, air blowers or compressors, valves, water and heat recovery (heat exchangers), sensors, system control strategies, electronics, and power conditioning. The system control strategies, sensors and power conditioning are particularly important for co-ordinating the fuel processing subsystems and fuel cell stacks in terms of mass, energy, and heat managements to ensure a rapid, smooth, efficient, and safe operation of the entire system. It is one of the major challenges for the researchers and engineers to optimize the BOP to make it more reliable, compact, and cost effective.

2.2. METRICS OF THE SIZE

Although the fuel cell stack is the heart of the fuel cell power generation system, often it is not the largest subsystem. For instance, as shown in Figure 2, in the Delphi's SPU 1B portable unit the two fuel stacks combined is smaller than the fuel processing system and the space occupied by the BOP.[7] The company made a great effort to reduce the size of the system while increased the efficiency. In the CFCL's NetGenä micro-CHP unit (2 kW), the planar fuel cell stack is ~70 l with insulation whereas the total volume of the unit is about 700 l (Figure 3).[8] Most of space is occupied by the insulation, heat exchangers, fuel processor, and BOP. One of the primary challenges for miniaturization of SOFC systems is how to reach the best compromise between the subsystems. Benchmark analyses on how the insulation impacts the efficiency of a SOFC module were conducted by Braun et al.[9] and Chung et al.[10] For a 1.5 kW SOFC module with a volume of 117 l and a surface area of 13,345 cm^2, a silica aerosol insulation with a thermal conductivity of 0.03 W m^{-1} and a thickness of 5 cm permits 9% energy loss. The energy loss is 5.5% when the thickness of the insulation is 10 cm. If the thickness of the insulation is 10 cm, the total volume of the insulation is 190 l, which is 1.6 time of the volume of the SOFC module. If the size of the SOFC module is reduced this ratio of the insulation volume to the SOFC module volume will increase because the area to volume ratio of the SOFC module will increase. It is worth of noting that the above analyses did not consider the impact of the heat emission on the electronics in the BOP. Much thicker insulation or additional space separation is required to reduce the heat received by the electronic devices that are in the package of the SOFC power source.

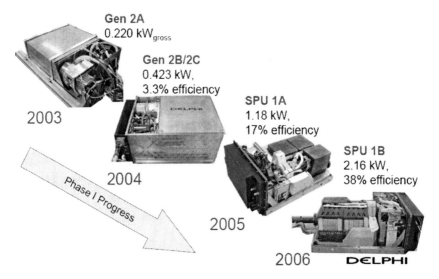

Figure 2. Progress of Delphi's SOFC APU technology.

Figure 3. CFCL's NetGenä micro-SOFC stacks.

2.3. SYSTEM EFFICIENCY

Large-scale hybrid SOFC and gas turbine (SOFC-GT) technology has been identified to be superior to other power generation technologies because of its high electric efficiency. Analyses[9, 10] indicate that a system electric efficiency ranging from 50% to 75% can be achieved by a hybrid SOFC-GT system depending on the actual values of the operation parameters. However, the highest electric efficiency has never been achieved by a real hybrid SOFC-GT system. Due to the space and mass constraints, the hybrid scheme can't be implemented in the miniaturized or micro-SOFCs. According to recent thermodynamic analyses,[9, 10] under a set of benchmark conditions a 1.5 kW CHP system can achieve a maximum electric efficiency of 40% and a total fuel efficiency of 80% when 100% internal reforming is realized and the exhaust gases are recycled. However, such an electric efficiency in a long operation period has not been reported for real micro-SOFC systems.

For a real system, the electric efficiency is always less than the maximum electric efficiency for a number of reasons as follows: (1) Some fuel (in the form of original fuel, not completely converted fuel, or reformate) is consumed for sustaining or maintaining the operation of the fuel processing system that is expected to deliver products with certain temperature, pressure, and flow rate (parasitic energy loss). (2) The fuel is not completely converted. (3) The flue gases from the burner must be hotter than the atmosphere. Thus, some heat is necessarily lost to the environment with the flue gases. (4) Real heat exchangers have less than ideal efficiencies because the temperature gradient is never zero. (5) The reactions exhibit irreversibility or non-ideality relating to heat transfer over a finite (not an infinitesimal) temperature difference, mixing of dissimilar substances, sudden phase changes, mass transport under finite concentration differences, free expansion, friction, etc. (6) Finally, there is always some heat loss in all components because insulation materials are not perfect.[11] According to Braun et al.[9] the primary loss of exergy that can be converted into electric power is the irreversibility but not the heat loss into the environment. In order to reduce the irreversibility, the linear flow velocity of the fluid has to be decreased or the rates of chemical reactions has to be increased (e.g. by catalysis) to approach to the equilibrium state. Clearly, the decrease of the linear velocity of the fluid in the fuel processor and in the SOFC stack means either increase of the system volume or decrease of the total power. Either of these will result in decrease of the power density or specific power of the system and not be acceptable for the micro-SOFC systems.

2.4. OTHER PROBLEMS

Another major problem for large-scale SOFCs and micro-SOFCs is the insufficient durability of the SOFC stacks. Carbon deposition, sulfur poisoning, high temperature oxidation, electrochemical corrosion, sintering, thermal fluctuations, and incompatible thermal expansions are main reasons for degradation of SOFC stacks.[1-5] In most cases, small fuel cell components are more vulnerable to these forms of degradations than lager fuel cell components.

3. Miniaturization of Fuel Processing System

Two reforming schemes have been implemented in the SOFC systems: external reforming and internal reforming (IR).[1-5] In the case of internal reforming, the reformation of the hydrocarbon fuels takes place in the same chamber of the SOFC anode.[12-15] The scheme of internal reforming is further categorized into indirect internal reforming (IIR) when the reforming reactions are separated spatially from the electrochemical reactions and direct internal reforming (DIR) when the reforming reactions take place on SOFC anode. Use of IR schemes eliminates the external reformer as a subsystem, reduces the total size and mass, facilitates heat transfer, and enhances the thermal efficiency. According to thermodynamic analyses,[9,10] a SOFC CHP that is fuelled with methane or natural gas and adopts IR and anode gas recycle schemes can have a little higher electric efficiency than a SOFC CHP fuelled with hydrogen rich gases on the basis of an equal power output. However, an external preprocessor or pre-reformer is necessary for processing hydrocarbon fuels, in particular, heavy hydrocarbon fuels. Direct interaction between the heavy hydrocarbons and anode may result in an extensive carbon deposition and fast degradation of the SOFC stack. Hydrocarbon fuels may contain a substantial amount of sulfur species. A desulfurizer should be included as an integrated part of the fuel processing system.[1-5, 9, 10]

3.1. REFORMING PROCESSES

Hydrocarbons or oxygenates can reformed using following processes:[16-25]

Steam reforming

$$C_mH_n + mH_2O = mCO + (m + n/2)H_2; \Delta H > 0 \qquad (1)$$

$$C_mH_nO_z + 2(m-z/2)H_2O = mCO_2 + 2(m-z/2+n/4)H_2; \Delta H>0 \quad (2)$$

Partial oxidation reforming

$$C_mH_n + (1/2)m(O_2+3.76N_2) = mCO + (1/2)nH_2 + 3.76(1/2)mN_2; \Delta H<0 \quad (3)$$

$$C_mH_nO_z + (m-z/2)(O_2+3.76N_2) = mCO_2 + (1/2)nH_2 + 3.76(m-z/2)N_2 \, \Delta H<0 \quad (4)$$

Autothermal reforming

$$C_mH_n + (1/2)mH_2O + (1/4)m(O_2+3.76N_2) = mCO + (1/2)(m+n)H_2$$
$$+ 3.76(1/4)mN_2; \Delta H<0 \quad (5)$$

$$C_mH_nO_z + mH_2O + (1/2)(m-z)(O_2+3.76N_2) = mCO_2 + (1/2)(n+2m)H_2$$
$$+ 3.76(1/2)(m-z)N_2; \Delta H>0 \quad (6)$$

Thus, the conversion of fuels to hydrogen and light hydrocarbons is carried out mainly via three techniques: (1) steam reforming (SR); (2) partial oxidation reforming (POXR); and (3) autothermal reforming (ATR). The SR reaction is strongly endothermic, whereas the POXR is exothermic. The SR process is very much suitable for long periods of steady-state operation and can deliver a high concentration of hydrogen (>70%). SR also allows reuse of heat and waste anode gas thereby resulting higher efficiency than those for POXR and ATR. However, because they need for indirect heat transfer (across a wall), the SR reactors are large, heavy, and responding slowly. POXR that does not need a water supply can generate heat and high temperatures but a relatively low hydrogen concentration and efficiency. ATR process combines the heat effects of POXR and SR to generate more hydrogen without an external heating source. In many fuel processing systems, ATR process is preferred because of its high efficiency and hydrogen content. In order to achieve a high efficiency and hydrogen content in reformate, it is essential to allow a long residence time and to homogenize the reforming process to avoid large gradients. This requirement for fine-tuning of mass, energy, and heat flows compromises the miniaturization of the fuel processing units.

The overall reactions for SR, POXR, and ATR processes can be summarized by:[18]

$$C_nH_mO_z + \frac{py}{2}(O_2 + 3.76N_2) + p(n - y - \frac{z}{p})H_2O \longrightarrow$$
$$nCO_p + [p(n - y - \frac{z}{p}) + \frac{m}{2}]H_2 + 3.76\frac{py}{2}N_2 + (\Delta H_r) \quad (7)$$

Where ΔH_r is heat of reaction, p can be 1 for CO or 2 for CO_2.

The theoretical maximum thermal efficiency of these reforming is given by:

$$\eta_{th} = \frac{-[p(n - y - z/p) + m/2]\Delta H_{f,steam} - n\Delta H_{f,CO_2} + n\Delta H_{f,CO_p}}{-n\Delta H_{f,CO_2} - (m/2)\Delta H_{f,steam} + \Delta H_{f,C_nH_mO_z}} \quad (8)$$

where $\Delta H_{f,steam}$, $\Delta H_{f,CO_2}$, $\Delta H_{f,CO}$, and $\Delta H_{f,C_nH_mO_z}$ are heats of formation for steam, CO_2, CO, $C_nH_mO_z$ respectively. The thermal efficiency varies with the composition of fuel but normally in the range between 87% and 97%.

3.2. CARBON DEPOSITION

Coking, coke formation, or carbon deposition is a phenomenon in which carbonaceous deposits form on the components of a reactor, in particular on the surface of reforming catalysts.[16, 18, 26] Carbon deposition can have detrimental effects on the reforming processes. First, carbon deposits cover the active sites and deactivate the catalysts. Second, carbon particles cause partial of total blockage of the paths for the reactants and products resulting slowing down or termination of the processes. Third, carbon deposition causes non-uniformity of the reactions and consequently "cold" and "hot" spots in the reactor. The high temperature at the hot spots may cause sintering of the catalysts and other reactor components and sometime "hot tube" phenomenon. Finally, the uneven flow distribution will cause self-accelerating situation with further overheating of the reactor. Therefore, carbon deposition can't be tolerated in reformers.

The parasitic carbon deposition reaction of hydrocarbons is as follows:[16]

$$C_mH_n = xC + C_{m-x}H_{n-2x} + xH_2 \quad (9)$$

$$2CO = C + CO_2 \quad (10)$$

$$CO + H_2 = C + H_2O \quad (11)$$

At temperatures above 650°C, higher hydrocarbons may react in parallel to reaction (9) via thermal cracking (pyrolysis) to form olefins that may easily form coke. As such, pre-reforming or fuel cracking is necessary for the heavy hydrocarbons. The cracking process for heavy hydrocarbons in cracking/pre-reforming unit (~450°C) is as follows:[16]

$$C_m H_n \rightarrow C_{mi} H_{n-k} + C_i H_k \qquad (12)$$

Very often, measures to eliminate of depress coke formation are more decisive for the process layout than for considerations such as the activity of the catalysts and reforming efficiency. A conservative design principle is simply to require that there should have no affinity for methane decomposition reaction at any position in the catalysts bed, or the reaction[16, 26] is impossible. If the actual gas above the catalyst shows affinity for carbon formation there will be a tendency towards carbon formation. This can be expressed by the carbon activity α_c^{eq} expressed by:[16, 26]

$$CH_4 + \ast \longrightarrow C - \ast \qquad \substack{\nearrow \text{gas} \\ \\ \searrow \text{carbon}} \qquad (13)$$

$$\alpha_c^{eq} = K_c \frac{p_{CH_4}}{p_{H_2}^2} < 1 \qquad (14)$$

where K_c is the equilibrium constant for the methane decomposition, p_{CH_4} are partial pressures of methane and hydrogen respectively. A safe, conservative design criterion would be to require $\alpha_c^{eq} < 1$ at any position in the reactor.[16, 26] However, although this may be realized in large-scale industrial process, it is difficult to be realized for small reforming systems with unit size and energy consumption constraints.

Thermodynamic calculations indicate that there is a minimum temperature above which coke formation is impossible for each of the reforming processes and the minimum temperature depends on the ratio of oxygen to carbon in the reactants.[18] At high O/C and H/C ratios the minimum temperature for coke-free conditions is lower. If the operating temperature is very low, the kinetics of coke formation is also low. As such coke formation becomes insignificant. Thus, two effective methods for coking suppression are to add excess steam and to supply hydrogen in the reactor. Hydrogen is an effective coke retarding reactant in pre-cracking (pre-reforming) and reforming processes, in essence by reducing the formation of olefin intermediates.[16, 26]

Catalysts modification is another approach for suppressing carbon deposition.[20] When using a catalyst with enhanced steam adsorption, the critical steam to carbon ratio and propensity of coke formation can be reduced. The rate of coking on acidic catalysts can be diminished by reducing the strength of the acidic sites or even neutralize them by addition of alkali. Use of shape sensitive catalysts such as ZSM-5 (zeolites) is effective in suppressing coking simply because deposition of coke cannot take place inside the zeolites.

3.3. DESULFURIZATION

Organic and inorganic compounds in fuels pose a major problem for all fuel processing systems and especially for catalysts. According to Cheekatamarla and Finnerty,[20] the heavy sulfur species, $C_xH_yO_zS_a$, in sulfur-containing fuels could be adsorbed on metallic catalysts and oxide support and cause reversible or irreversible poisoning. The poisoning process is a complex process that may involve adsorption, diffusion in the lattice, transfer of the sulfur between the catalysts and catalyst support, reduction and oxidation. The dead end of the process may be the adsorbed sulfur, S, on the catalysts and combination of the sulfur in the catalyst support in the form of sulfates. As such, it is required that the sulfur concentrations in the hydrocarbons feed to the reformer to be in the range of 1–10 ppm.

The removal of sulfur from liquid hydrocarbons is difficult and requires severe process conditions. Desulfurization may be carried out both downstream or upsteam of reforming step. In practice, to avoid too much trouble caused by sulfur, desulfurization at a mild temperature upstream of reforming step is very popular. A new integrated process concept, so-called SARS-HDSCS, has been developed at Pennsylvania State University.[27, 28] The selective adsorption SARS process can be applied as organic sulfur trap for sulfur removal from fuels before the reformer for fuel cells on-board or on-site, and it may be applied in a periodically replaceable form such as a cartridge. Further improvement in adsorption capacity is desired. Another existing process, hydrodesulfurization (HDS) at high temperature (320–380 °C) and high pressure (3–7 MPa) over sulfided CoMo or NiMo catalysts is used at the upstream of pre-reforming step (480°C, 2.5 MPa) to reduce the sulfur in hydrocarbon fuels.[29–32] The hydrogen sulfide produced in the process is scrubbed using ZnO pellets as the absorbent. The major sulfur compounds existing in current liquid hydrocarbon fuels are thiophenic compounds and their alkyl-substituted derivatives. Some of them have been considered to be the refractory sulfur compounds in the fuels due to the steric hindrance of the alkyl groups in HDS. Consequently, it is difficult or very costly to use the existing HDS technology to reduce the sulfur in the fuels to less than 10 ppmw.

Some catalysts such as pyrochlore-based materials, $Gd_2Ti_{1.4}Mo_{0.6}O_7$, have high resistance to poisoning by hydrogen sulfide but are not tolerant to the heavy sulfur species in logistic fuel.[20] One of the possible mechanisms of accumulation of adsorbed sulfur on the catalysts is illustrated in Figure 4. Other materials, $La_{1-x}Sr_xVO_3$ (LSV), $La_{1-x}Sr_xCr_{1-x}Mn_xO_{3-\delta}$ (LSCM), and $La_{1-x}Sr_xBO_3$ (LSB)[33–35] also exhibit good resistance to H_2S.

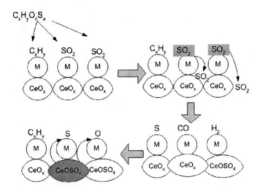

Figure 4. A possible mechanism with which accumulation of S on the catalyst surface. M is referred as a catalyst which could be conventional Ni/Cu/YSZ or the new pyrochlore-based materials, $Gd_2Ti_{1.4}Mo_{0.6}O_7$, that are resistant to H_2S.

3.4. LARGE-SCALE INDUSTRIAL REFORMING SYSTEMS

Production of syngas or hydrogen rich gases from natural gas and other light or heavy hydrocarbons is one of the major processes in the chemical industry. The large-scale reforming systems are designed and built with much less compromises and constrains.[16] The reforming reactors of choice are fixed bed tubular steam reformers. The large dimension of the systems allows very detailed process monitoring, control, improvements, and optimization. For example, the heat produced in the downstream of a production process can be recovered for sustaining the steam reforming (i.e. heat exchange reforming) and for generating extra electricity using a steam turbine. Process monitoring data indicate that in the entire tubular reformer, the gas composition is very close to the equilibrium gas composition and a high thermal efficiency of the entire system approaching 95%.[16] Kinetic and thermodynamic limits or criteria are considered in the designs and operation schemes to avoid carbon deposition at many points in the entire reforming unit. A desulfurization unit may be in place to mitigate sulfur poisoning of the catalysts.[16]

Autothermal reforming (ATR) for industrial syngas production has been used since late 1950 and has been improved since 1990.[16–18] The new developments have been operation at low steam to carbon ratios and new burner designs ensuring safe operation. The burner provides proper mixing of the feed streams in a turbulent diffusion flame to avoid soot formation.

As alternative reforming routes, both non-catalytic partial oxidation (POX) and catalytic partial oxidation (CPOX) are used for industrial production of syngas.[16–18] The POX needs a burner and high temperatures (ca. 1,400°C) to ensure complete conversion and to reduce soot formation. In CPOX, the reactants are premixed and all conversion takes place in a catalytic reactor without burner at lower temperatures (ca. 800°C). The hydrogen selectivity and conversion can be above 90% and above 95% respectively.

In summary, in large-scale industrial reforming systems detailed process monitoring and control methods are in place to achieve a condition as close to the equilibrium condition as possible. The system efficiency is very close to the theoretical value given by Eq. (8).

3.5. CONSTRAINTS IN A REFORMING UNIT FOR FUEL CELL APPLICATIONS

A fuel processor used in a micro-SOFC system must meet specific technical and marketing demands that allow the entire fuel cell system to be competitive technically and economically. The fuel processor must be small, lightweight, processing feeds of varying composition that may include fuel cell poisons. The unit must also be sufficiently rugged to withstand frequent shut-downs and cold start-ups and must operate for many years, with minimum service. Above all, the reforming unit must have a high energy efficiency that permits a high electric efficiency of the entire fuel cell system. Meeting these constraints requires a number of design trade-offs and for this reason, every fuel processing unit is the result of a series of compromises. Depending on the technical requirements of the fuel cell system, the designers have to make a sets design compromises, resulting in fuel cell processor designs. However, in general, the energy or thermal efficiency of the reforming units in a fuel cell system is much lower than that of a large-scale industrial reforming system.[19–25]

Fuel processing units in SOFCs were developed with the following considerations[6]

- Air, water, fuel, raffinate, and waste streams have to be properly arranged to facilitate the processing steps and utilize the energies carried by the streams as much as possible.

- Several heat recuperators and heat exchangers may be used to enable high heat exchange efficiency between exothermal and endothermal processes and reduce heat loss to the environment.
- Desulfurization must be carried out at least at upstream of the reforming steps or preferentially at both upstream and downstream.
- A cracking or pre-reforming unit may be set up prior to the main reforming unit for avoiding carbon deposition.
- These controls and processes have to be performed under stringent space and weight limits.

3.6. STEAM REFORMING PROCESSORS

Many SR units have been developed by commercial organizations.[6] A few of them have been described in literature, including IdaTech's heat exchange steam reformer and IFC's plate assembly steam reactor.

In IdaTech's annular heater exchange design, desulfurized gas fuel and vaporized water are allowed to pass through a pre-reformer into a steam reformer.[11] The design enables use of unreacted fuel, raffinate from a fuel cell outlet in a central burner to provide heat for the endothermal SR in the primary reforming in the inner catalysts bed. The flue gas from burning the un-reacted fuel is utilized to enable pre-reforming in the outer catalysts bed. For the Idatech steam reforming processor, the fuel conversion and the thermal efficiency are estimated to be 65–83% and 75–83% respectively on natural gas fuel. The hydrogen production rate was 25–80 slm.

To enhance the efficiency and to make it easier for scaling up, IFC proposed a design,[38] in which the SR system is an assembly of repeating units, each of which has its own burner, reformer, regenerator–heat exchanger sections. Because the adjacent reformer and the burner passages share a common wall, both sides of which are covered with a catalyzed alumina coating (only the walls of the regenerator sections of the assemblage are not catalyzed), the heat released due to exothermic oxidation reaction can be effectively and locally transferred to the SR side. The catalysts on the combustion side can be nickel or nickel synergetic with noble metal catalysts such as Pt, Pd, Rh. The catalysts on the SR side can be copper-based materials if methanol is used as a fuel or noble metal catalysts if hydrocarbon is to be used.

3.7. AUTOTHERMAL REFORMING AND PARTIAL OXIDATION REFORMING (ATR AND POXR) PROCESSORS

Argon National Laboratory developed an annular integrated reformer design that used monolithic catalysts.[37] The ATR catalysts were 15.24 cm (6 in.) long monoliths coated with a proprietary catalyst. The fuel processor contains 1–2 inner ATR zones and 1–2 outer reforming zones, several cooling zones, and a sulfur removal zone. The heat exchanging was enabled with highly efficient microchannel heat exchanger. The volume of the processor (excluding BOP and insulation) was 17 l. The power was between 2.5 and 16 kW. The efficiency for the reformer was a function of power and between 77% and 90% on methane. Argon National Laboratory also developed a cylindrical ATR containing four catalyst monoliths. The efficiency was between 40% and 80% depending on the ratio of oxygen to carbon.

Lenz and Aicher[19] studied ATR of kerosene, a heavy liquid hydrocarbon fuel using 15 kW cylindrical reformer containing monolithic catalysts. They found that the oxidation and the steam reforming in hot part (close to entrance) are mass transfer controlled while the steam reforming reaction in the colder part is reaction kinetics controlled. The reforming efficiency is a function of air to fuel ratio and between 57% and 77%.

3.8. LIMITATION OF HEAT AND MASS TRANSFER

Heat transfer between the reforming and combustion reactors and mass transfer on the catalysts surface dictate the design and performance of a steam reformer. According to a transport and kinetic analysis by Zalc and Loffler,[38] in order to produce 31 mol/min or 50 kW on the lower heating value of hydrogen, a tubular reforming reactor of 1 inch diameter containing fixed bed pellet catalysts has to be at least 7.75 l. If the reforming reactor adopts a plate design, the minimum volume can be 2.2 l. Yet, if the catalysts are directly deposited on the wall of the plate reactor, the minimum volume can be 1.2 l. If catalysts are directly deposited on both sides of shared wall of combustion reforming channels, the total volume for a reforming unit can be as small as 2.4 l. However, it is worth of pointing out that although the small reforming systems can satisfy the power output requirement with a relatively small reactor volume, the performance of the reforming units are below that of the large-scale industrial reforming reactors.

3.9. SHORT CONTACT TIME REFORMING PROCESS

In order to overcome the limitation of heat and mass transfer processes, a short contact time (SCT) technology was developed.[39-45] Compared with the existing homogeneous processes, the essence of SCT catalysts allows to considerably shorten the reaction times 5–200 milliseconds), enable high sensitivity, conversion and efficiency, decrease the volumes, and lower the operating temperatures, thus enabling the use of air instead of pure O_2 as the oxidant, and inhibiting carbon deposition. The SCT reforming process was realized initially by using Rh-Pt alloy based catalysts that enabled fast reforming reactions that quickly established thermodynamic equilibria. The catalysts allow building small CPOX or ATR reactors based on a Microlith substrate. The Microlith substrate consists of a series of ultrashort channel length, low thermal mass, catalytically coated metal meshes with very small channel diameters. For a conventional long channel honeycomb monolith a fully developed boundary layer is present over a considerable length of the catalytic surface, limiting the rate of reactant transport to the active sites. The ultra short channel length in Microlith substrate minimizes boundary layer buildup and results in remarkably high heat and mass transfer coefficients compared to conventional monolith substrates. The geometry of the substrate provides about three times higher geometric surface area (for supporting catalyst) over conventional reactors (e.g. monoliths) with equivalent volume and open frontal area.

An autothermal reforming (ATR) reactor based on the Microlith catalyst substrate technology was developed by Precision Combustion Inc. using the concept of the short contact time (SCT) reactor and tested for reforming isooctane into hydrogen.[42-45] The reactor was made of stacked Microlith screens coated with Lastabilized alumina washcoat and a precious metal based catalyst. The reactor diameter was ~4 cm and length ~1.2 cm, which corresponds to a volume of about 15 cm^3 (~1 $in.^3$) and a mass of ~12 g for the catalyst and substrate. The reactor operated at 700–800°C and yielded a conversion of 70–100% and hydrogen sensitivity of ~90%. The isooctane feed rate corresponded to about 3.4 kW of thermal energy input and water and airflow were regulated to provide the specified steam to carbon and oxygen to carbon ratios. This implies a power density of 160 kW/l and 200 kW/kg. The thermal efficiency of the reactor was claimed to be 70% on isooctane.

3.10. MICROCHANNEL FUEL PROCESSOR

Microchannel fuel processor technology is another approach that helps overcome the limitation of heat and mass transfer. The microchannel fuel processor technology is originated from the concept of microchemistry that *"deals with the development, correlation, and systematization of the methods for handling small quantity of materials..."* [46] Availability of the latest micro-electro-mechanical systems (MEMS) enables a transition of microchemistry from fundamental research to micro-reactor technology.

One of the major areas of the development is the quest for miniature fuel processors on microchips for portable power sources. Micro-scale reaction systems provide significantly higher surface area to volume ratios compared to conventional reactors and this can effectively facilitate novel reaction paths by suppressing unwanted homogeneous phase side reactions, not possible in the conventional reactors. Pacific Northwest National Laboratory (PNNL)[47-51] is the first institute who designed and tested first microchannel combustor, heater exchanger, and integrated reformer. In the steam reforming reactor, the combustor and steam reformer share the same reactor wall with both side catalyzed with combustion and reforming catalysts respectively. This eliminates the stationary boundary layer and suppresses the mass transfer and heat transfer resistance thereby significantly increasing the reaction rates and reducing the residence time and startup time. An integrated steam reforming system was designed in accordance with Second Law of Thermodynamics principles. Thus, the heat transfer between the reformers, combustors, recuperators, vaporators, heat exchangers is driven by minimum temperature gradients. They claimed that the system had a capacity of 10–20 kW with a total volume of ~7 l. The efficiency based on LHV was claimed at the level of 70–80% on hydrocarbon fuels. Microchannel fuel processors based on the same basic design have also been reported.[47] Ryi et al.[52] reported a microchannel reforming reactor had a capacity of 26 W with a volume of 0.064 l. The efficiency based on LHVs was initially 25.6–37.7% on hydrocarbon fuels and later was increased to 69% by using new type of catalysts and reactor materials. However, these reactors have to use hydrogen as the fuel in the combustor.

Holladay et al.[48-51] developed a self-sustained miniature fuel processor using the microchannel design. The volume of the reactor is only 0.002l. The capacity of the fuel processor is 0.2 W with an efficiency (based on LHV) of 9% on methanol fuel. Tanaka et al. also developed a MEMS-based miniature fuel reformer. The capacity is 0.2 W with an efficiency of 6%. The low efficiencies of these systems are attributed to the ineffective thermal insulation and lack of heat and fuel reuses in the system.

Microchannel fuel processors are promising to meet the power density requirement for the miniaturization of the fuel cell power systems.[47–59] However, several issues need to be addressed for the microchannel systems. First, the interconnection between the microstructure and macrostructures need to be improved because the microstructure has to be interfaced with macrostructures. Second, the pre-reformer and desulfurization units need to be miniaturized and incorporated into the microstructure units to avert coke and sulfur-poisoning. Otherwise, either the volume of the entire system is limited by the pre-reformer and desulfurization units or the microchannel fuel processor can only operate on fuels free of sulfur and having low coking tendency (e.g. pure methanol). Third, in order to achieve a high processing rate and efficiency, noble metal catalysts have to be used, result in high costs. Fourth, the thermal insulation technology has to be intensified to significantly reduce the loss of heat. Finally, the efficiency and durability of the microchannel fuel processors have to be significantly improved.

4. Miniaturization of Solid Oxide Fuel Cell

4.1. PLANAR SOFCS

Planar-type SOFCs or pSOFCs adopt the basic structure and configuration of PEMFCs thereby having a high volumetric power density. A pSOFC stack is a series of planar units that consist of an electrolyte layer, anode and cathode layers, current collecting layers on the electrodes, interconnect layers that provide electric conductance between two neighboring units and flow channels. There are two basic designs of planar-type SOFCs: electrolyte-supported and electrode-supported.[4] The electrolyte-supported SOFC can be made using a rather simple fabrication procedure. However, the relatively high ohmic resistance due to the thick electrolyte layer (about 150–200 μm for yttria stabilized zirconia or YSZ) requires a high operating temperature of 1,000°C or above. Although the high operating temperature is favorable for a heat and power co-generation system, it causes sever degradation of the constituent materials and components at the temperature and deficient stability, reliability, and durability of a cell stack.

In the anode-supported pSOFCs, the anode/electrolyte/cathode assemblies are supported with a thick, porous anode layer while the electrolyte can be made very thin (10–100 μm). As such, the internal resistance of the SOFCs can be reduced thereby permitting operation at lower temperatures (700–800°C).[1,4] Wen et al.[60] developed an electrolyte-supported pSOFC. The electrolyte layer was yttria-stabilized zirconia (YSZ) strengthened with Al_2O_3 fibers and a thickness of 150 μm. The performance at 1,000°C can enable a power density of ~70 W/l. Jung et al.[61] developed an anode-supported pSOFC with an YSZ electrolyte of 8 μm and a $(La_{0.7}Sr_{0.3})_{0.95}CoO_3$ (LSCo) current collecting layer of ~30 μm.[62] At 750°C the volumetric power density can be 200–1,000 W/l. The authors claimed that the electric efficiency of the pSOFC was 65% for hydrogen fuel and oxygen.

Fabrication of pSOFC requires a huge investment on equipment. Unlike the metallic and polymeric components for polymer electrolyte fuel cells, manufacture of ceramic components is complex, tedious, time-consuming, and difficult for mass production due to large quality variations and low repeatability.[4,5] Sealing in the pSOFCs is very important because any leakage of fuel and/or oxidant will result in significant reduction of performance.[4,5,63] In general, glass-based materials are highly attractive for sealing purposes because of their structural deformability for gas tightness at high temperatures. Unfortunately, serious technical problems with glass-based sealants in terms of chemical and thermo-mechanical stability are still to be solved.

Durability of the pSOFCs is one of the major concerns and technical hindrances for commercialization of pSOFCs. The thin electrolyte layer of ~10 μm is particularly vulnerable to thermal stress and shocks while operating at high current density at 800–1,000°C. The ceramic and metallic components are also subjected to high temperature oxidation, reduction, sintering, alloying, and de-alloying at the high temperatures. Thermal insulation from environment and efficient heat transfer are critical factors for achieving a high efficiency. For a SOFC stack operating at ~1,000°C, the insulation is larger than the stack itself.

4.2. MICRO-TUBULAR SOFCS

Tubular SOFCs were initially developed for large-scale power plants.[1–4] The diameter of the SOFC tubes is typically 2.5 cm and the length is 100–200 cm. The volumetric power density of these SOFCs is ~110 W/l (excluding reaction chamber, burner, supporters, insulation, etc.). Use of small-scale tubular SOFCs enables design of cell stacks with high volumetric power density due to increased electrode area in a unit volume. Micro-tubular SOFC that is 1–15 mm in diameter can achieve higher mass specific active surface area, lower internal resistance and higher performance or same performance at lower temperatures in comparison with large tubular SOFC. Moreover, in comparison to thin film pSOFCs, the sealing for micro-tubular SOFCs can be easer, more reliable, and more durable because the sealing can be placed in the low temperature regions. Sammes et al.[64] reported a 100 W anode supported micro-tubular SOFC stack. The stack consists of 40 tubular SOFCs with a diameter of 1.32 cm and a length of 11 cm. The SOFCs are connected with current connectors made from a Ni alloy. The connection between the SOFCs and connectors are made using a brazing technique. The area power density of the stack is 0.059 W/cm^2. Suzuki et al.[65–68] have published a series of papers on development of micro-tubular SOFCs. The diameter of their micro-tubular SOFCs is 1–2.5 mm. The inner layer is anode while the outer layer is electrolyte made from Gd doped ceria (GDC) that has a high conductivity at low temperatures (500–650°C). In one of their designs the tubes were assembled in cathode matrices that form a cube bundle.[68] The matrices were made from a cathode material, $La_{0.6}Sr_{0.4}Co_{0.2}Fe_{0.8}O_{3-y}$ (LSCF). Ag wires, stripes, and layers are applied on the SOFC surfaces in order to reduce the resistance along the axis of the tube. The volumetric power density of the stack at 550°C was reported as being about 2,000 W/l.

4.3. SINGLE-CHAMBER SOFCS

A 'single-chamber SOFC (SC-SOFC)' is a SOFC wherein both the anode and cathode are exposed to the same mixture of fuel and oxidant gas.[69] The primary advantage single-chamber SOFCs or SC-SOFCs is their simplified structure. Because the design of SC-SOFC does not require separation between anode and cathode chambers, sealing between the anode and cathode chambers is not required and a dense, gastight electrolyte layer between the anode and cathode is not necessary. The electrolyte layer can be porous and very thin. The very thin electrolyte layer enable lowering the operating temperature of the SOFCs to the level of 500°C at which ferric stainless steel can be used for reactor walls and for interconnectors.

The potential for rapid start-up can also be enhanced because the operating temperature is reduced and the reaction between the fuel and oxidant generates heat. In addition, carbon deposition is less of a problem due to the presence of a large amount of oxygen in the mixture. The SC-SOFCs have several benefits[69]: (1) enhanced durability; (2) fast start-up; (3) low operating temperature and low manufacture costs; and (4) simplification and miniaturization of the fuel cell stacks.

There are three types of possible geometries for SC-SOFCs.[69] The A- and B-type geometries position two electrodes on opposite sides of an electrolyte, which is the same arrangement as that found in dual-chamber fuel cells. In the A-type geometry, a mixture of fuel and oxidant gases is first delivered to the cathode, which reduces the oxidant gas. The mixture is then passed through the anode oxidizing the fuel. In the B-type geometry, a mixture of fuel and oxidant gases is simultaneously supplied to both the anode and cathode. On the other hand, the C-type geometry positions two electrodes on the same side of an electrolyte, which is also regarded as a 'surface migration fuel cell'.

In all three geometries, the two electrodes should meet the following criteria: (1) one electrode (anode) has to be electrochemically active for oxidation of the fuel but should be inert to oxygen reduction; (2) the other electrode (cathode) has to show the opposite properties. Consequently, the SOFC generates the maximum open-circuit voltage (OCV) when each electrode is ideally selective to the corresponding electrode reaction. One factor that lowers the OCV is when the target electrode reaction cannot overwhelm the counter electrode reaction in rate at the electrode, decreasing the mixed potential for the target electrode reaction. Another factor is the direct chemical reaction of the fuel with the oxidant in the gas phase or on the electrode surface. This also causes an energy loss in the fuel cell.

The concept of SC-SOFCs was first demonstrated by Hibino and Iwahara in 1993.[69] They used yttria-stabilized zirconia (YSZ), Ni–YSZ cermet (or Pt), and Au as the electrolyte, anode and cathode, respectively. The performance of the SC-SOFC was 0.0023 W cm^{-2} at 0.35 V with a methane–air mixture of 2:1 and at 950°C. One breakthrough in SC-SOFCs was made by Hibino et al. in 1999.[69] They applied a conventional Ni/YSZ/La$_{0.8}$Sr$_{0.2}$MnO$_3$ (LSM) cell to an SC-SOFC. The resulting fuel cell generated an OCV of 795 mV and a peak power density of 0.121 W cm^{-2} in a methane–air mixture feed at 950°C. Another interesting finding was that the C-type fuel cell showed an increased power density with decreasing gap between the two electrodes, suggesting that this geometry allows for a reduction in the ohmic resistance. The best performance in their reports to date is a maximum power density of 0.66 W cm^{-2} and open circuit voltage (OCV) of 0.78 V obtained in a configuration of a porous YSZ electrolyte layer (20 μm), Ni–YSZ anode, and LSCF cathode in methane CH$_4$-air mixture at a furnace temperature of 606°C. It was found that the real temperature at the electrodes was usually 150–200°C greater than the furnace temperature due to the extensive exothermal reactions on the electrode surface. Using a similar design, Shao et al.[70] demonstrated that SC-SOFC could generate a peak power of 0.78 W cm^{-2} at 790°C.

These results demonstrate that the area specific power density of a SC-SOFC can be comparable to those of a state-of-the-art pSOFCs and micro- tubular SOFCs. The electrolyte layer for the SC-SOFC can be porous and gas-tight electrolyte layer for gas separation is not necessary. In fact, electrochemical measurements indicate that the resistance of the porous YSZ layer is 2–3 times lower than that of a dense YSZ layer with the same macroscopic geometry.[71] Fleig et al.[72] proposed to make SC-SOFC on a chip with alternating anode and cathode stripes (comb-like) on the same side of an electrolyte sheet using the conventional semiconductor manufacture techniques. Their theoretical analyses indicate that the separation between the anode and cathode can be in the micrometer level with a minimum resistance between the electrodes.

Although SC-SOFC is a very attractive concept for make SOFC on chip, there are several technical barriers: (1) all the fuel passed on the cathode side is in principle exhausted from the SOFC; (2) a part of the fuel is subjected to deep oxidation due to the low selectivity of the anode and cathode; (3) the selectivity of the anode and cathode becomes lower as the fuel utilization increases; (4) there has been no report on development of a SC-SOFC based power generation system in literature.

4.4. INTERNAL REFORMING IN SOFCS

Application of internal reforming offers several advantages compared with external reforming:[12, 13]

1. The system volume, mass, and costs are reduced because the separate steam reformer unit is not needed.
2. With DIR less steam is required (the anode reaction produces steam).
3. There is a more evenly distributed load of hydrogen in a DIR cell, which may result in a more uniform temperature distribution.
4. There is a higher methane conversion.

There are however some disadvantages:

1. It may be necessary to incorporate catalyst, which may require modification of stack hardware.
2. Catalyst can become deactivated by carbon deposition particularly with heavy hydrocarbon fuels.
3. Conventional catalysts may be poisoned by impurities (e.g. sulphur compounds) in the feed fuel, or by sintering of the active metal and/or support.
4. Although the reforming and electrochemical reactions may be synergistic, integration of the two functions may reduce the flexibility of operation of the fuel cell.

The key requirements for IR catalysts are:

1. Stability under operating conditions (oxidizing and reducing).
2. High electronic conductivity.
3. Minimum chemical reactivity and diffusivity with other cell components.
4. Thermal expansion similar to the electrolyte.

5. Stable pore structure.
6. Stable reactivity for steam reforming and electrocatalysis.

Ni-YSZ anodes have high reforming activity but degrade at high steam/methane ration. Significant improvement in performance by using Ni supported on CeO_2-ZrO_2 or on CeO_2-YSZ mixture was reported. Noble metal containing catalysts such as Rh-YSZ, Ru/YSZ, and Pt/YSZ have high resistance to sintering and high reforming activity. A maximum power density of 4.55 W cm^{-2} was observed. It has been found that using mixed ionic and electronic conductors such as Sm doped Ceria (SDC) and TiO_2 doped YSZ have beneficial effects on internal reforming. The reason is that the hydrogen as the product of reforming can be oxidized in the entire surface of the catalysts not only at the "triple phase boundary" (TPB) sites.

Carbon deposition is a serious concern for the use of internal reforming.[73] The likelihood of carbon deposition is a function of the nature of catalysts and supports.[73-75] Anodes based on ceria have been found to effective in oxidation of CH_4 without significant carbon deposition.[12, 75] Doping the anode with Mo also can mitigate carbon deposition.[12, 13]

Gradual internal reforming of methane is a new concept of DIR[76, 77] This technology is to associate the reforming catalysts and electro-oxidation catalysts in the anode. The electro-oxidation of hydrogen produces water that sustains steam reforming of hydrocarbons. This method may enable a DIR that does not require steam.

IIR has also demonstrated effective in improving the performance of SOFCs on hydrocarbon fuels. Zhan and Barnett[78] used a partially stabilized zirconia (PSZ) porous layer to separate a Ni-YSZ anode layer and Ru-CeO_2 reforming catalyst layer. PSZ layer allow diffusion of the syngas produced from the reforming reaction to the anode, enables an efficient heat transfer between the reforming and anode, and effective separation of the anode from the oxygen. The maximum power density was 0.45 W cm^{-2} in 300 sccm of a 10.7% C_3H_8-18.7% O_2-70.6% Ar mixture at 750°C. The IIR in the SOFC allows SR, ATR, and POXR that does not require steam.

4.5. ELECTROLYTE MATERIALS FOR LOW TEMPERATURE SOFCS

Low temperature SOFC is commonly referred to SOFCs that operate at temperature below 650°C.[4] At the temperature range of 450–650°C, many metallic and ceramic materials can sustain for a long period of time. Reducing the operating temperature will result in less complexity in design and fabrication, enhanced flexibility in materials selection, enhanced reliability and durability, less insulation, higher thermal efficiency, and less stack volume and mass. The key of reducing the operating temperature is development of electrolytes that have sufficiently high oxygen ion conductivity at low temperatures. YSZ that has a fluorite structure is the most commonly used electrolyte material. YSZ has a sufficiently high conductivity at temperatures above 800°C (0.02 S cm^{-1}). YSZ is also chemically stable in both reducing and oxidizing condition, mechanically strong at elevated temperatures, and has a low thermal expansion rate. However, YSZ has a low conductivity at 450–650°C (10^{-5}–10^{-3} S cm^{-1}).

Doped ceria materials including Gd doped ceria (GDC), Y doped ceria (YDC), and Sm doped ceria (SDC) are promising electrolyte materials for low temperature SOFCs.[79-84] Among them, YDC has the highest conductivity of 0.0344 S cm^{-1} at 600°C but it is not mechanically strong. GDC is the most commonly used electrolyte materials for low temperature SOFCs. GDC has a little higher thermal expansion rate than YSZ. If it is used at low temperature regime the mismatching of the thermal expansions in the GDC based SOFCs is not significant. GDC is both electronic and oxygen ion conductors. In a reducing environment, the electronic conductivity may be greater than the oxygen ion conductivity, resulting in current leakage.

High oxygen ion conductivity can be achieved by substituting Lathanum with alkaline earth elements and/or incorporating divalent metal cations into gallium sublattice in order to increase the oxygen vacancy density.[79, 83, 84] The commonly used electrolyte material in this series is $La_{1-x}Sr_xGa_{1-y}Mg_yO_{3-\delta}$ (LSGM). The maximum conductivity is achieved at x = 0.10–0.20 and y = 0.15–0.20. Disadvantages of the material include possible reduction and volatization of gallium oxide, formation of stable secondary phases in the course of processing, the relatively high cost of gallium, and significant reactivity with perovskite electrodes under oxidizing conditions and with anode under reducing conditions.

$Bi_4V_2O_{11}$ (BIMEVOX) is particularly interesting due to its high ionic conductivity with respect to other types of electrolytes.[79, 84, 85] The conductivity of this series of materials can be 0.1 S cm^{-1} at 600°C. Disadvantages of the materials include thermodynamic instability in reducing environment, volatilization of bismuth oxide at moderate temperatures, a high corrosion activity, and low mechanical strength. The materials also have a high thermal expansion rate. Solution to the problem is the use of multiplayer cells having a layer of Bi_2O_3-based ionic conductor applied on another ionic conductor that also act as mechanical support and protection against reduction.

5. Summary

Without insulation the power densities of both the best fuel processor (SCT or microchannel reformer) and SOFC stack (pSOFC or micro-tubular SOFC) can be 1,000–2,000 W/l. Including the insulation, the power densities can be 380–760 W/l. If the BOP and packaging is 2.5 times of the insulated SOFC stack, the power density of the total system can be 84–168 W/l. The power density of a SOFC system is much less than that of an IC engine (ca. 250 W/l) even if the system is well optimized. If the 100% internal reforming is realized, the power density of the SOFC system can increase to 108–216 W/l. However, 100% of internal reforming is not realistic because the commercial fuels including natural gas, liquefied petroleum gas, gasoline, diesel, bio-diesel, and bio-ethanol, contains certain amount of impurities including sulfur and cyclic hydrocarbons.

Carbon deposition and sulfur poisoning will cause degradation of the catalysts and reduce the durability of the SOFC system to an unacceptable level. Thus, at least, a pre-cracking unit is necessary for light hydrocarbons. For heavy hydrocarbon fuels, an external reformer is required.

If 100% internal reforming is realized and the exhaust gases from the SOFC anode and cathode are recycled, the electric efficiency of the high temperature (1,000°C) SOFC system can reach 40%. With other operation schemes, the highest efficiency can only be about 35%. The primary loss of exergy is the loss of irreversibility of the processes but not the heat loss or fuel loss in flue gases, which can be effectively controlled by insulation and recycling schemes respectively. Reduction of the irreversibility require enhancement of mass transfer, heat transfer, and reaction kinetics via development highly effective catalysts.

Lowering the operating temperature of SOFC to the range from 450 to 650°C enables reduction of system size, heat loss, and manufacture costs and enhancement of power density and stack durability. The major disadvantage of the low operating temperature is that the heat generated and rejected from the SOFC stack can't be easily reused for fuel reforming because the operating temperature of existing effective catalytic reforming processes is 700–850°C. More fuel will have to be consumed for sustaining the fuel reforming process. The efficiency of the system will be reduced to below 30%. Nevertheless, lower temperature SOFC power sources are suitable for applications of the small-scale (1–500 W) power sources for portable electronic devices for which high electric efficiency is not the primary technical requirement.

There are seemly two approaches to enhance the efficiency of low temperature SOFC power sources to the level of 40%. The first is to identify catalysts for reformation of hydrocarbon fuels at lower temperature range from 450°C to 550°C. This allows efficient reuse of the rejected heat from low temperature SOFC stack for reformation of fuel. The second is to identify more active catalysts and electrolyte materials that have high conductivity in the low temperature range. In the past decades, the milestones of advancement of SOFC technology have been the inventions of new stack configurations including pSOFC, microtubular SOFC, and anode supported electrode assembly, and new fuel processing technologies including SCT and microchannel reformers. It is difficult to or not likely to have further breakthrough in these areas. The next "quantum jump" will be in the development of low temperature electrolytes and catalysts.

References

1. K. Kendall, N.Q. Minh, and S.C. Singhal, in: *High-Temperature Solid Oxide Fuel Cells-Fundamentals, Design and Applications*, edited by S. Singhal and K. Kendall (Elsevier Science, Oxford, England, 2003), pp. 197–229.
2. M.A. Khaleel and J.R. Selman, in: *High-Temperature Solid Oxide Fuel Cells-Fundamentals, Design and Applications*, edited by S. Singhal and K. Kendall (Elsevier Science, Oxford, England, 2003), pp. 293–332.
3. C. Song, Fuel processing for low-temperature and high-temperature fuel cells Challenges, and opportunities for sustainable development in the 21st century. Catalysis Today 77, 17–49 (2002).
4. EG&G Technical Services, Inc., Fuel Cell Handbook, Seventh Edition. U.S. Department of Energy, Office of Fossil Energy, National Energy Technology Laboratory, Morgantown, West Virginia, 2004.
5. M.C. Williams, J.P. Strakey, and SC. Singhal, U.S. distributed generation fuel cell program. Journal of Power Sources 131, 79–85 (2004).
6. Aidu Qi, Brant Peppley, and Kunal Karan, Integrated fuel processors for fuel cell application: A review. Fuel Processing Technology 88, 3–22 (2007).
7. Steven Shaffer, Update on Delphi's Development of a Solid Oxide Fuel Cell Power System, Honolulu, Hawaii, 2006 Fuel Cell Seminar, 2006.
8. Julian Dinsdale, Karl Föger, Raj Ratnaraj, Jonathan Love, and Alison Washusen Ceramic Fuel Cells Limited (CFCL), COMMERCIALISATION OF CFCL'S ALL-CERAMIC STACK TECHNOLOGY Paper presented at the Fuel Cell Seminar, Miami, November 2003.
9. R.J. Braun, S.A. Klein, and D.T. Reindl, Evaluation of system configurations for solid oxide fuel cell-based micro-combined heat and power generators in residential applications. Journal of Power Sources 158, 1290–1305 (2006).
10. Tsang-Dong Chung, Wen-Tang Hong, Yau-Pin Chyou, Dong-Di Yu, Kin-Fu Lin, and Chien-Hsiung Lee, Efficiency analyses of solid oxide fuel cell power plant systems. Applied Thermal Engineering. Publication proof, 2007.
11. Daniel G. Löffler, Kyle Taylor, and Dylan Mason, A light hydrocarbon fuel processor producing high-purity hydrogen. Journal of Power Sources 117, 84–91 (2003).
12. A.L. Clerk, Advances in catalysts for internal reforming in high temperature fuel cells. Journal of Power Sources 71, 111–121 (1998).

13. Stephen H. Clarke, Andrew L. Dicks, Kevin Pointon, Thomas A. Smith, and Angie Swann, Catalytic aspects of the steam reforming of hydrocarbons in internal reforming fuel cells. Catalysis Today 38, 411–423 (1997).
14. P. Aguiar, C.S. Adjiman, and N.P. Brandon, Anode-supported intermediate-temperature direct internal reforming solid oxide fuel cell II. Model-based dynamic performance and control. Journal of Power Sources 147, 136–147 (2005).
15. P. Aguiar, D. Chadwick, and L. Kershenbaum, Modelling of an indirect internal reforming solid oxide fuel cell. Chemical Engineering Science 57, 1665–1677 (2002).
16. J.R. Rostrup-Nielsen, Production of synthesis gas. Catalysis Today 18, 305–324 (1993).
17. Jens R. Rostrup-Nielsen, New aspects of syngas production and use. Catalysis Today 63, 159–164 (2000).
18. S. Ahmed and M. Krumpelt, Hydrogen from hydrocarbon fuels for fuel cells. International Journal of Hydrogen Energy 26, 291–301 (2001).
19. Bettina Lenz, and Thomas Aicher, Catalytic autothermal reforming of Jet fuel. Journal of Power Sources. Journal of Power Sources 149, 44–52 (2005).
20. Praveen K. Cheekatamarla, and C.M. Finnerty, Reforming catalysts for hydrogen generation in fuel cell applications. Journal of Power Sources 160, 490–499 (2006).
21. Meng Ni, Dennis Y.C. Leung, and Michael K.H. Leung, A review on reforming bio-ethanol for hydrogen production. International Journal of Hydrogen Energy, publication proof, 2007.
22. Praveen K. Cheekatamarla, and Alan M. Lane, Catalytic autothermal reforming of diesel fuel for hydrogen generation in fuel cells I. Activity tests and sulfur poisoning. Journal of Power Sources 152, 256–263 (2005).
23. S. Krummricha, B. Tuinstra, G. Kraaij, J. Roes d, and H. Olgun, Diesel fuel processing for fuel cells – DESIRE. Journal of Power Sources 160, 500–504 (2006).
24. A. Docter and A. Lamm, Gasoline fuel cell systems. Journal of Power Sources 84, 194–200 (1999).
25. Inyong Kang, Joongmyeon Bae, and Gyujong Bae, Performance comparison of autothermal reforming for liquid hydrocarbons, gasoline and diesel for fuel cell applications. Journal of Power Sources 163, 538–546 (2006).
26. Jens R. Rostrup-Nielsen, Industrial relevance of coking. Catalysis Today 37, 225–232 (1997).
27. Xiaoliang Ma, Subramani Velu, Jae Hyung Kim, Chunshan Song, Deep desul-furization of gasoline by selective adsorption over solid adsorbents and impact of analytical methods on ppm-level sulfur quantification for fuel cell applications. Applied Catalysis B: Environmental 56, 137–147 (2005).
28. C. Song, An overview of new approaches to deep desulfurization for ultra-clean gasoline, diesel fuel and jet fuel. Catalysis Today 86, 211–263 (2003).
29. Yixin Lu and Laura Schaefer, A solid oxide fuel cell system fed with hydrogen sulfide and natural gas. Journal of Power Sources 135, 184–191 (2004).
30. I.V. Babich and J.A. Moulijn, Science and technology of novel processes for deep desulfurization of oil refinery streams: a review. Fuel 82, 607–631 (2003).
31. Sylvette Brunet, Damien Mey, Guy Pe´rot, Christophe Bouchy, and Fabrice Diehl, On the hydrodesulfurization of FCC gasoline: a review. Applied Catalysis A: General 278, 143–172 (2005).
32. G. Murali Dhar, B.N. Srinivas, M.S. Rana, Manoj Kumar, and S.K. Maity, Mixed oxide supported hydrodesulfurization catalysts – a review. Catalysis Today 86, 45–60 (2003).

33. Shaowu Zha, Philip Tsang, Zhe Cheng, and Meilin Liu, Electrical properties and sulfur tolerance of La$_{0.75}$Sr$_{0.25}$Cr$_{1-x}$Mn$_x$O$_3$. Journal of Solid State Chemistry 178, 1844–1850 (2005).
34. Zhe Cheng, Shaowu Zha, Luis Aguilar, and Meilin Liu, Chemical, electrical, and thermal properties of strontium doped lanthanum vanadate. Solid State Ionics 176, 1921–1928 (2005).
35. Luis Aguilar, Shaowu Zha, Zhe Cheng, Jack Winnick, Meilin Liu, A solid oxide fuel cell operating on hydrogen sulfide (H$_2$S) and sulfur-containing fuels. Journal of Power Sources 135, 17–24 (2004).
36. R.R. Lesieur, T.J. Corrigan, US6203587B1, Mar. 20, 2001.
37. S. Ahmed, S.H.D. Lee, J.D. Carter, and M. Drumpelt, US6713040B2, Mar. 30 2004.
38. J.M. Zalc and D.G. Löffler, Fuel processing for PEM fuel cells: transport and kinetic issues of system design. Journal of Power Sources 111, 58–64 (2002).
39. L. Bobrova, I. Zolotarsky, V. Sadykov, and V. Sobyanin, Hydrogen-rich gas production from gasoline in a short contact time catalytic reactor. International Journal of Hydrogen Energy, publication proof, 2006.
40. D.A. Hickman and L.D. Schmidt, Synthesis gas formation by direct oxidation of methane over pt monoliths. Journal of Catalysis 138, 267–282 (1992).
41. S.S. Bharadwaj and L.D. Schmidt, Catalytic partial oxidation of natural gas to syngas. Fuel Processing Technology 42, 109–127 (1995).
42. Jakob J. Krummenacher, Kevin N. West, and Lanny D. Schmidt, Catalytic partial oxidation of higher hydrocarbons at millisecond contact times: decane, hexadecane, and diesel fuel. Journal of Catalysis 215, 332–343 (2003).
43. Jakob J. Krummenacher and Lanny D. Schmidt, High yields of olefins and hydrogen from decane in short contact time reactors: rhodium versus platinum. Journal of Catalysis 222, 429–438 (2004).
44. Marco Castaldi, Maxim Lyubovsky, Rene LaPierre, and William C. Pfefferle and Subir Roychoudhury, Performance of Microlith Based Catalytic Reactors for an Isooctane Reforming System. Precision Combustion, Inc., SAE International, 2003-01-1366.
45. Subir Roychoudhury, Marco Castaldi, Maxim Lyubovsky, Rene LaPierre, and Shabbir Ahmed, Microlith catalytic reactors for reforming iso-octane-based fuels into hydrogen. Journal of Power Sources 152, 75–86 (2005).
46. M.V. Kothare, Dynamics and control of integrated microchemical systems with application to micro-scale fuel processing. Computers and Chemical Engineering 30, 1725–1734 (2006).
47. By Robert S. Wegeng, Larry R. Pederson, Ward E. TeGrotenhuis, and Greg A. Whyatt, Compact fuel processors for fuel cell powered automobiles based on microchannel technology. Fuel Cell Bulletin 28, 2001.
48. A.Y. Tonkovich, J.L. Zilka, M.J. LaMont, Y. Wang, and R.S. Wegeng, Microchannel reactors for fuel processing applications. I. Water gas shift reactor. Chemical Engineering Science 54, 2947–2951 (1999).
49. Jamelyn D. Holladay, Evan O. Jones, Max Phelps, and Jianli Hu, Microfuel processor for use in a miniature power supply. Journal of Power Sources 108, 21–27 (2002).
50. J.D. Holladay, J.S. Wainright, E.O. Jones, and S.R. Gano. Power generation using a mesoscale fuel cell integrated with a microscale fuel processor. Journal of Power Sources 130, 111–118 (2004).
51. J.D. Holladay, E.O. Jones, R.A. Dagle, G.G. Xia, C. Cao, and Y. Wang, High efficiency and low carbon monoxide micro-scale methanol processors. Journal of Power Sources 131, 69–72 (2004).

52. Shin-Kun Ryi, Jong-Soo Park, Seung-Hoon Choi, Sung-Ho Cho, and Sung-Hyun Kim, Novel micro fuel processor for PEMFCs with heat generation by catalytic combustion. Chemical Engineering Journal 113, 47–53 (2005).
53. Yeena Shin, Okyoun Kim, Jong-Chul Hong, Jeong-Hoon Oh, Woo-Jae Kim, Seungjoo Haam, and Chan-Hwa Chung, The development of micro-fuel processor using low temperature co-fired ceramic (LTCC). International Journal of Hydrogen Energy 31, 1925–1933 (2006).
54. E.R. Delsman, M.H.J.M. De Croon, A. Pierik, G.J. Kramer, P.D. Cobden, Ch. Hofmann, V. Cominos, and J.C. Schouten, Design and operation of a preferential oxidation microdevice for a portable fuel processor. Chemical Engineering Science 59, 4795–4802 (2004).
55. Shuji Tanaka, Kuei-Sung Chang, Kyong-Bok Min, Daisuke Satoh, Kazushi Yoshida, and Masayoshi Esashi, MEMS-based components of a miniature fuel cell/fuel reformer system. Chemical Engineering Journal 101, 143–149 (2004).
56. Taegyu Kim and Sejin Kwon, Catalyst preparation for fabrication of a MEMS fuel reformer. Chemical Engineering Journal 123, 93–102 (2006).
57. Arunabha Kundu, Jae Hyuk Jang, Hong Ryul Lee, Sung-Han Kim, Jae Hyoung Gil, Chang Ryul Jung, Yong Soo Oh, MEMS-based micro-fuel processor for application in a cell phone. Journal of Power Sources 162, 572–578 (2006).
58. Jae Hyoung Gil, Chang Ryul Jung, and Yong Soo Oh, MEMS-based micro-fuel processor for application in a cell phone. Journal of Power Sources 162, 572–578 (2006).
59. Oh Joong Kwon, Sun-Mi Hwang, Je Hyun Chae, Moo Seong Kang, Jae Jeong Kim, Performance of a miniaturized silicon reformer-PrOx-fuel cell system. Journal of Power Sources 165, 342–346 (2007).
60. T.-L. Wen, D. Wang, M. Chen, H. Tu, Z. Lu, Z. Zhang, H. Nie, and W. Huang, Material research for planar SOFC stack. Solid State Ionics 148, 513–519 (2002).
61. H.Y. Jung, S.-H. Choi, H. Kim, J.-W. Son, J. Kim, H.-W. Lee, and J.-H. Lee, Fabrication and performance evaluation of 3-cell SOFC stack based on planar 10 cm × 10 cm anode-supported cells. Journal of Power Sources 159, 478–483 (2006).
62. Zhenrong Wang, Jiqin Qian, Jiadi Cao, Shaorong Wan, and Tinglian Wen, A study of multilayer tape casting method for anode-supported planar type solid oxide fuel cells (SOFCs). Journal of Alloys and Compounds, publication proof, 2006.
63. Sonja M. Gross, Thomas Koppitz, Josef Remmel, Jean-Bernard Bouche, and Uwe Reisgen, Joining properties of a composite glass-ceramic sealant. Fuel Cell Bulletin, 12–16 (2006).
64. N.M. Sammes, and Y. Du, R. Bove, Design and fabrication of a 100W anode supported micro-tubular SOFC stack. Journal of Power Sources 145, 428–434 (2005).
65. T. Suzuki, Y. Funahashi, T. Yamaguchi, Y. Fujishiro, and M. Awano, Fabrication and characterization of micro tubular SOFCs for advanced ceramic reactors. Journal of Alloys and Compounds, publication proof, 2007.
66. Toshio Suzuki, Toshiaki Yamaguchi, Yoshinobu Fujishiro, and Masanobu Awano Current collecting efficiency of micro tubular SOFCs. Journal of Power Sources 163, 737–742 (2007).
67. Toshio Suzuki, Yoshihiro Funahashi, Toshiaki Yamaguchi, Yoshinobu Fujishiro, and Masanobu Awano, Anode-supported micro tubular SOFCs for advanced ceramic reactor system. Journal of Power Sources, publication proof, 2007.

68. Yoshihiro Funahashi, Toru Shimamori, and Toshio Suzuki, Fabrication and characterization of components for cube shaped micro tubular SOFC bundle. Journal of Power Sources 163, 731–736 (2007)
69. Masaya Yano, Atsuko Tomita, Mitsuru Sano, and Takashi Hibino, Recent advances in single-chamber solid oxide fuel cells: a review. Solid State Ionics 177, 3351–3359 (2007).
70. Zongping Shao 1, Jennifer Mederos, William C. Chueh, and Sossina.M. Haile, High power-density single-chamber fuel cells operated on methane. Journal of Power Sources 162, 589–596 (2006).
71. Toshio Suzuki, Piotr Jasinski, Vladimir Petrovsky, Harlan U. Anderson, and Fatih Dogan, Performance of a porous electrolyte in single-chamber SOFCs. Journal of The Electrochemical Society, 152(3) A527–A531 (2005).
72. J. Fleig, H.L. Tuller, and J. Maier, Electrodes and electrolytes in micro-SOFCs: a discussion of geometrical constraints. Solid State Ionics 174, 261–270 (2004).
73. Tatsuya Iida, Mitsunobu Kawano, Toshiaki Matsui, Ryuji Kikuchi, and Koichi Eguchia, Internal reforming of SOFCs: carbon deposition on fuel electrode and subsequent deterioration of cell. Journal of the Electrochemical Society, 154 (2), B234–B241 (2007).
74. S. Assabumrungrat, N. Laosiripojana, and P. Piroonlerkgul, Determination of the boundary of carbon formation for dry reforming of methane in a solid oxide fuel cell. Journal of Power Sources 159, 1274–1282 (2006).
75. N. Laosiripojanaa and S. Assabumrungrat, The effect of specific surface area on the activity of nano-scale ceria catalysts for methanol decomposition with and without steam at SOFC operating temperatures. Chemical Engineering Science 61, 2540 – 2549 (2006).
76. Samuel Georges, Gaelle Parrour, Marc Henault, and Jacques Fouletier, Gradual internal reforming of methane: a demonstration. Solid State Ionics 177, 2109–2112 (2006).
77. J.-M. Kleina,Y. Bultel, S. Georges, and M. Pons, Modeling of a SOFC fuelled by methane: from direct internal reforming to gradual internal reforming. Chemical Engineering Science 62, 1636–1649 (2007).
78. Zhongliang Zhan and Scott A. Barnett, Use of a catalyst layer for propane partial oxidation in solid oxide fuel cells. Solid State Ionics 176, 871–879 (2005).
79. Stephen J. Skinner and John A. Kilner, Oxygen ion conductors. Materials Today, 30–37 (2003).
80. W.Z. Zhu and S.C. Deevi, A review on the status of anode materials for solid oxide fuel cells, Materials Science and Engineering A362, 228–239 (2003).
81. Da Yu Wang, D.S. Park, J. Griffith, and A.S. Nowick, Oxygen ion conductivity and defect interactions in yttria-doped ceria. Solid State Ionics 2, 95–105 (1981).
82. N.M. Sarnrnes, and Zhihong Cai, Ionic conductivity of ceria/yttria stabilized zirconia electrolyte materials. Solid State Ionics 100, 39–44 (1997).
83. Jeffrey W. Fergus, Electrolytes for solid oxide fuel cells. Journal of Power Sources 162, 30–40 (2006).
84. V.V. Kharton, F.M.B. Marques, and A. Atkinson, Transport properties of solid oxide electrolyte ceramics: a brief review. Solid State Ionics 174, 135– 149 (2004).
85. N.M. Sammes, G.A. Tompsett, H. NaÈfe, and F. Aldinger, Bismuth based oxide electrolytes: structure and ionic conductivity. Journal of the European Ceramic Society 19, 1801–1826 (1999).

PLANAR SOLID OXIDE FUEL CELLS: FROM MATERIALS TO SYSTEMS

YASUNOBU MIZUTANI
Fundamental Research Department, Technical Research Institute,
Toho Gas Co., Ltd., Tokai-city, Aichi, JAPAN

1. Introduction

In this article, recent technology updates regarding planar type solid oxide fuel cells (SOFCs) are reviewed. Intermediate-temperature operation of SOFCs is a recent R&D trend because of such advantages as cost effectiveness due to the use of low-cost metallic parts and thermal cycle ability. Materials are related to cell performance and operating temperatures, and stack designs are related to hot modules and system design. Also, "reforming" is an important issue for the achievement of high-efficiency SOFC systems.

Based on Toho's wide range of experience from materials to 1 kW systems for small-scale stationary applications, a case example is suggested for developing mini-or micro-fuel cells. Topics of discussion include reducing the operating temperature of an electrolyte-supported SOFC 1,000°C to 800°C with a newly developed ScSZ electrolyte; the development of compact SOFC stacks with metallic interconnect plates; the examination of a catalytic partial oxidation (CPOx) reformer and steam reforming fuel processor for pipeline city gas fuel; and important issues for the realization of high-efficiency and cost-effective SOFC systems.

2. Market Strategy and Requirements

The marketing of co-generation systems, i.e. CHP (Combined Heat and Power) systems, has been growing over the past 20 years. In Japan, 11,610 CHP systems with a total output of 8,786 MW are stock capacity in 2007.[1] In particular, small-scale stationary CHP systems are showing remarkable growth, and systems with electric power outputs around 1 kW are attractive application for residential use. CHP systems driven by an IC (Internal combustion) type gas engine have already been commercialized, and 20,000 units were installed in FY2006. Also, PEFC (Polymer Electrolyte Type

_____
E-mail: master@tohogas.co.jp

Fuel Cell) systems have been developed for residential use, and over 1,000 units/year are currently being field-tested.[2] The requirements for stationary-application SOFCs, in contrast with those of CHP systems, include high electric generating efficiency, long-term reliability and durability, compactness, and lower costs. Moreover, their operating patterns should be selected to obtain maximum energy economy based on their primary characteristics and their electricity/heat ratios, as shown in Table 1.

Table 1. Characteristics of generators and their operating procedures for residential CHP.

Type of generator	Efficiency (LHV)		Operating procedure	
	Electricity (%)	Heat (%)	Operating style	Operating pattern
IC Engine	20	60	Daily start and stop	Full load
PEFC CHP	35	45	Daily start and stop	Full and partial load
SOFC CHP	45	30	Non-stop	Load compliance

3. Materials and Cells

It is well known that the types of SOFC cells are categorized as "tubular" and "planar" according to their shape and structure. Planar SOFCs have the advantage of high (volumetric) power densities and lower fabrication cost, but gas sealing is a technical hurdle. Also, thick dense electrolyte-supported cells and thin electrolyte porous electrode-supported cells have been proposed.[3-5] Thin electrolyte cells have the advantage of reduced operating temperature, but ensuring their mechanical reliability is a technical hurdle because of porous ceramics structures and RedOx (Reduced and Oxidation Atmosphere) stability.

Electrolyte-supported (thick electrolyte) cells are conventional but have features such as mechanical robustness, low-cost manufacturing process, good stability for reduced and oxidized (RedOx) cycles, and gas-diffusible thin electrodes. To reduce operating temperature from 1,000°C to around 800°C, an alternative electrolyte made of scandium-doped zirconia (ScSZ), which has a higher ionic conductivity than yttrium-doped zirconia (YSZ), was applied in Toho's SOFC (Figure 2, Table 2).[6-8] The configuration and mechanical strength of cells should be considered from the stack design. In general, mechanical reliability and electrochemical performance have a trade-off relationship. Electrode materials should be selected from the viewpoint of initial performance, durability, chemical stability, productivity and material costs. Multi-layered electrodes are one of the solution, and manufacturing process is important for the realization of mass-produced, cost-effective cells. Tape casting, screen printing and firing (sintering) processes are applicable and well established.

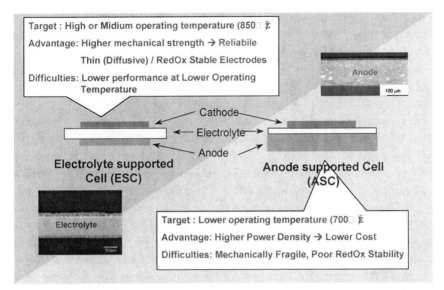

Figure 1. Structure of two types of planar SOFC cells in cross section. Electrolyte supported cell (ESC) and anode supported cell (ASC).

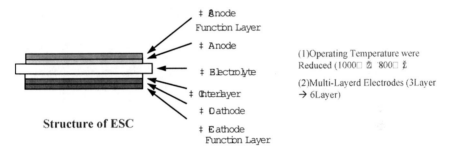

Figure 2. Multi-layered structure of electrolyte supported cell developed by Toho Gas.

TABLE 2. History of materials used for electrolyte-supported cell developed by Toho Gas.

Generation	1G	2G	3G
Temperature	1,000°C	950°C	800°C
(1) Anode FL	–	Ni	Ni-composite
(2) Anode	Ni-YSZ	Ni-ScSZ	Ni(Ti)-ScSZ
(3) Electrolyte	11ScSZ	4ScSZ	4ScSZ
(4) Interlayer	–	–	Doped CeO_2
(5) Cathode	La(Sr)MnO_3	La(Sr)MnO_3	La(Sr)Co(Fe)O_3
(6) Cathode FL	–	La(Sr)CoO_3	LSC-composite

Durability of cells (materials) is quite a complex subject because of high-temperature operation and varying atmospheric conditions. Chemical stability with neighbor materials, sintering at high temperatures, reduced/oxidation cycles, and electrochemical mass transport, etc. are considerable issues. Also, there are many degradation phenomena related to thermodynamics, and describes in phase diagrams.[9]

Cell reliability is always a weak point of SOFCs because they are composed of fragile and brittle ceramics. Stress applied to cells during SOFC operation, mechanical testing of SOFC cells, and cell design (material, configuration, shape) are considerable issues. For example, applied stresses in planar SOFC cells are described in Table 3. As for the mechanical testing of thin ceramic sheets, the piston-on-ring test is applicable to the measurement of cells under actual production. Fracture strength, fracture toughness, failure probability and thermal/mechanical properties during actual operation (from RT to high temperatures, and atmosphere) should be considered. In order to understand cell fracture, the questions "when does the fracture happen?" and "where does it originate?" need to be answered.

TABLE 3. Applied stresses in planar SOFC cells during fabrication and system operation.

Mechanical stresses	Warped or waved cells (must be made flat for stacking) → Bending stress
	Inclusions and foreign (large) particles → Bending stress and stress concentration (large stress)
Simple thermal stresses	Stable operation (thermal distribution) → stable thermal stress
	Start-up and shut-down → non-stable thermal stress
	Endothermic reaction (reforming) → anode intake will be quenched
Thermal stress caused by TEC mismatch	Mismatched in thermal expansion coefficients (TEC)
	Between each materials in cell (PEN) → internal stress
	Between each parts (cell and interconnector) → tensile stress
Other considerable issues	High temperature strength, fatigue fracture, failure probability, size effect, biaxial stress, proof testing, etc.

TEC (Thermal Expansion Coefficient) mismatch often causes large internal (residual) stress, as shown in Figure 3. In this case, the internal stress was reduced by increasing the temperature, and the stress was reversible. The cell then became stress-free and safer in the SOFC (high-temperature) operating conditions. Non-destructive stress measurement is an effective technique for measuring internal stress in SOFC cells.[10]

RedOx (Reduced and Oxidized Atmosphere) cycles also present severe stress conditions for an ASC (Anode-Supported Cell), because of Ni–NiO changes cause a large amount of shrinkage/expansion, as shown in Figure 4.

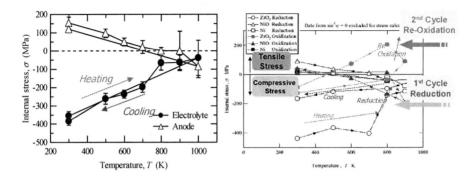

Figure 3. History of internal stress in case of thermal cycling.

Figure 4. History of internal stress in case of thermal and Red-Ox cycling.

In this case, irreversible stress was occurred in the re-oxidation process, and fatal tensile stress was applied to the ZrO_2 electrolyte.

4. Stacks

There are several difficulties in "stacking" with ceramic cells and metallic interconnect plates, including mismatched of thermal expansion coefficients, gas sealing, lossless electrical connections, and proof of thermal and mechanical stresses, etc. Steady and non-steady thermal stresses are applied to the cells while the SOFC system is in operation. Thermal stress can be estimated from temperature distributions, and it is determined in stack designs as the gas flow configuration. Stress concentration should be avoided in the cell design. In a planar type stack, the material cost and machining cost of the metallic interconnect plates is the most significant portion of the total cost. However, interconnect plate accuracy is strongly needed to obtain a uniform flow of process gas across a single cell plane and between cells respectively. Non-uniform gas flow causes partial lack of fuel gas, lower fuel utilization, lower efficiency, and degradation of cells. Ferritic stainless steels are applicable to interconnect plates, and several candidate materials have been developed, such as Crofer 22APU or ZMG232; these, however, require a protective coating to

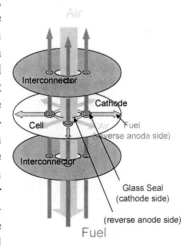

Figure 5. Design and configuration of stack with planar cell developed by Toho Gas.

ensure durability. The use of perovskites and the spinels for materials and flame spray coating, slurry coating, and electro-plating technique for the cathode-side coating has been reported. Requirements, related issues, and Toho's solutions for a planar SOFC stack are described in Table 4.[11]

TABLE 4. Requirements, related issues, and solutions for planar SOFC stack as a case example.

Requirements	Relating issues	Solutions for our stack
(1) Higher electrical efficiency	Tight gas seal	Less sealing area (only around the center manifolds) and small pressure difference between anode and cathode manifolds
	Current collecting	Fine gas flow pattern by etching process
	Higher fuel utilization	Uniform gas distributions
(2) Long lifetime	Thermal cycle stability	Unrestricted cells with interconnect plates
	Degradation of cells	Avoid Cr poisoning by protective coatings
	Oxidation of metallic interconnect plates	Selected special metals (Crofer 22 APU, ZMG232)
(3) Lower cost	High power density	Sc-TZP cells, Reduce stacking loss
	Lower cost materials	Ferritic stainless steel interconnect plates
	Lower cost parts	Inexpensive machining process
	Easy to stacking	Less number of parts

5. System Design

A considerable issue for the design of an SOFC system is "reforming" of the fuel. In the case of a polymer electrolyte fuel cell (PEFC) system, complex multi-stage reforming systems are required because of the limited photonic conductivity of the electrolyte and platinum base electrodes, and they occupy a large portion of the cost of a BOP (Balance of Plant) system. On the other hand, many types of reforming options, i.e. steam reforming, partial oxidation, autothermal reforming, and dry reforming, are applicable to SOFC systems (Figure 6, Table 5), because the carbon monoxide can be used as the fuel for the SOFC.

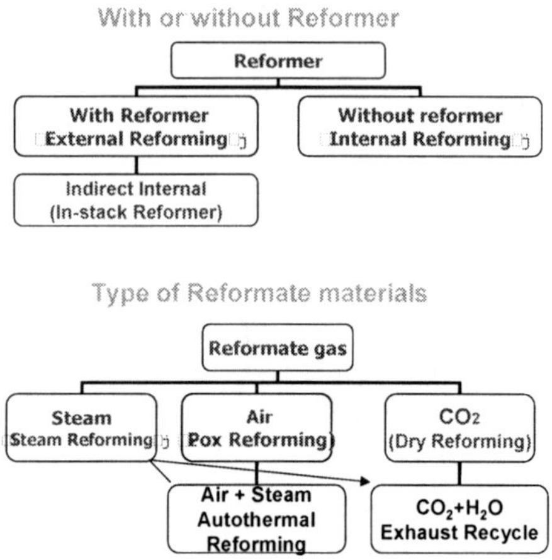

Figure 6. Type of reforming system in SOFC.

TABLE 5. Type of reforming system and their advantage/disadvantage.

Type	Catalytic Partial Oxidation	External Steam Reforming	Internal Reforming with exhaust gas recycling	Direct Methane Generation (Without Reforming)
System	Fuel → Reformer → SOFC; Air → Reformer	Fuel → Reformer → SOFC; Water → Steamer/Water Cleaner → Reformer	Fuel → Mixer → SOFC; Gas Recycling ← Exhaust Gas	Fuel → SOFC
Reaction	$CH_4 + 1/2 O_2$ → $CO + 2H_2$ (exothermic)	$CH_4 + H_2O$ → $CO + 3H_2$; $CO + H_2O$ → $CO_2 + H_2$ (endothermic)	$CH_4 + H_2O$ → $CO + 3H_2$; $CH_4 + CO_2$ → $2CO + 2H_2$; $CO + H_2O$ → $CO_2 + H_2$	$CH_4 + 4O^{2-}$ → $CO_2 + 2H_2O + 8e^-$
BOP	■Catalyst reformer	■Water cleaner ■Steamer ■Catalyst reformer	■Exhaust gas recycling instrument	■None
Efficiency	◇	○	○	○○
Point	It realizes simple systems	A lot of BOPs complicate systems	Recycling devices Startup Shutdown procedure	It needs durability against carbon deposition

Figure 7 shows a simplified SOFC system with a compact CPOx reformer for the fuel reforming of natural gas. Advantages and disadvantages of the CPOx reforming system should be discussed. The CPOx reforming systems are simple, cost effective, and easy to start, and supplying them with reforming aids (air) is easy. CPOx reforming consumes a part of the fuel energy (exothermic reaction), and high efficiency in an SOFC system is only achieved when the excess exhaust heat energy of the SOFC is utilized for endothermic reforming. A desulfurizer is required to avoid fuel poisoning of the reforming catalyst and Ni anode.

Of course, the reforming efficiency of a steam reforming system is higher than that of a partial oxidation reforming system. Thermal balance and mass balance are important for controlling SOFC systems to obtain high efficiency and long life, and the A/F (Air Fuel Ratio) in CPOx reforming and S/C (Steam Carbon Ratio) should be considered in the equilibrium calculation to avoid the carbon deposition. Moreover, Ru or Ni catalysts are often used for steam reforming; however, carbon deposition tendencies are changed by the type of catalyst material. Also, steam-supplying methods are important because of their BOP costs and durability (impurities) issues.

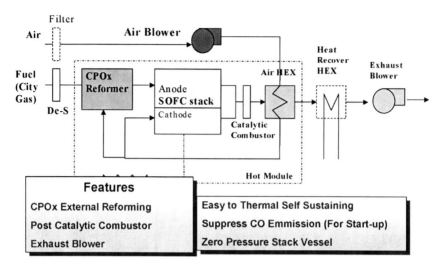

Figure 7. Flow diagram of simplified SOFC system. Toho's first 1 kW SOFC system.

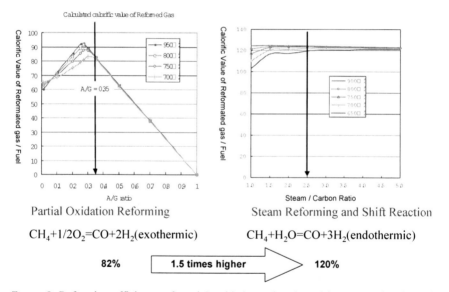

Figure 8. Reforming efficiency of partial oxidation reforming and stream reforming with various air/fuel ratios and steam/carbon ratios.

When an SOFC stack is thermally combined with a steam reformer, or when internal reforming is applied, an extremely high electric conversion efficiency is easily obtained.[12] Over 55% efficiency was obtained in Toho's 1 kW SOFC modules (Figure 9).

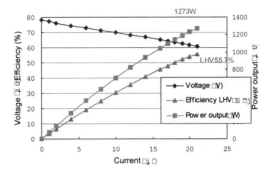

Figure 9. Output power and efficiency of steam reforming SOFC module.

6. Future Perspectives

High electrical efficiency of SOFC systems has already been demonstrated, but the path toward a low-cost, reliable system has not yet been clarified. From the viewpoint of a simple stack and controllable system, mechanically robust and chemically stable cells are required. From the viewpoint of cost-effective cells, on the other hand, a mechanically soft type of stack and perfect process controls in BOP are required for the system. Discovering new materials and making steady progress in developing cells and stacks could solve these antinomic subjects in the future.

Acknowlegements

The authors would like to acknowledge and thank colleague in Toho Gas and Professor S. Kakac, University of Miami.

References

1. Japan Cogeneration Center, JAPAN, http://www.cgc-japan.com/english/e_top.html
2. New Energy Foundation, JAPAN, http://www.NEF.or.jp
3. L.G.J de Haart, K. Mayer, U. Stimming, and U. Vinke, Operation of anode-supported thin electrolyte film solid oxide fuel cells at 800°C and below, *Journal of Power Sources*, 71, 302–305(1998).
4. Y. Mizutani, M. Kawai, K. Nomura, and Y. Nakamura, Characteristics of substrate type SOFC using Sc-doped zirconia electrolytes, *Solid Oxide Fuel Cells VI*, S.C. Singhal and M. Dokiya eds., Electrochemical Society PV99-19, Pennington, USA (1999) pp. 185–192.

5. Y. Baba, K. Ogasawara, H. Yakabe, Y. Matsuzaki, and T. Sakurai, Development of anode-supported SOFC with metallic interconnectors, *Solid Oxide Fuel Cells VIII*, S.C. Singhal and M. Dokiya (eds.), Electrochemical Society PV2003-07, Pennington, USA (2003) pp. 119–126.
6. Y. Mizutani, M. Tamura, M. Kawai, and O. Yamamoto, "Development of high-performance electrolyte in SOFC," *Solid State Ionics*, 72 (1994) pp. 271–275.
7. Y. Mizutani, M. Tamura, M. Kawai, K. Nomura, Y. Nakamura, and O. Yamamoto, "Characterization of the Sc_2O_3–ZrO_2 system and its application as the electrolyte in planar SOFC", *Solid Oxide Fuel Cells IV*, M. Dokiya, O. Yamamoto, H. Tagawa, and S.C. Singhal (eds.), Electrochemical Society PV95-1, Pennington, USA (1995) pp. 301–309.
8. Y. Mizutani, K. Hisada, K. Ukai, H. Sumi, M. Yokoyama, Y. Nakamura, and O. Yamamoto, From rare earth doped zirconia to 1 kW solid oxide fuel cell system, *Journal of Alloys and Compounds*, 408–412, 518–524 (2006).
9. H. Yokokawa, T. Horita, N. Sakai, M. Dokiya, and T. Kawada, Thermodynamic representation of nonstoichiometric lanthanum manganite, *Solid State Ionics*, 86–88, 1161–65 (1996).
10. H. Sumi, K. Ukai, M. Yokoyama, Y. Mizutani, Y. Doi, S. Machiya, Y. Akiniwa, and K. Tanaka, Changes of internal stress in solid-oxide fuel cell during Red-Ox cycle evaluated by in situ measurement with synchrotron radiation, *Journal of Fuel Cell Science and Technology*, 3, 68 (2006).
11. Y. Mizutani, K. Hisada, K. Ukai, M. Yokoyama, and H. Sumi, "Experimences with the first Japanese-made solid oxide fuel cell system", *Journal of Fuel Science and Technology*, 2, 179–185 (2005).
12. J. Shimano, H. Yamazaki, Y. Mizutani, K. Hisada, K. Ukai, M. Yokoyama, K. Nagai, S. Kashima, H. Orishima, S. Nakatsuka, H. Uwani, and M. Hirakawa, Development status of a planar type of 1 kW class SOFC system. *Solid Oxide Fuel Cells 10 (SOFC-X)*, K. Eguchi, S.C. Singhal, H. Yokokawa, and J. Mizusaki (eds.), Electrochemical Society transactions Vol. 7. No. 1 (2007) pp. 141–148.

MATHEMATICAL ANALYSIS OF PLANAR SOLID OXIDE FUEL CELLS

A. PRAMUANJAROENKIJ[1*], S. KAKAÇ[2], AND X. Y. ZHOU[1]
[1]*Department of Mechanical and Aerospace Engineering*
University of Miami, Coral Gables, FL 33124
[2]*TOBB University of Economics and Technology*
Sogutozu, Ankara, Turkey

1. Introduction

Fuel cells are electrochemical devices that convert the chemical energy stored in a fuel directly into electrical power. The main attractive features of fuel cell systems are quiet operation and low emissions. However, some issues in reducing manufacturing and material costs, improving stack performances, their reliabilities and lifetimes, must be solved to enable commercialization. Solid Oxide Fuel Cell (SOFC) is considered as one of the most promising energy conversion device and as an alternative of existing power generation systems. SOFCs operate at high temperatures from 600°C to 1,000°C to ensure sufficient ion conductivity through their electrolytes which are nonconductive to electrons. Main SOFC components include air channel, cathode, electrolyte, anode, fuel channel, and interconnects.

Tubular, planar, and monolithic structures are primary SOFC structures. Until recent years the integrated planar solid oxide fuel cell (IP-SOFC) developed by Rolls-Royce has gained a lot of attention. Tubular and planar SOFCs are generally manufactured in different designs as electrolyte-supported, anode-supported, cathode-supported, and metal-supported SOFCs. The planar configuration has received great attentions continuously because it is naturally more compact than the tubular configuration and it provides higher volume specific power (W/cm^3), however, sealing around cell edges and controlling temperature gradients are among the major issues to be resolved.[1] Gas flow configurations in planar solid oxide fuel cells can be

---

*Department of Mechanical Engineering, Faculty of Science and Engineering, Kasetsart University, Chalermprakiat Sakonnakhon Province Campus, 59 Moo.1, Chiang Krer, Amphur Maung, Sakonnakhon 47000 Thailand, e-mail: anchasa@gmail.com

arranged to be cross-flow, or co-flow, or counter-flow as basic designs (Figure 1). The flow configuration has significant effects on temperature and current distribution within the stack.

Mathematical models of SOFCs help to understand, examine, and design various components and operation parameters of SOFCs, there by playing important roles in SOFC developments. The modeling results can be used to optimize designs and select optimal operation conditions. Some models were performed by using mathematical models coupled with experimental data. Because not all variables can be obtained from testing, estimated values of physical properties and/or reaction kinetic parameters have to be derived from fitting fuel cell polarization data on laboratory-scale SOFCs into a mathematical model.[2]

The weakness of a mathematical model is that there is no guarantee for simulating the exact processes in fuel cells, especially when some guessed and arbitrary unknowns are incorporated into the model. Because of the approximate nature of the numerical models, there are also some errors in simulation results. Nevertheless, careful comparison between numerical and experimental data can validate and improve the model as a predictive tool. Numerical results obtained from commercial modeling packages and in-house programs can acclaim an acceptable accuracy via this comparison tuning and validation approach.

Categorized according to the phenomena, SOFCs can be examined as a heat and mass exchanger in a viewpoint of fluid dynamics and transport phenomena, as an electrochemical generator in a viewpoint of electrochemical modeling at a continuum level (suitable for integration into modeling of full scale stacks), as a chemical reactor in viewpoints of chemical reactions depending on fuel composition and heat effects associated with their electrochemical conversion, and as a system component by combining stack

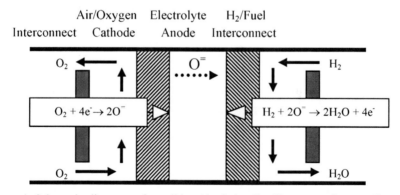

Figure 1. Schematic diagrams of a solid oxide fuel cell with counter-flow configuration showing reactions at interfaces.

models with system models such as reformer, contaminant removal unit, compressors, etc.[2] When SOFC is modeled in a viewpoint of fluid dynamics and transport phenomena, the model includes evaluation of flow and temperature distributions using the conservation laws in fluid mechanics and heat transfer such as the conservation of mass, momentum, and energy. When SOFC is modeled in a viewpoint of electro-chemical modeling at a continuum level, the model includes a charge potential balance and current-voltage relationship. When SOFC is modeled in viewpoints of chemical reactions, the model includes rate equations for mole changes of all considered chemical reactions. Some interesting review papers and models are cited in references.[1-24]

The objective of the present work is to develop a mathematical model to analyze a planar solid oxide fuel cell performance. The analysis uses an in-house program which can help developers to understand all parameters and effects which affect the solutions of governing equations considered. For this purpose, a transport model of the planar SOFC coupled with electro-chemical reaction is developed by the use of different materials for components namely an electrolyte made from yttria-stabilized zirconia (YSZ) and gadolinia-doped ceria (CGO), the anode made from nickel-zirconia cermet, the cathode made from strontium-doped lanthanum manganite, and the ceramic interconnect made from strontium-doped lanthanum chromite. The first objective of this work is to predict the transportations in Ceramatec[7] SOFC (Figures 2 and 3) operated with hydrogen/oxygen gases and to show the development of a generalized actual SOFC voltage versus current density relationship with modified Nernst equation. Performance comparisons between two electrolyte materials, YSZ and CGO are obtained. The governing equations are written in discretized forms and solved

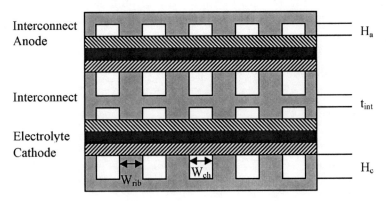

Figure 2. The planar solid oxide fuel cell is designed by Ceramatec (not to scale).[7]

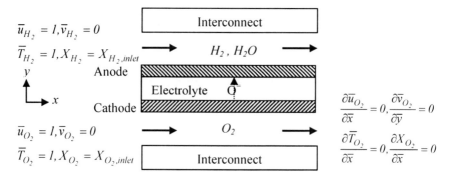

Figure 3. Model boundaries of a solid oxide fuel cell with co-flow configuration.

numerically by an in-house computer program based on a finite volume method. Because of SOFC operating temperatures at 500°C, 600°C, 800°C, and 1,000°C, the modified Nernst equation is used to obtain a reversible cell voltage. The concentration polarization in this work is calculated in an electrochemical point of views. Furthermore, effects of different anode/electrolyte thicknesses to power density and actual voltage of the typical SOFCs are shown.

To facilitate a global view of the limitations of this work, the following assumptions are taken into account in the analysis: The gas mixtures are assumed to be perfect gases, incompressible and single-phase flows with constant physical properties like viscosity, thermal conduction, and diffusion coefficient. The fluid flow is steady fully-developed laminar flow everywhere. The co-flow is considered in gas channels. The temperature of the gas phase inside the porous medium pores and the temperature of the solid matrix equalize instantaneously and the porous medium is assumed to be isotropic, homogenous, and rigid. The electrochemical reactions take place at the electrode/electrolyte interfaces. Details used to calculate the reactive surface area per unit volume (A_v) are obtained from Hussain et al.[17] and Costamagna et al.[31] The reactive surface area per unit volume is used to calculate the source terms of species.

2. Mathematical Model

SOFCs are more flexible for the use of fuels than low temperature fuel cells; when using hydrogen as fuel, the electrochemical reactions of the SOFC at the anode, cathode, and overall cell reactions, respectively, are

$$H_2 + O^{2-} \rightarrow H_2O + 2e^- \tag{1}$$

$$1/2 O_2 + 2e^- \rightarrow O^{2-} \tag{2}$$

$$1/2 O_2 + H_2 \rightarrow H_2O \tag{3}$$

When using carbon monoxide fuel, the electrochemical reactions of the SOFC:

At the anode:

$$2CO + 2O^{2-} \rightarrow 2CO_2 + 4e^- \tag{4}$$

At the cathode:

$$O_2 + 4e^- \rightarrow 2O^{2-} \tag{2}$$

And the overall cell reaction is

$$2CO + O_2 \rightarrow 2CO_2 \tag{5}$$

At high temperature, carbon monoxide can be oxidized by water to form carbon dioxide and hydrogen gas as the water-gas-shift reaction,

$$CO + H_2O \rightarrow CO_2 + H_2 \tag{6}$$

When using hydrocarbons ($C_nH_mO_p$) as fuels like propane, the electrochemical reactions of the SOFC at the anode, cathode, and overall cell reactions are:

At the anode:

$$C_nH_m + (2n+0.5m)O^{2-} \rightarrow nCO_2 + (0.5m)H_2O + (4n+m)e^- ;$$
$$p = 2n + 0.5m \tag{7}$$

At the cathode:

$$(n+0.25m)O_2 + (4n+m)e^- \rightarrow (2n+0.5m)O^{2-} \tag{8}$$

The overall cell reaction is

$$C_nH_m + (n+0.25m)O_2 \rightarrow nCO_2 + (0.5m)H_2O \tag{9}$$

The hydrocarbon-reforming reaction can be described as

$$C_nH_m + nH_2O \rightarrow (n+0.5m)H_2 + nCO \tag{10}$$

A mathematical SOFC model involving electrochemical, chemical, fluid dynamic, and thermodynamic processes must consider different domains corresponding to the components of SOFC. Although the steady-state operation is a good basic assumption in most of the model analysis, in some model studies transient processes are also considered.

2.1. ELECTROCHEMICAL MODEL

The actual solid oxide fuel cell voltage is decreased from its open circuit voltage because of irreversible losses, as shown in Figure 4. Multiple phenomena contribute to irreversible losses in an actual output voltage. The major losses which are called polarization, or overpotential, or overvoltage (η) in the solid oxide fuel cell are: activation polarization (η_{act}); ohmic polarization (η_{ohm}); and concentration polarization (η_{conc}). The actual cell voltage (V_{ac}) is less than its open circuit voltage, E, and is calculated by,

$$V_{ac} = E - \text{Losses} = E - \eta_{total} \tag{11a}$$

$$V_{ac} = E - \eta_{act} - \eta_{ohm} - \eta_{conc} \tag{11b}$$

The open circuit voltage, E, varies as a function of substance properties, such as species concentration which can be obtained from the Nernst equation:

$$E = -\frac{\Delta \bar{g}_f}{n_e F} \quad \text{and} \quad \Delta \bar{g}_f = \Delta \bar{g}_f^0 - RT \ln \frac{\prod a_{products}^{v_i}}{\prod a_{reactants}^{v_i}} \tag{12a}$$

$$E = E^0 - \frac{RT}{n_e F} \ln \frac{\prod a_{products}^{v_i}}{\prod a_{reactants}^{v_i}} \tag{12b}$$

where

$$E^0 = -\frac{\Delta \bar{g}_f^0}{n_e F} = \text{the standard-state reversible voltage (in Volt)}.$$

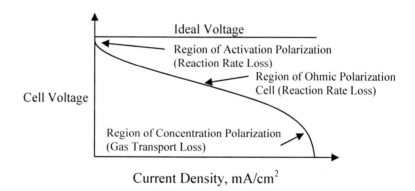

Figure 4. The typical actual fuel cell voltage/current characteristic.

Although temperature enters into the Nenst equation as a variable, the Nernst equation does not fully account for how the reversible voltage varies with temperature. When fuel cells operate at an arbitrary temperature, $T \neq T_0$, the Nernst equation must be modified to obtain the reversible open circuit voltage of hydrogen fuel cell as[4]

$$E = E_T - \frac{RT}{n_e F} \ln \frac{\prod a_{products}^{v_i}}{\prod a_{reactants}^{v_i}} \qquad (13)$$

$$E_T = E^0 + \frac{\Delta \bar{s}}{n_e F}(T - T_0), \qquad (14)$$

where $\Delta \bar{s}$ represents the entropy change of substances.

In low- and medium-temperature fuel cells, the activation polarization plays the most important role on total losses causing voltage drop from ideal voltage; or on the other hand, the activation polarization in high temperature fuel cell is less significant. The activation polarization of electrolyte-supported SOFC occurs in both electrodes; the activation polarization of the cathode is obviously higher than that of the anode due to its lower exchange current density. The activation polarization increases steeply at low current density but gradually at higher current density.[5] The activation polarization can be determined from Butler-Volmer equation[26] which is used to relate the surface overpotential to the rate of reaction as follows:[27]

$$j_a = j_{0a}\left[\exp\left(\frac{\beta n_e F \eta_{act,a}}{RT}\right) - \exp\left(-\frac{(1-\beta)n_e F \eta_{act,a}}{RT}\right)\right] \qquad (15a)$$

$$j_c = j_{0c}\left[\exp\left(\frac{\beta n_e F \eta_{act,c}}{RT}\right) - \exp\left(-\frac{(1-\beta)n_e F \eta_{act,c}}{RT}\right)\right] \qquad (15b)$$

where $j_{0,a}$ and $j_{0,a}$ are anode and cathode exchange current densities, respectively. The exchange current density is the current densities for the forward and reverse reaction at equilibrium, β is a symmetry factor representing the fraction of the applied potential that promotes the reaction. Frequently, it is assumed that β should have the value of 0.5.[27]

When $\beta = 0.5$, from $\sinh u = 1/2(e^u - e^{-u})$, Eq. (15a) can be arranged as

$$\frac{j_a}{2 j_{0a}} = \sinh\left(\frac{n_e F \eta_{act,a}}{2RT}\right) \qquad (16a)$$

The activation polarizations for both electrodes, anode and cathode, can be written as

$$\eta_{act,a} = \frac{2RT}{n_e F}\sinh^{-1}\left(\frac{j_a}{2j_{0a}}\right) \tag{16b}$$

$$\eta_{act,c} = \frac{2RT}{n_e F}\sinh^{-1}\left(\frac{j_c}{2j_{0c}}\right) \tag{16c}$$

The Ohmic polarization is caused by the electrical resistances of the electrodes and resistance of ion transport in the electrolyte. Three general methods of reducing this polarization are the use of high conductivity electrodes, the use of appropriate interconnect materials and design, and the use of thin electrolyte. The Ohmic polarization can be calculated by the following expression:[7–9]

$$\eta_{ohm} = \left(\mathfrak{R}_{contact} + \sum \aleph_{com}\delta_{com}\right)j \tag{17}$$

The loss due to reductions of reactant gas concentrations that will cause reductions in the partial pressures of the reactant gases is the mass transport or concentration polarization. It is more significant in cases where the reactant gases are not pure gases, for an example, hydrogen supplied from a reformer reaction. The activation loss of the high temperature fuel cell is less than that of the low temperature fuel cell since the high operating temperature allows for a lower polarization.[26] In electrochemical point of views, the concentration polarization can be written in term of species concentrations in each electrode[10, 27]

$$\eta_{conc} = \pm\frac{RT}{n_e F}\sum_i s_i \ln\left(\frac{\omega_{i\infty}}{\omega_{i0}}\right) \tag{18}$$

with a relationship between s_i and n_e as

$$\sum_i z_i s_i = -n_e \tag{19}$$

The sign in Eq. (18) is positive for products and negative for reactants. Ivers-Tiffée and Virkar[11] noted that the concentration polarization is a function of several physically measurable parameters and they suggested the following expressions for the concentration polarization:

$$\eta_{conc,a} = -\frac{RT}{2F}\ln\left(1-\frac{j}{j_{as}}\right) + \frac{RT}{2F}\ln\left(1+\frac{p_{H_2,a}\, j}{p_{H_2O,a}\, j_{as}}\right) \tag{20a}$$

$$\eta_{conc,c} = -\frac{RT}{4F}\ln\left(1-\frac{j}{j_{cs}}\right) \tag{20b}$$

where j_{as} and j_{cs} are described as the current densities at which the partial pressure of gas is near zero at the electrode/electrolyte interfaces, and can be defined as

$$j_{as} = \frac{2F p_{H_2,a} D_a^{eff}}{RT\delta_a} \quad (21a)$$

$$j_{cs} = \frac{4F p_{O_2,c} D_c^{eff}}{\left(\frac{p_c - p_{O_2,c}}{p_c}\right) RT\delta_c} \quad (21b)$$

where δ_a and δ_c are anode and cathode thicknesses, respectively, and D_a^{eff} and D_c^{eff} are anode and cathode effective diffusion coefficients, respectively. $p_{H_2,a}$ is the hydrogen partial pressure at anode, $p_{H_2O,a}$ is the water vapor partial pressure at anode, and $p_{O_2,c}$ is the oxygen partial pressure at cathode.

2.2. TRANSPORT IN THE GAS CHANNELS AND ELECTRODES

There are only two phases, gas and solid, which are involved in solid oxide fuel cells. In the air and fuel channels, generally, it is assumed that the consumption of oxygen and hydrogen will not greatly reduce the air and fuel densities, and the air flow is assumed to be incompressible, the fluid properties are constant and no chemical reactions occur in the air and fuel channels.

The transport in the porous electrodes consists of those in both the gas and solid phases are incorporated with the electrochemical reactions. In the most of the model analysis, the reaction is assumed to occur only in the gas phase. The microscopic gas flows in pore space with the chemical reaction must be written. Energy equation includes heat generation term as a result of chemical reaction rate. Concentration equation will also include the chemical reaction rate.

The four equations for conservation of mass, momentum, energy, and concentration must be written. By setting the correct value for porosity in each domain (i.e. $\varepsilon = 1$ for the flow channels), these four equations are valid in all gaseous transportation domains.[4] Assuming fluid properties are constant, air flow is incompressible and no chemical reaction occur in the air and fuel channels, one can write the governing equations as follow.[2, 4, 10, 12–14, 22, 27]

The conservation of mass:

$$\frac{\partial}{\partial t}(\rho\varepsilon) + \nabla\cdot(\rho\varepsilon\mathbf{V}) = S_{m,i} \qquad (22)$$

The conservation of momentum:

$$\frac{\partial}{\partial t}(\rho\varepsilon\mathbf{V}) + \nabla\cdot(\rho\varepsilon\mathbf{V}\mathbf{V}) = -\varepsilon\nabla p + \nabla\cdot(\varepsilon\zeta) + \frac{\varepsilon^2\mu\mathbf{V}}{\kappa} \qquad (23)$$

The last term on the right-hand side of Eq. (23) is known as Darcy's law, which quantifies the viscous drag of fluids in porous media.[4] The species balance equation can be written as

$$\frac{\partial}{\partial t}(\rho\varepsilon c_i) + \nabla\cdot(\rho\varepsilon\mathbf{V}c_i) = -\nabla\cdot\varepsilon\vec{J}_i + S_{s,i} \qquad (24)$$

where $S_{s,i}$ is the additional species sources of species i or the rate of production or consumption of species i, which can be neglected in the bulk of the solution.[27] The source term, $S_{s,i}$, in the air and fuel passages due to the electrochemical reactions at the electrodes can be obtained from

$$S_{m,i} = \pm\frac{M_i j}{n_e F} \qquad (25)$$

The sign in Eq. (25) is positive for products and negative for reactants.

For flows of low velocity in porous electrodes, it is anticipated that the transport process is dominated by diffusion. The electrodes usually are the porous cermet; various investigators solved the mass transport in the porous medium by using one of three approaches:

1. Fick's Law is the simplest diffusion model and used in dilute or binary systems,[13, 14, 22, 26, 28]

$$\vec{J}_i = -\rho_i D_i^{eff}\nabla c_i \qquad (26a)$$

$$\vec{N}_i = -\omega_i D_i^{eff}\nabla X_i \qquad (26b)$$

where D_i^{eff} is the diffusion coefficient of species i which combining the effective molecular diffusion, $D_{m,i}^{eff}$, and effective Knudsen diffusion, $D_{Kn,i}^{eff}$, as[15]

$$D_i^{eff} = \left(\frac{1}{D_{m,i}^{eff}} + \frac{1}{D_{Kn,i}^{eff}}\right)^{-1} \qquad (27)$$

2. The Stefan-Maxwell model is more commonly used in multi-component systems. This model for an ideal gas mixture can be written as,[4, 10, 13, 15, 16, 22, 28]

$$-\omega_t \nabla X_i = \sum_{j=1, j\neq i}^{n} \frac{X_j \overset{\omega}{N_i} - X_i \overset{\omega}{N_j}}{D_{i,j}} \quad (28)$$

3. The Dusty Gas model (called as the extended Stefan-Maxwell equation) is the Stefan-Maxwell equation incorporating Knudsen diffusion for multi-component systems[13, 15, 16, 22] and can be defined as

$$-\omega_t \nabla X_i = \frac{\overset{\omega}{N_i}}{D_{Kn,i}} + \sum_{j=1, j\neq i}^{n} \frac{X_j \overset{\omega}{N_i} - X_i \overset{\omega}{N_j}}{D_{i,j}} \quad (29a)$$

$$-\omega_t \nabla X_i = \sum_{j=1, j\neq i}^{n} \frac{X_j \overset{\omega}{N_i} - X_i \overset{\omega}{N_j}}{\wp_{i,j}^{eff}} \quad (29b)$$

Hussain[17] introduced their effective diffusion coefficient, $\wp_{i,j}^{eff}$, as

$$\wp_{i,j}^{eff} = \frac{\varepsilon}{\tau} \left(\frac{D_{i,j} D_{Kn,i}}{D_{i,j} + D_{Kn,i}} \right) \quad (30)$$

The energy conservation equation can be written as follows:

$$\frac{\partial}{\partial t}(\rho \varepsilon e) - \varepsilon \frac{\partial p}{\partial t} + \nabla \cdot (\rho \varepsilon \mathbf{V} e) = \nabla \cdot \varepsilon (-k_{eff} \nabla T) + S_e + S_{rad} \quad (31)$$

Dagan[29] introduced k_{eff} term:

$$k_{eff} = -2k_{solid} + \frac{1}{\frac{\varepsilon}{2k_{solid} + k_g} + \frac{1-\varepsilon}{3k_{solid}}} \quad (32)$$

and Beale[10] and VanderSteen[13] presented k_{eff} term as

$$k_{eff} = \varepsilon k_g + (1-\varepsilon) k_{solid} \quad (33)$$

S_e represents heat generation by the electrochemical reactions, chemical reactions, and Joule heating.[2] Beale[10] shows the Joule heating as

$$S_{e,joule} = \frac{(E - V_{ac}) j}{\delta_e} \quad (34)$$

Beale[10] noted that the relative importance of radiation in SOFCs is the subject of debate[18–21] some simulations exclude the radiative heat source, and implied the surface-to-surface radiation in the micro-channels accounted-for by the network method as

$$Q_{rad} = \frac{\sigma T_k^4 - q_{0k}}{(1-\Im_k)/\Im_k A_k} = \sum_{j=1}^{N} \frac{q_{0j} - q_{0k}}{1/f_{k-j} A_k} \quad (35)$$

Equation (31) can be written in terms of temperature, T, as follows:

$$\frac{\partial}{\partial t}(\rho \varepsilon c_p T) - \varepsilon \frac{\partial p}{\partial t} + \nabla \cdot (\rho \varepsilon \mathbf{V} c_p T) = \nabla \cdot \varepsilon (-k_{eff} \nabla T) + S_e + S_{rad} \quad (36)$$

2.3. TRANSPORT IN THE SOLID ELECTROLYTE

The solid electrolyte serves as both an electronic insulator and an ionic conductor. It also insulates the air and fuel electrodes, while letting only oxygen ions to pass through. No mass transfer occurs through the electrolyte since it is normally made from nonporous materials. The only energy balance can be considered, then from Eq. (36), reduces to[2, 14]

$$\nabla \cdot (-k_e \nabla T) + S_e + S_{rad} = 0 \quad (37)$$

$O^=$ transport in the electrolyte is described by considering the ion transport from the conservation of charge as:[4, 10, 13, 14, 16, 22, 27]

$$\nabla \cdot \mathbf{i} = 0 = \nabla \cdot \mathbf{i}_{io} + \nabla \cdot \mathbf{i}_{el} \quad (38a)$$

$$-\nabla \cdot \mathbf{i}_{io} = \nabla \cdot \mathbf{i}_{el} \quad (38b)$$

and Ohm's Law as

$$\mathbf{i}_{io} = -\sigma_{io}^{eff} \nabla \phi_{io} \quad (39)$$

$$\nabla \cdot (\nabla \phi_{io}) = \begin{cases} \pm \dfrac{j'_{FC}}{\sigma_{io}^{eff}} & \text{at the interfaces} \\ 0 & \text{elsewhere} \end{cases} \quad (40)$$

where $\mathbf{i}_{io}$ and $\mathbf{i}_{el}$ are charge fluxes, ions and electrons respectively and ϕ_{io} is ionic potential in the electrolyte (the ion conductor). The sign in Eq. (40) is positive for cathode/electrolyte interface and negative for anode/electrolyte interface.[17] At the electrolyte interfaces, a simplified Butler-Volmer equation can be used to obtain the volumetric current density, j'_{FC}, which incorporates a reactive surface area per-unit volume of the porous electrode structure, A_{ac}, as[4, 16, 17, 27]

$$j'_{FC} = A_{ac}j_{0,electrode}\left(\frac{\omega_i}{\omega_i^0}\right)^{v_i}\left\{\exp\left[\frac{\beta n_e F\Delta\phi}{RT}\right] - \exp\left[-\frac{(1-\beta)n_e F\Delta\phi}{RT}\right]\right\} \quad (41)$$

$$\Delta\phi = \begin{cases} \phi_{io} - \phi_{el} = \eta_{act,a} & \text{at the anode/electrolyte interface} \\ \phi_{el} - \phi_{io} = \eta_{act,c} & \text{at the cathode/electrolyte interface} \end{cases} \quad (42)$$

The reactive surface area represents a small fraction of the electrode structure which is accessible by reactant species[30] and this value is one of the electrode morphology effects which also strongly influence the electrode resistance.[31] j'_{FC} can be used to calculate the source term in the species balance equation (Eq. (25)) by substituting it into current density, j. Beale[10] mentioned that the Nernst potential can be considered to be the sum of the potential differences across the anode $\Delta\phi_a$ and cathode $\Delta\phi_c$ as

$$E = \Delta\phi_a + \Delta\phi_c \quad (43)$$

$$\Delta\phi_c = \Delta\phi_c^0 - \frac{RT_c}{4F}\left(\ln p_c + \ln X_{O_2}\right) - \eta_{act,c} \quad (44)$$

$$\Delta\phi_a = \Delta\phi_a^0 + \frac{RT_a}{2F}\left(\ln X_{H_2} - \ln X_{H_2O}\right) - \eta_{act,a} \quad (45)$$

where $\Delta\phi_c^0$ and $\Delta\phi_a^0$ are the arbitrary potential differences across the anode and cathode and T_c and T_a are the cathode and anode side temperatures. He also indicated that if the electrodes are sufficiently thin, both potential differences may be applied as sources of equal and opposite magnitude on either side of the electrode boundary, sources of elsewhere in the electrolyte and the interconnect are zero.

2.4. TRANSPORT IN THE INTERCONNECTS

The interconnect is assumed to be impermeable for planar SOFC material and used to collect the current from the SOFC; therefore only energy balance can be considered and from Eq. (36) reduces to the following[2, 13, 14]

$$\nabla \cdot (-k_{int}\nabla T) + S_e + S_{rad} = 0 \quad (37)$$

The electron transport can be considered from the conservation of charge at the interfaces between the interconnects and the electrodes as

$$\nabla \mathbf{i}_{el,electrode} = -\nabla \mathbf{i}_{el,interconnect} \quad (46)$$

2.5. GEOMETRICAL COMPONENTS

A sample planar solid oxide fuel cell is designed by Ceramatec[7] as shown in Figure 2, is considered as the basic configuration in this present work.

TABLE 1. Typical solid oxide fuel cell dimensions.[7]

Element	Description	Size (mm)
Ha	Anode channel height	0.8
Hc	Cathode channel height	1.5
Wrib	Rib width	1.0
Wch	Channel width	2.0
tint	Interconnect thickness	0.5
ta	Anode electrode thickness	25×10^{-3}
tc	Cathode electrode thickness	25×10^{-3}
te	Electrolyte thickness	150×10^{-3}

TABLE 2. The operating conditions and electrochemical and transport properties of the typical solid oxide fuel cell at 1,000°C. [6, 7]

Description	Unit	Value
Anode inlet pressure	Atm	1.0
Cathode inlet pressure	Atm	1.0
Cell temperature	°C	1,000
Hydrogen inlet mole fraction		0.9
Water vapor anode-inlet mole fraction		0.1
Oxygen inlet mole fraction		0.7
Water vapor anode-inlet mole fraction		0.3
Anode reference exchange current density[a]	A/m^2	1.67×10^8
Cathode reference exchange current density[a]	A/m^2	5.51×10^9
Anode electrode porosity		0.375
Cathode electrode porosity		0.375
D_a^{eff} [a]	m^2/s	3.5×10^{-5}
D_c^{eff} [a]	m^2/s	7.3×10^{-6}
YSZ electrolyte conductivity[b]	S m^{-1}	$\left(85 \times 10^3\right) \exp\left(\dfrac{-11 \times 10^3}{T}\right)$
Interconnect conductivity[b]	S m^{-1}	$\dfrac{\left(9.3 \times 10^6\right)}{T} \exp\left(\dfrac{-1.1 \times 10^3}{T}\right)$

[a]The information from Costamagna and Honegger.[6]
[b]The information from Bossel.[7]

The typical information is given in Table 1 which shows solid oxide fuel cell dimensions[7] and Table 2 gives typical operating conditions and the assumed electrochemical and transport properties of the solid oxide fuel cell components and gases used. To compare performances of the YSZ-electrolyte and CGO-electrolyte solid oxide fuel cells, the operating conditions and electrochemical and transport properties given in Tables 3 and 4 are used.

3. Results and Discussions

3.1. FLOW FIELDS

Figures 5 and 6 show flow fields at the center of cathode and anode domains (electrode and channel) in the YSZ electrolyte-supported cell operated at 1,000°C, respectively. The results show that the macroscopic velocities in the porous mediums drop very fast from the interfaces with channel, in a thin layer of the order of the pore size as mentioned by Dagan.[29]

3.2. TEMPERATURE DISTRIBUTIONS IN CATHODE AND ANODE DOMAINS

Figures 7 and 8 show temperature distributions at the center of cathode and anode domains (electrode and channel) in the YSZ electrolyte-supported cell operated at 1,000°C, respectively; these results indicate that gas temperatures are affected by operating temperature. Temperature does not only affect the cell performance but also affect gas properties within the cathode and anode domains in cases that the gas is fed to the cell at a different temperature from the operating temperature.[32] Both gas temperatures are affected by operating temperature; the lowest temperatures of both gases are different with the same inlet temperature, 25°C.

TABLE 3. The operating conditions and electrochemical and transport properties of the YSZ-electrolyte and CGO-electrolyte solid oxide fuel cell at 800°C.[5, 15, 25]

Description	Unit	Value
Cell temperature	°C	800
Hydrogen inlet mole fraction		0.95
Water vapor anode-inlet mole fraction		0.05
Oxygen inlet mole fraction		0.21
Nitrogen inlet mole fraction		0.79
Anode exchange current density[a]	A/m^2	5,300
Cathode exchange current density[a]	A/m^2	2,000
Porosity of cathode and anode[b]		0.375
Tortuosity of cathode and anode[b]		2.75
$D_{H_2-H_2O}$[b]	m^2/s	7.68×10^{-4}
$D_{O_2-N_2}$[b]	m^2/s	1.83×10^{-4}
CGO Electrolyte conductivity[c]	S m^{-1}	$\left(2.706 \times 10^6\right) \dfrac{\exp\left(-\dfrac{0.64}{8.6173 \times 10^{-5} T}\right)}{T}$

[a]The information from Chan and Xia.[5]
[b]The information from Suwanwarangkul et al.[15]
[c]The information from Leah et al.[25]

TABLE 4. The operating conditions and electrochemical and transport properties of the YSZ-electrolyte and CGO-electrolyte solid oxide fuel cell at 600°C and 500°C.[25]

Description	Unit	600°C	500°C
Hydrogen inlet mole fraction		0.97	0.97
Water vapor anode-inlet mole fraction		0.03	0.03
Oxygen inlet mole fraction		0.21	0.21
Water vapor anode-inlet mole fraction		0.79	0.79
Anode reference exchange current density[a]	A/m^2	3.2×10^{13}	3.2×10^{13}
Cathode reference exchange current density[a]	A/m^2	7.0×10^{11}	7.0×10^{11}
$D_{H_2-H_2O}$	m^2/s	5.38×10^{-4}	4.36×10^{-4}
$D_{O_2-N_2}$	m^2/s	1.27×10^{-4}	1.03×10^{-4}

[a]The information from Leah et al.[25]

MATHEMATICAL ANALYSIS OF PSOFCS

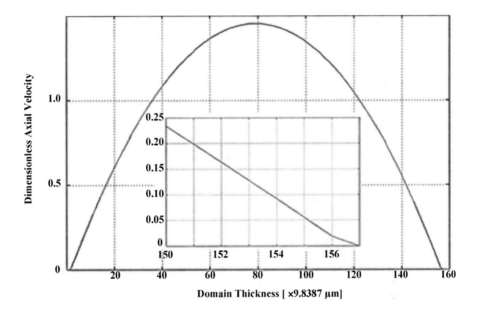

Figure 5. The axial velocity profile across (a) the cathode domain and (b) electrode.

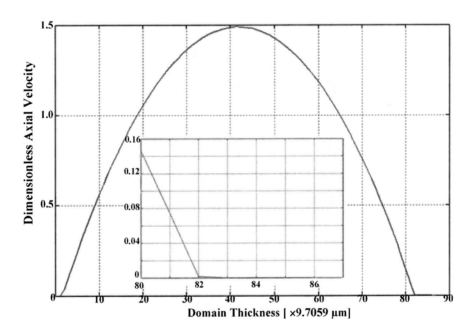

Figure 6. The axial velocity profile across (a) the anode domain and (b) electrode.

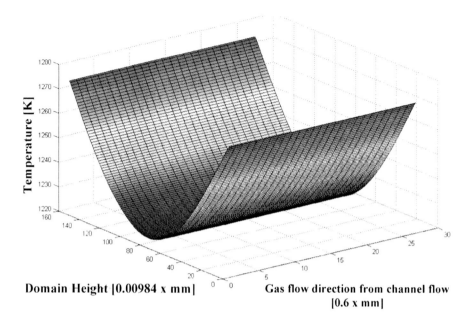

Figure 7. Temperature distribution of the cathode domain.

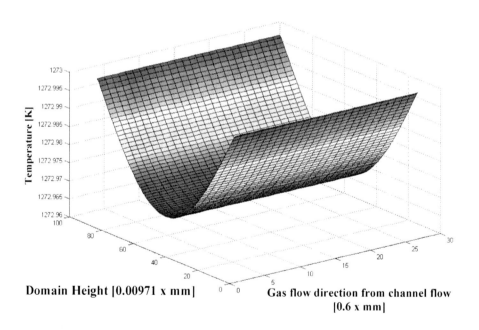

Figure 8. Temperature distribution of the anode domain.

3.3. PERFORMANCE PREDICTION

Figures 9 and 10 show the typical SOFC performance predictions of the actual voltage and power density of the YSZ electrolyte-supported cell operated at 1,000°C influenced by all overpotentials, concentration loss, ohmic loss, and activation loss. As shown in Figure 10, the concentration losses in both electrodes are very low, almost the same line. The power density reaches the highest value of 2,636 W/m^2, when the current density is 8,750 A/m^2 and the actual voltage is 0.3065 V (Figure 9). At high temperature, total ohmic losses of YSZ, nickel-zirconia cermet, strontium-doped lanthanum manganite, and strontium-doped lanthanum chromite do not play an important role. When the current density closes to 20,000 A/m^2, the actual voltage approaches to zero. From the results, the concave curvature of the actual voltage is due to both activation and concentration losses as mentioned by Chan and Xia.[5] Figure 11 shows the total losses within cathode and anode domains which are obtained from the present modeling; as it can be seen, the significant loss is from the anode side.

To determine more performance analysis of SOFCs at lower temperatures, two different electrolyte materials namely, YSZ and CGO, are simulated at 800°C by using the modified electrolyte-supported modeling to compare these electrolyte materials (Figure 12). Both SOFCs are calculated under conditions as shown in Table 3 by using the dimensions from the Ceramatec design.[7] The YSZ-electrolyte conductivity is the same as indicated in Table 2. The power density of the YSZ electrolyte-supported

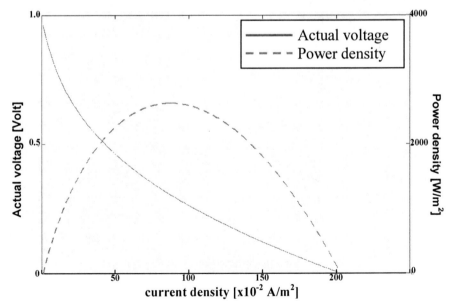

Figure 9. Actual voltage prediction of electrolyte-supported Ceramatec SOFC at 1,000°C.

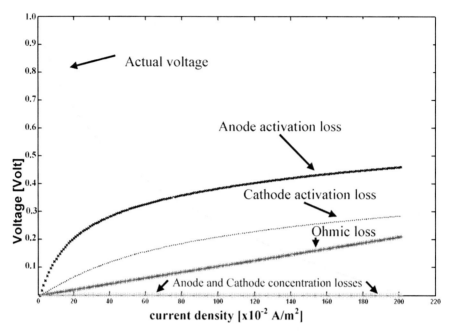

Figure 10. All overpotentials prediction of electrolyte-supported Ceramatec SOFC at 1,000°C.

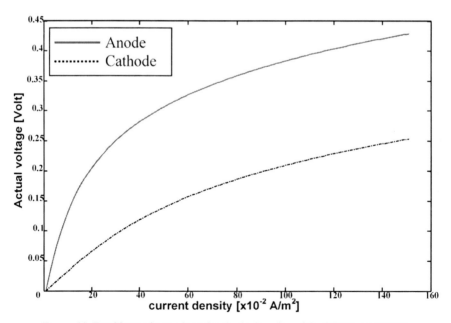

Figure 11. Total losses in anode and cathode domains of the SOFC at 1,000°C.

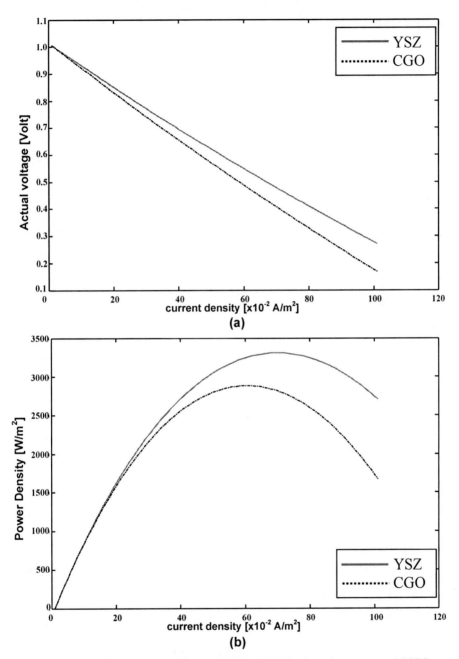

Figure 12. The performance comparisons of YSZ and CGO electrolyte-supported SOFCs at 800°C (a) actual voltage and (b) power density.

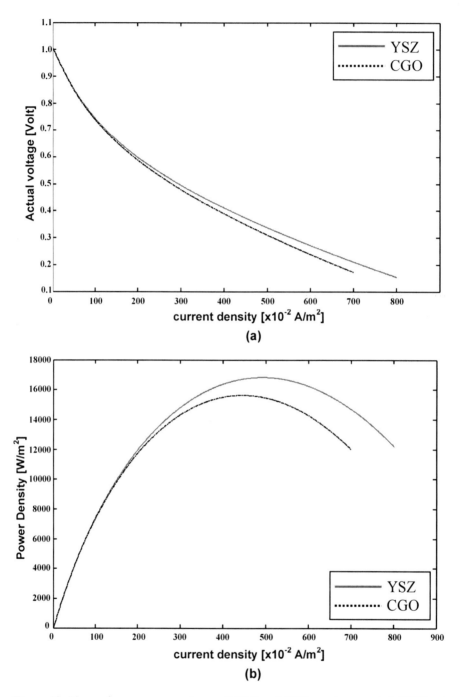

Figure 13. The performance comparisons of YSZ and CGO anode-supported SOFCs at 800°C (a) actual voltage and (b) power density.

SOFC reaches the highest value of 3,307 W/m^2, when the current density is 7,000 A/m^2 and the actual voltage is 0.4793 V. The power density of the CGO electrolyte-supported SOFC reaches the highest value of 2,883.3 W/m^2, when the current density is 6,050 A/m^2 and the actual voltage is 0.4846 V. As shown in Figure 12, the YSZ-electrolyte SOFC gives higher performance than the CGO-electrolyte SOFC with the same operating conditions, the same Ceramatec dimensions, and the same electrode materials.

The performance comparisons of both electrolyte materials have been continued by considering anode-supported SOFCs using both materials (Figure 13). The anode thickness, the cathode thickness, and the electrolyte thickness of both anode-supported SOFCs are 1.02 mm, 20 μm, and 8 μm, respectively. The anode-supported SOFCs are also calculated under conditions as shown in Table 3. The power density of the YSZ anode-supported SOFC reaches the highest value of 16,834 W/m^2, when the current density is 49,200 A/m^2 and the actual voltage is 0.3429 V. The power density of the CGO anode-supported SOFC reaches the highest value of 15,634 W/m^2, when the current density is 44,550 A/m^2 and the actual voltage is 0.3517 V. The CGO based SOFCs have lower performance than the YSZ based SOFCs mainly because CGO is less conductive than YSZ at 800°C (Figure 14). The conductivity values of CGO are higher than that of YSZ temperatures lower than 750°C.

The electrolyte- and anode-supported SOFCs using conditions shown in Table 4 are modeled at 600°C to study the SOFC performances at the lower temperature (Figures 15 and 16). The power density of the YSZ electrolyte-supported SOFC reaches the highest value of 454 W/m^2, when the current density is 1,000 A/m^2 and the actual voltage is 0.50444 V. The power density of the CGO electrolyte-supported SOFC reaches the highest value of 896.3 W/m^2, when the current density is 1,900 A/m^2 and the actual voltage is 0.49796 V (Figure 15). The power density of the YSZ anode-supported SOFC reaches the highest value of 5,174.6 W/m^2, when the current density is 12,000 A/m^2 and the actual voltage is 0.43484 V. The power density of the CGO anode-supported SOFC reaches the highest value of 9,188.1 W/m^2, when the current density is 22,500 A/m^2 and the actual voltage is 0.41018 V (Figure 16). The CGO based SOFCs have higher performance than the YSZ based SOFCs mainly because CGO is more conductive than YSZ at 600°C (Figure 14).

It can be seen that the reduction of performance with decreasing temperature is not proportional to reduction of conductivity. The performance of a SOFC is determined by electrolyte conductivity and electrode activity overpotentials. This complex problem can only be analyzed with a numerical model considering all controlling factors.

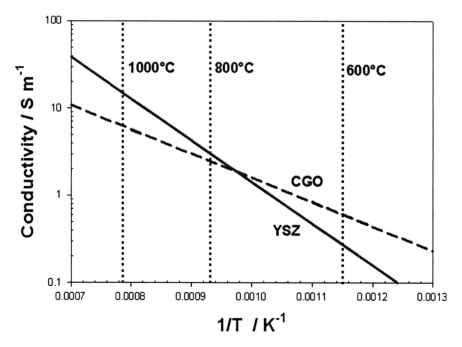

Figure 14. The electrolyte conductivity comparisons of YSZ and CGO.

Kharton et al.[33] compared the electronic and the ionic conductivities of CGO and indicated that in reducing environment the electronic conductivity at the anode side will be greater than the ionic conductivity for temperatures greater than about 500°C. The CGO electrolyte- and anode-supported SOFCs using conditions shown in Table 4 are modeled at 500°C to study the SOFC performances at the lower temperature (Figures 17). The power density of the CGO electrolyte-supported SOFC reaches the highest value of 293.4474 W/m^2, when the current density is 700 A/m^2 and the actual voltage is 0.4891 V. The power density of the CGO anode-supported SOFC reaches the highest value of 2,705.6109 W/m^2, when the current density is 7,300 A/m^2 and the actual voltage is 0.3758 V.

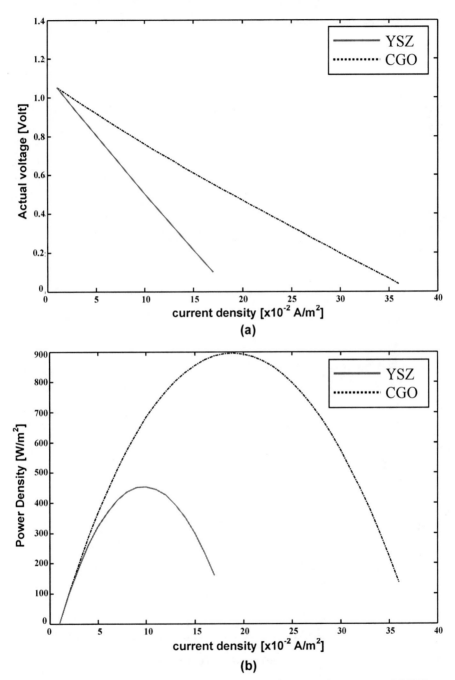

Figure 15. The performance comparisons of YSZ and CGO electrolyte-supported SOFCs at 600°C (a) actual voltage and (b) power density.

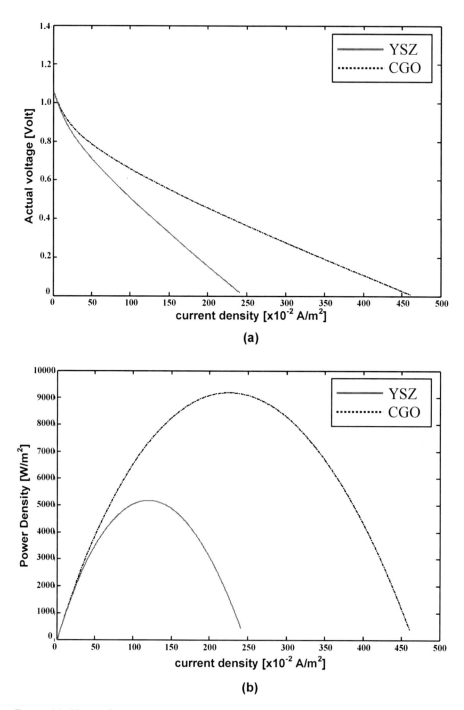

Figure 16. The performance comparisons of YSZ and CGO anode-supported SOFCs at 600°C (a) actual voltage and (b) power density.

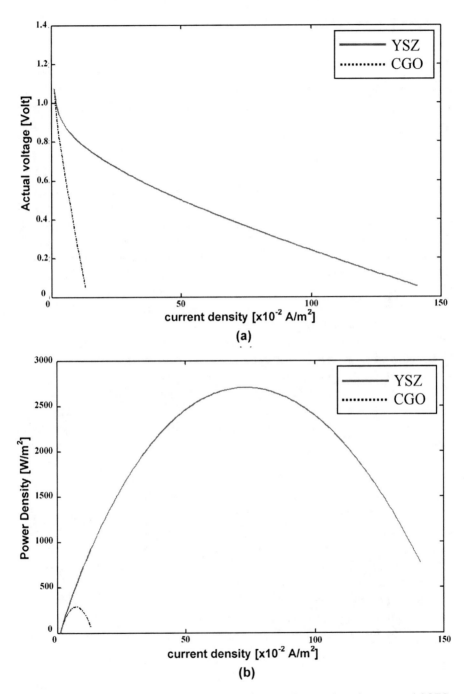

Figure 17. The performance comparisons of CGO electrolyte- and anode-supported SOFCs at 500°C (a) actual voltage and (b) power density.

4. Conclusions

The mathematical analysis has been developed by using finite volume method, experimental data from literature, and solving the governing equations numerically to predict solid oxide fuel cell performances with different operating conditions and with different material properties. This study shows that the transportations of velocity and temperature in electrode domains are different. The performance of an anode-supported planar SOFC is better than that of an electrolyte-supported planar SOFC for the same material used, under the same electrode electrochemical considerations, and same operating conditions. From this study, there are available current density ranges for each type of solid oxide fuel cells, electrolyte- and anode-supported solid oxide fuel cells with different electrolyte materials and different component thicknesses. The anode-supported solid oxide fuel cells can be used to give the high power density in the higher current density range than the electrolyte-supported solid oxide fuel cells. Even though the electrolyte-supported solid oxide fuel cells give the lower power density and can operate in the lower current density range but they can be used as a small power generator which is portable and provide low power. On the other hand, the electrolyte- and anode-supported solid oxide fuel cells can be considered to be used in different applications depending on power density, current density, and actual voltage requirements.

Furthermore, the effects of the electrolyte materials and operating temperatures play important roles to the performance predictions. This should be noted that performance comparisons are obtained by using the same electrode materials. The YSZ-electrolyte solid oxide fuel cell in this work shows higher power density than the CGO-electrolyte solid oxide fuel cell at the higher temperatures than 750°C. The performance of the CGO-electrolyte solid oxide fuel cell is higher at temperatures lower than 750°C, the CGO can be considered as the electrolyte material for the intermediate- and low- temperature SOFCs. Since the conductivity values of the CGO and YSZ in this work depend on operating temperatures and the effects of different electrode materials, they can be implied that conductivity values of electrolyte and electrode materials can help to improve SOFC overall performances at suitable operating temperatures, a lot of scientists have been working on solving the conductivity values by developing electrolyte and electrode material properties and introducing new electrolyte and electrode materials.

NOMENCLATURE

a_i	activity values of substances where i represents each species in the reactions
A_{ac}	actual reaction surface area per unit volume, m^{-1}
c_i	mass fraction of species i
D	diffusion coefficient, m^2/s
$D_{i,j}$	binary diffusion coefficient of species i in j, m^2/s
$D_{m,i}^{eff}$	effective molecular diffusion of species i, m^2/s
$D_{Kn,i}^{eff}$	effective Knudsen diffusion of species i, m^2/s
$D_{Kn,i}$	Knudsen diffusion coefficient of species i, m^2/s
e	gas internal energy, Joule
E	reversible electrochemical cell voltage, Volt
E_T	open circuit voltage at an arbitrary temperature, $T \neq T_0$, Volt
F	Faraday's constant, 96,487 coulombs/mol
f_{k-j}	radiation configuration or view factor
$\Delta \overline{g}_f^0$	the standard free-energy change of fluid substances in the reaction at 1 atm, J/mol
i	charge flux, A/m^3
j	current density, A/m^2
j_0	exchange current density, A/m^2
$j_{0,electrode}$	reference exchange current density for each electrode, A/m^2
j_{as}	anode limiting current density, A/m^2
j_{cs}	cathode limiting current density, A/m^2
j'_{FC}	volumetric current density, A/m^3
$\overset{\nu}{J}_i$	mass diffusion flux of species i, kg/(s m^2)
k	thermal conductivity, W/m K
M_i	molecular weight of species i gas k, kg/mol
n_e	number of electrons participating in the reaction
$\overset{\omega}{N}_i$	molar diffusion flux of species i, mol/(s m^2)
p	fluid pressure, atm or Pa
q_{0k}	radiosity (the total emitted and reflected radiation), W
Q_{rad}	radiative heat source, W
R	gas constant, 8.314472 J/K mol
$\mathfrak{R}_{contact}$	contact resistance of the solid oxide fuel cell, Ω m^2
s_i	stoichiometric coefficient of species i in the electrode reaction
$\Delta \overline{s}$	entropy change of substances in the reaction, J/(mol K)
S_e	volumetric heat source, W/m^3
$S_{e,joule}$	joule heating, W/m^3
$S_{m,k}$	additional mass sources in gas k, kg/(s m^3)
S_{rad}	volumetric radiative heat source, W/m^3
$S_{s,i}$	additional species sources of species i, kg/(s m^3)

T	reaction temperature, K	σ	Stefan-Boltzmann constant, 5.67×10^{-8} W/m² K⁴
T_0	standard temperature, K		
V	velocity vector, m/s	$\mathfrak{I}_k$	emissivity of the kth element
V_{ac}	actual cell voltage, Volt		
X	molar fraction	σ_{io}^{eff}	ionic conductivity of the purely ionic conducting material
z_i	charge number of species i		

Greek symbols

α	charge transfer coefficients	σ_{el}^{eff}	electrical conductivity of the purely electrical conducting material
β	symmetry factor		
η_{total}	total polarization, Volt	ϕ	charge potential
η_{act}	activation polarization, Volt		

Subscripts

η_{conc}	concentration polarization, Volt	a	anode
		c	cathode
		com	component
η_{ohm}	ohmic polarization, Volt	e	electrolyte
υ_i	stoichiometric coefficients where i represents each species in the reactions	eff	effective
		el	electron or electrical
		g	gas properties of electrodes
		H_2	hydrogen
v_i	reaction order for each gas reaction	H_2O	water
		i	species i in the electrode reaction
δ	thickness, m		
ω	molar concentration, mol/m³	int	interconnect
		io	ion or ionic
ω_i^0	reference molar concentration of species i, mol/m³	$i0$	property of species i at the electrode surface
		$i\infty$	property of species i in the bulk solution
$\wp$	diffusion coefficient		
ρ	fluid density, kg/m³	O_2	oxygen
$\aleph_{com}$	resistivity of each component in the solid oxide fuel cell, Ω m	k	kth considered element
		$products$	product substances
		$reactants$	reactant substances
τ	tortuosity	$solid$	solid properties of electrodes
μ	fluid dynamic viscosity, Pa s or kg/(m s)		
		t	total
ε	porosity	**Superscripts**	
ζ	shear stress tensor of the fluid, N/m²	eff	effective transport coefficients
κ	permeability, m²		

References

1. K. Kendall, N.Q. Minh, and S.C. Singhal, in: *High-Temperature Solid Oxide Fuel Cells-Fundamentals, Design and Applications*, edited by S. Singhal and K. Kendall (Elsevier Science, Oxford, England, 2003), pp. 197–229.
2. M.A. Khaleel and J.R. Selman, in: *High-Temperature Solid Oxide Fuel Cells-Fundamentals, Design and Applications*, edited by S. Singhal and K. Kendall (Elsevier Science, Oxford, England, 2003), pp. 293–332.
3. Z.F. Zhou, C. Gallo, M.B. Pague, H. Schobert, and S.N. Lvov, Direct Oxidation of Jet Fuels and Pennsylvania Crude Oil in a Solid Oxide Fuel Cell, *Journal of Power Sources* 133, 181–187 (2004).
4. R. O'Hayre, S.W. Cha, W. Colella, and F. Prinz, *Fuel Cell Fundamentals* (Wiley, New Jersey, 2005).
5. S.H. Chan and Z.T. Xia, Polarization Effects in Electrolyte Electrode-Supported Solid Oxide Fuel Cells, *Journal of Applied Electrochemistry* 32, 339–347 (2002).
6. P. Costamagna and K. Honegger, Modeling of Solid Oxide Heat Exchanger Integrated Stacks and Simulation at High Fuel Utilization, *Journal of the Electrochemical Society* 145(11), A3995–A4007 (1998).
7. U.G. Bossel, *Performance Potentials of Solid Oxide Fuel Cell Configurations* (Electric Power Research Institute, Inc., California, 1992)
8. U.G. Bossel, *Facts and Figures, Energy agency final report* (Swiss Federal Office of Energy, Bern, Switzerland 1992)
9. R. Bove, P. Lunghi, and N.G. Sammes, SOFC Mathematical Model for Systems Simulations. Part One: from a Micro-Detailed to Macro-Black-Box Model, *International Journal of Hydrogen Energy* 30, 181–187 (2005).
10. S.B. Beale, in: *Transport Phenomena in Fuel Cells*, edited by B. Sunden and M. Faghri (WIT Press, Massachusetts, 2005), pp. 43–82.
11. E. Ivers-Tiffée and A.V. Virkar, in: *High-Temperature Solid Oxide Fuel Cells-Fundamentals, Design and Applications*, edited by S. Singhal and K. Kendall (Elsevier Science, Oxford, England, 2003), pp. 230–260.
12. U. Pasaogullari and C.Y. Wang, Computational Fluid Dynamics Modeling of Solid Oxide Fuel Cells, *Electrochemical Society Proceedings* 2003–07, 1403–1412 (2003).
13. J.D.J. VanderSteen, B. Kenney, J.G. Pharoah, and K. Karan, Mathematical Modeling of the Transport Phenomena and the Chemical/Electrochemical Reactions in Solid Oxide Fuel Cells: A Review, *Proceedings of Canadian Hydrogen and Fuel Cells 2004*, Toronto, Canada (CD-ROM) (2004).
14. R. Bove, and S. Ubertini, Modeling Solid Oxide Fuel Cell Operation: Approaches, Techniques and Results, *Journal of Power Sources* 159, 543–559 (2006).
15. R. Suwanwarangkul, E. Croiset, M.W. Fowler, P.L. Douglas, E. Entchev, and M.A. Douglas, Performance Comparison of Fick's, Dusty-gas and Stefan-Maxwell Models to Predict the Concentration Overpotential of a SOFC Anode, *Journal of Power Sources* 122, 9–18 (2003).
16. M.M. Hussain, X. Li, and I. Dincer, Mathematical Modeling of Transport Phenomena in Porous SOFC Anodes, *International Journal of Thermal Sciences* 46, 48–56 (2007).
17. M.M. Hussain, X. Li, and I. Dincer, Mathematical Modeling of Planar Solid Oxide Fuel Cells, *Journal of Power Sources* 161, 1012–1022 (2006).
18. D.L. Damm and A.G. Fedorov, Radiation Heat Transfer in SOFC Materials and Components, *Journal of Power Sources* 143, 158–165 (2005).
19. D.L. Damm and A.G. Fedorov, Spectral Rdiative Heat Transfer Analysis of the Planar SOFC, *Journal of Fuel Cell Science and Technology* 2, 258–262 (2005).

20. J.D.J. VanderSteen and J.G. Pharoah, Modeling Radiation Heat Transfer with Participating Media in Solid Oxide Fuel Cells, *Journal of Fuel Cell Science and Technology* 3, 62–67 (2006).
21. K.J. Daun, S.B. Beale, F. Liu, and G.J. Smallwood, Radiation Heat Transfer in Planar SOFC Electrolytes, *Journal of Power Sources* 157, 302–310 (2006).
22. S. Kakac, A. Pramuanjaroenkij, and X.Y. Zhou, A Review of Numerical Modeling of Solid Oxide Fuel Cells, *International Journal of Hydrogen Energy* 32(7), 761–786 (2006).
23. K. Sudaprasert, R.P. Travis, and R.F. Martinez-Botas, A Computational Fluid Dynamics Model of a Solid Oxide Fuel Cell, *Journal of Power and Energy* 219, 159–167 (2005).
24. R. Suwanwarangkul, E. Croiset, E. Entchev, S. Charojrochkul, M.D. Pritzker, M.W. Fowler, P.L. Douglas, S. Chewathanakup, and H. Mahuadom, Experimental and Modeling Study of Solid Oxide Fuel Cell Operating with Syngas Fuel, *Journal of Power Sources* 161, 308–322 (2006).
25. R.T. Leah, N.P. Brandon, and P. Aguiar, Modelling of Cells, Stacks, and Systems Based Around Metal Supported Planar IT-SOFC Cells with CGO Electrolytes Operating at 500–600°C, *Journal of Power Sources* 145, 336–352 (2005).
26. J. Larminie and A. Dicks, *Fuel Cell Systems Explained* (Wiley, West Sussex, England 2003).
27. J. Newman and K. Thomas-Alyea, *Electrochemical Systems* (Wiley, New Jersey, 2004).
28. R.B. Bird, W.E. Stewart, and E.N. Lightfoot, *Transport Phenomena* (Wiley, New York, 1962).
29. G. Dagan, *Flow and Transport in Porous Formations* (Springer-Verlag, New York, 1989).
30. X. Li, *Principles of Fuel Cells* (Taylor & Francis, New York, 2006).
31. P. Costamagna, P. Costa, and V. Antonucci, Micro-modeling of solid oxide fuel cell electrodes, *Electrochimica Acta* 43, 375–394 (1998).
32. J. Yuan and B. Sunden, Analysis of Intermediate Temperature Solid Oxide Fuel Cell Transport Processes and Performance, *Journal of Heat Transfer* 127, 1380–1390 (2005).
33. V.V. Kharton, F.M.B. Marques, and A. Atkinson, Transport properties of solid oxide electrolyte ceramics: a brief review, *Solid State Ionics* 174, 135–149 (2004).

THE PROPERTIES AND PERFORMANCE OF MICRO-TUBULAR (LESS THAN 1 MM OD) ANODE SUPPORTED SOFC FOR APU-APPLICATIONS

NIGEL SAMMES[1,2]*, ALEVTINA SMIRNOVA[2], ALIDAD MOHAMMADI[2], FAZIL SERINCAN[2], ZHANG XIAOYU[2], MASANOBU AWANO[3], TOSHIO SUZUKI[3], TOSHIAKI YAMAGUCHI[3], YOSHINOBU FUJISHIRO[3], AND YOSHIHIRO FUNAHASHI[4]

[1]*Department of Metallurgical and Materials Engineering Colorado School of Mines 1500 Illinois Street Golden, Colorado, USA*
[2]*Connecticut Global Fuel Cell Center, University of Connecticut, Storrs, CT 06269, USA*
[3]*Institute of Advanced Industrial Science and Technology (AIST), Nagoya, Japan*
[4]*Fine Ceramics Research Association, Nagoya, Japan*

1. Introduction

Solid Oxide Fuel Cells (SOFCs) consisting of durable ceramic components can be considered nowadays as the most promising fuel cell systems for alternative sources of energy due to their high electrical conversion efficiency, superior environmental performance, and fuel flexibility.[1–3] In comparison to conventional SOFCs, micro tubular SOFCs[4–6] have many advantages such as high resistance to thermal shock and higher power densities reaching 0.3 W cm^{-1} at 550°C.[7] Recent improvements of the mechanical and electrochemical properties at reduced operating temperatures allows the use of cost-effective materials for interconnects and balance of plant.[8,9]

Traditionally, yttria stabilized zirconia (YSZ) and NiO-YSZ cermets have been used as electrolyte and anode materials in SOFC systems, respectively. Electrochemical and mechanical properties of these materials under different operational conditions have been extensively studied[10,11]

_____
*E-mail: nsammes@mines.edu

and optimized by changing the cermet structure and particle size distribution.[12-14] To reduce the operating temperature, and increase cell performance, alternative materials for the cell components have been developed, such as perovskite single phase and composite anode and cathode materials and electrolytes operating in the intermediate temperature range.[15-17]

The present paper is focused on modeling, mechanical, and electrochemical properties of the micro-tubular SOFCs and SOFC cube stacks for potential use in an APU-system, which requires quick start-up. High volumetric power density, and ease of fabrication. To predict microtubular SOFC performance, a two-dimensional (2D) model was considered, followed by the results of mechanical and electrochemical testing. The attention was paid to the development of materials for electrodes and electrolytes that lower the operating temperatures to below 650°C, as well as to the development of manufacturing processes for sub-millimeter tubular SOFC cells for arrangement and integration at the micro level.

The results of a micro-tubular SOFC, and a cube-type cell stack module, fabricated and tested to check its applicability for commercialization of an auxiliary power unit (APU) and a stationary small power source system are presented. The micro-tubular SOFCs with high energy efficiency, quick start-up and shut-down performance and low production cost are described in addition to the mechanical properties and fabrication technology data for the SOFC materials operating in the temperature range of 450–650°C.

2. Mathematical Model

The modeled geometry is based upon micro tubular cells fabricated as a part of the Advanced Ceramic Reactor Project (ACRP) initiated by New Energy and Industrial Technology Development Organization (NEDO) of Japan. The single cell tests are assumed to be carried out in a firebrick furnace in which the heat is supplied by the wires surrounding the cell so that there is an even peripheral heat distribution. The modeled domain covers only a portion of the test geometry for a couple of reasons. Firstly, due to the high aspect ratio of the cell length to the electrolyte thickness, the modeled length is taken as quarter of the actual cell length. Secondly, for the sake of simplicity while handling the radiation heat transfer, the model domain does not include the heating elements. Since the test geometry employs axial symmetry, contiguous rectangles which revolve around an axis are used to represent the cell.

While defining the geometry, parameters (Table 1) are determined either by direct measurements or from the SEM images provided in the work of Suzuki et al.[18]

The presented model takes into account the steady state heat, mass, momentum and charge transfer phenomena. Heat equation is used with both convection and conduction mode in all the domains except the electrolyte for which only the conduction mode heat transfer is modeled. For the anode, cathode and current collector regions, an effective heat conduction coefficient is suggested to combine the heat transfer in solid and fluid phases in a single equation. Fick's law is used to govern mass transfer. Momentum transfer is governed by the Navier-Stokes equation in the channels, whereas Darcy's law is used for porous regions such as the anode, cathode and current collectors. Finally, electronic and ionic charge transfers are modeled by Ohm's law.

The heat, mass, momentum, and charge transfer governing equations are solved using a commercial computational fluid dynamics package COMSOL, which is equipped with predefined partial differential equations. Along with a direct UMFPACK linear system solver, nonlinear stationary solvers are used.[19] With 7,878 triangular mesh elements, the solution becomes satisfactory in the error tolerance band of 10^{-4} considering 70,126 degrees of freedom in the domain. Weak formulation is used for solving the model, which generates the exact Jacobian for the fast convergence of nonlinear models.

TABLE 1. Geometry parameters.

Description	Value	Reference
Interior radius of the anode tube, r_i	1.0 mm	Direct measurement
Exterior radius of the anode tube, r_o	1.3 mm	Direct measurement
Electrolyte Thickness, t_e	25 µm	[18]
Cathode thickness, t_c	50 µm	[18]
Cathode current collector thickness, t_{cc}	50 µm	[18]
Length of the actual cell, l	1.0 cm	Assumed
Air channel thickness, t_{ac}	2.0 mm	Assumed

The model presented in the preceding sections is validated by the experimental data of Suzuki et al.[18] The operational parameters and material properties used in the model are taken to be the same as the actual ones for comparison. The ones that are not given or not found in the literature are chosen with reasonable assumptions. In Figure 1a, polarization curves obtained from the model and the experiments are compared. It can be seen that the model results correlate reasonably well with experimental data. The comparison of the power density curves as shown in Figure 1b is also promising. However, the model overestimates the actual results in mid-range current densities.

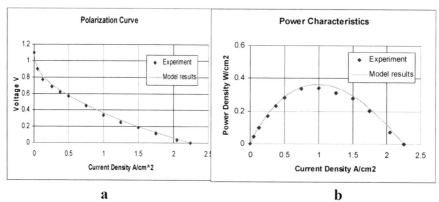

Figure 1. Comparison of (a) polarization curves and (b) power characteristics obtained from the model (solid line) and acquired from experiments of Suzuki et al.[18]

It can be seen from the polarization curve that at higher current densities the curve has a concave up characteristic which is not an usual result for fuel cell applications. For an ordinary fuel cell, at higher current densities, concentration losses are dominant which is reflected as a concave down curve. As is discussed in the introduction section, the concentration losses in micro scaled fuel cells would not be as notable as those observed in macro scaled ones due to significantly smaller concentration gradients. In addition, at higher current densities, the temperature of the fuel cell increases because of ohmic resistance.

The increase in temperature improves the cell performance due to the fact that conductivity of the electrolyte increases with increasing temperature. In Figure 2a the temperature distribution inside the cell for different operating voltages, marked on the vertical axis, can be seen. Corresponding ionic conductivities of the gadolinia-doped ceria (GDC) electrolyte can be found in Figure 2b. Ionic conductivity of GDC as a function of temperature was obtained from.[20]

The temperature distribution inside the modeled domain is shown in Figure 3a. The effect of the radiation is seen as the temperature profiles evolve from the regions closer to the most right boundary. As discussed in the previous sections some portion of the energy irradiated by the wires is absorbed by the cell while the rest is given to the fluid flow.

This consequence is reflected on the temperature distribution as the temperature profiles have a shape of a bell pointing to the outlet of the channel, which explains that fluid regions closer to the heating elements acquire energy by radiation, whereas the regions closer to the cell acquire energy by convection. Since these two distinct heat transfer mechanisms transmit different amounts of energy to the fluid, the bell shaped profile is inclined towards the cell meaning flow gains more energy from radiation. Hereby, it could be implied that preheating of the gases before entering the furnace would result in an increase in cell performance because in the case of preheated gases, less energy would be released from the cell to the flow.

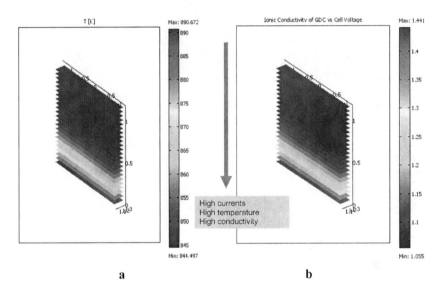

Figure 2. (a) Temperature distribution inside the cell for different operating voltages indicated in vertical y-axes and (b) corresponding values of the GDC electrolyte ionic conductivity.

Corresponding temperature distribution inside the cell is plotted in a different figure (Figure 3b) to distinguish the small temperature difference compared to the overall distribution. As discussed in the previous sections, the temperature difference inside the cell is significantly small for a micro scaled fuel cell. Only less than 3°C of a difference is observed. For a more meaningful measurement, the longitudinal temperature difference for a unit length is 12°C/cm and radial temperature difference for unit length is 3°C/cm. These values are consistent with the order of magnitude reported in the previous work of our group.[21]

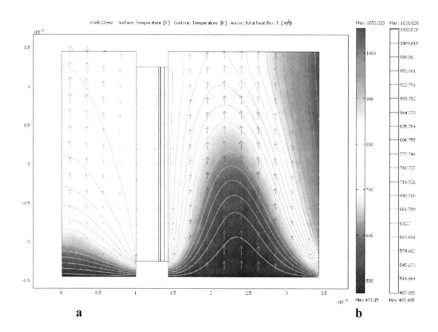

Figure 3. (a) Temperature distribution inside the computational domain at the operating voltage of 0.7 V (min 473 K, max 1,053 K), contours show the temperature profiles and (b) temperature distribution inside the cell at the same operating point (min 850.63 K, max 853.48 K).

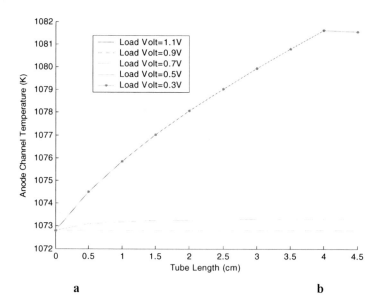

Figure 4. Temperature difference in the radial direction when the dimensions are scaled up ten times.

For the quantitative evaluation of the micro-tubular SOFC, the model geometry was scaled up 10 times in the radial direction, and 10 times in the longitudinal direction. With the same amount of energy flux given to the system and the same flow rates, the results are plotted (Figure 4) in terms of temperature differences as functions of normalized cell thickness and length. In both cases, temperature differences increase notably when cell geometry is enlarged. Enlarging the geometry 10 times resulted in a temperature difference of 20.44 times more than that found for the original dimensions, whereas longitudinal scaling resulted in a temperature difference of 14.75 times higher.

In the context of miniaturization it should be stated that for a micro scaled fuel cell the concentration gradients are small enough to decrease mass transport losses notably. Figure 5 shows the hydrogen and oxygen molar concentrations for the operating voltages of 0.8, 0.45 and 0.1 V. Comparing 0.8 V and 0.1 V it can be deduced that there is 39% decrease in H_2 concentration whereas the decrease is only 9% for O_2. This result is noteworthy because it is known that for macro-scaled fuel cells either, SOFC or PEMFC, at higher current densities severe mass transport limitations are predominant. In some cases, with the diminishing concentration of the reactants, starvation of the cell takes place. The consequence of the small gradients in miniaturized cells is shown in the polarization curve characteristics, as discussed previously.

The effect of scaling on hydrogen concentration can be seen in Figure 6. For the cell operating at 0.7 V, the hydrogen concentration profiles along a vertical line in the anode are compared for the original dimensions and dimensions 10 times scaled in the radial direction. The drop in the concentration along the profile is three times more than for the cell with larger dimensions.

3. Mechanical Testing

The micro-hardness results for different anode and electrolyte materials are shown in Table 2. The GDC powder from Anan Kazei in comparison to the United Ceramics Limited YSZ powder has almost the same mechanical properties. However, the mechanical properties of GDC from DKKK with coarser particle size were even better. Since all different powders were fired under the same conditions, this can be related to the particle size of powders and the effect of grain boundaries impurities known to change sintering properties of ceramic materials. The YSZ nano-powder from TOSOH probably activates another sintering mechanism due to the nano-powder properties well known for their lower sintering temperatures. In this case a strong bonding in the grain boundary regions as well as inside of the grain areas improves the hardness of the pellets (Table 3).

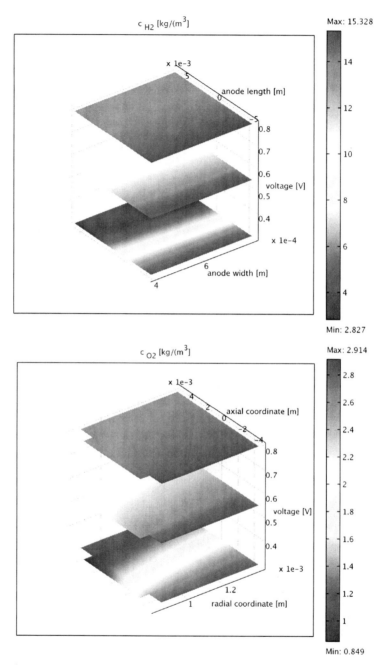

Figure 5. (Top) Hydrogen concentration distribution for operating voltages of 0.8, 0.45 and 0.1 V. (min 15.02 mol/m^3, max 25.75 mol/m^3), contours show the temperature profiles; (bottom) oxygen concentration distribution for the same operating voltages (min 4.952 mol/m^3, max 5.408 mol/m^3).

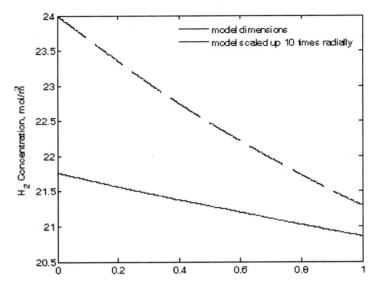

Figure 6. Comparison of the hydrogen concentration profiles in the anode for the original cell (solid) and the scaled up cell (dashed).

TABLE 2. Average micro-hardness (300 g load) and coefficient of variation.

Material	Hardness (VHN)	Coefficient of variation (%)
GDC (Anan Kazei)	673.8	10.2
GDC (DKKK)	899.3	7.1
YSZ (TOSOH) nano-crystalline powder	1,379.2	4.9
YSZ (United Ceramics Limited)	690.5	3.0
NiO-GDC	1,358.3	4.6
Screen Printed GDC on the anode substrate	775.7	4.9

TABLE 3. Average nano-hardness (5* and 50**g load), elastic modulus, and standard deviation.

Material	Hardness (GPa)	Modulus (GPa)	Standard deviation
GDC** (Anan Kazei) pellet	4.8	129.9	4.3
GDC** (DKKK) pellet	7.7	143.4	0.7
YSZ**(United Ceramics Limited) pellet	4.9	131.2	4.5
GDC* (Anan Kazei) screen-printed layer on the anode substrate	1.9	27.2	2.3

Nano-indentation results for different samples (Table 2) confirm that the mechanical properties of GDC powder is in the same range or better than those of the United Ceramics YSZ powder. For the GDC layer screen printed on the anode support, application of the lower load reduces the depth of penetration of the indenter, which minimizes the effect of substrate on the hardness of the coating. As expected, unlike the micro-hardness results, the hardness and modulus of screen printed GDC is lower than those of the GDC pellet (Figure 7). Different GDC and YSZ pellets were tested using ASTM F 349-78 (1996) standard, however, the data collection from delaminated GDC pellets, made from the coarser powder, was a challenge since they broke before any reliable data collection. Chipped edges, due to high contact stress under the loaded balls, also occurred for some specimens. This issue was solved by applying a thin plastic layer underneath the specimen to reduce the local stress at the contact points. The calculated biaxial flexure strength for GDC was $\sigma = 20$ MPa, however, since the pellet was delaminated this number is much less then the true strength value of GDC pellets. The modulus of rupture of the pellets made from the YSZ nano powder was about 114 MPa. These calculations were based on the peak loads of 60.2N and 297.0N for GDC and YSZ pellets, respectively.

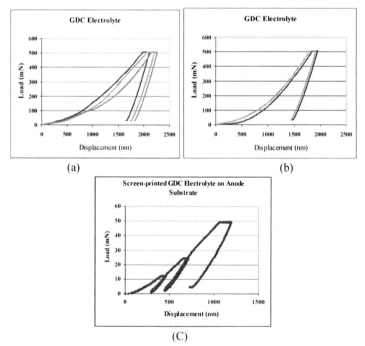

Figure 7. Nano-hardness tests on: (a) Anan Kazei GDC pellets; (b) DKKK GDC pellets; (c) Screen-printed GDC on NiO-GDC anode. The higher displacement in Anan Kazei specimen represents lower hardness of the substrate.

The measured fracture strength of NiO-YSZ powder after different redox cycles are reported in Table 4. It has been shown that there is no significant increase in the fracture strength between the non-reduced and reduced anode. It has also been shown that after the first and third redox cycles the maximum strength resulted in 43% and 77% increase, compared to that of the reduced anode, respectively. Further redox cycles reduced the maximum strength values. However, those values still remained higher than those of the reduced anodes.

TABLE 4. Average fracture strength, standard deviation, and weibull parameter of the C-ring anodes.

Cycle	$\sigma_{average}$ (MPa)	Standard deviation	Weibull parameter
Non-reduced anode	65.1	3.8	16
Reduced anode	69.0	6.7	10
First redox	98.6	12.5	8
Second redox	118.0	23.0	5
Third redox	122.4	5.5	22
Fourth redox	105.3	9.8	10
Fifth redox	105.3	15.8	6

The morphology and structure of the anode pellets after indentation, as well as micro- and nano- indentation marks shown in Figure 8, clearly demonstrates the cracks propagated from the micro-indentation tips.

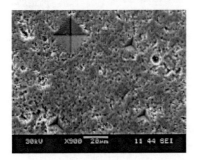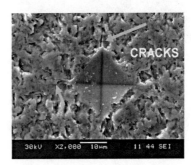

Figure 8. The SEM micrographs of an anode pellet used for measuring the fracture toughness of the material (a) micro- and nano-indentation marks and (b) propagated cracks.

4. SOFC Manufacturing and Testing

For the development of material technology towards lower operating temperatures, the target power density was set to 0.5 W/cm^2 at 650°C. The high performance electrodes and electrolyte were investigated by using advanced internal structure control on nano-and micro-scale level and thin film process technology, as well as improving cell components by interfacial control of electrode, electrolyte and interconnect. Fabrication technology of materials and components for the next stage were examined. As a result, the electrode materials, which reduced energy loss to 1/10, were obtained.

To achieve an output performance of 2 kW/l required for portable or stationary type power sources, development of innovative manufacturing process technologies for the high performance SOFCs and modules has been undertaken. By optimizing the structures of the ceramic electrodes, a cell performance of a power density of 1 W/cm^2 at 570°C was achieved (Figure 10). Successful fabrication processes for the cube consisting of 36 cells in 1 cm^3 was established (Figure 9). In addition, a Honey-comb type micro-tubular SOFCs, which consist of sub-millimeter cells, were demonstrated (Figure 11) and promising results regarding cell performance of over 2 W per 1 cm^3 at 650°C and quick start-up (five minutes starting from room temperature) were obtained.

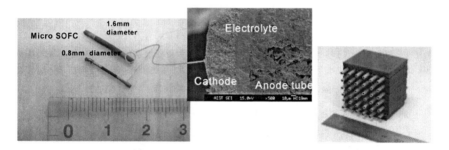

Figure 9. Micro-tubular SOFC with 0.8 and 1.8 mm diameter, SEM image of the fuel cell cross-section, and a 36 cell stack.

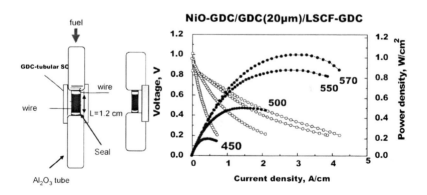

Figure 10. Micro tubular SOFCs configuration and performance characteristics.

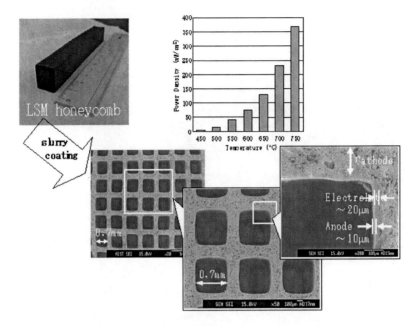

Figure 11. Honeycomb type cathode-supported micro-SOFC integration process and electrochemical properties at different temperatures of operation.

The model cube (27 cm^3) with 36 micro tubular SOFCs (2 mm in diameter) was fabricated and is shown in Figure 12. The module demonstrated a high output performance of about 15 W at 650°C and 13.5 W below 650°C. Recent work, however, on a microbundle of nine-tubular cells (2 mm in diameter; see Figure 13) in a cube-bundle has produced 2.1 W cm^{-3} (equivalent to over 2 kW/l) at 550°C (see Figure 14). A 25-cell micro-bundle (each tube 0.8 mm in diameter) has also been fabricated and is now the subject of future work (Figure 15). This will be the building block for the APU unit to be studied.

Figure 12. Module assembly and evaluation of prototype module for various applications.

Figure 13. *Nine*-cell stack module and assembly using 2 mm diameter tubes.

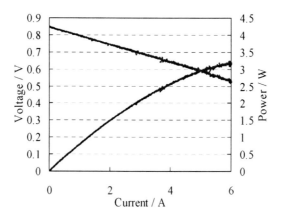

Figure 14. Performance of bundle in Figure 16, tested on hydrogen at 550°C.

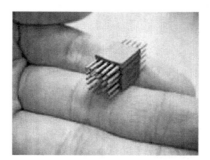

Figure 15. *Twenty-five*-cell stack module and assembly using 0.8 mm diameter tubes.

Acknowledgments

Financial funding was provided by the New Energy and Industrial Technology Development Organization (NEDO) in the frame of the Advanced Ceramic Reactor Project Contract Number AG060145.

References

1. M.C. Williams, J.P. Strakey, W.A. Surdoval, and L.C. Wilson, Solid oxide fuel cell technology development in the US, *Solid State Ionics*, 177, 2039–2044 (2006).
2. M. Masashi, H. Yoshiko, and N.M. Sammes, Sintering behavior of Ca- or Sr-doped $LaCrO_3$ perovskites including second phase of $AECrO_4$ (AE = Sr, Ca) in air, *Solid State Ionics*, 135, 743–748 (2000).
3. Y. Yi, A. Rao, J. Brouwer, and G. S. Samuelsen, Fuel flexibility of an integrated 25 kW SOFC reformer system, *J. Power Sources*, 144 (1), 67–76 (2005).
4. J. Turner, M.C. Williams, and K. Rajeshwar, Hydrogen economy based on renewable energy sources, *Electrochem. Soc. Interface*, 13(3), 24–31 (2004).
5. K. Kendall and M. Palin, A small solid oxide fuel cell demonstrator for microelectronic applications, *J. Power Sources*, 71, 268–270 (1998).
6. A. Smirnova, G. Crosbie, and K. Ellwood, Application of fourier-based transforms to impedance spectra of small-diameter tubular solid oxide fuel cells, *J. Electrochem. Soc.*, 148, A610–A615 (2001).
7. Y. Funahashi, T. Shimamori, T. Suzuki, Y. Fujishiro, and M. Awano, Fabrication and characterization of components for cube shaped micro tubular SOFC bundle, *J. Power Sources*, 163, 731–736 (2007).
8. X. Zhou, J. Ma, F. Deng, G. Meng, and X. Liu, Preparation and properties of ceramic interconnecting materials, $La_{0.7}Ca_{0.3}CrO_{3-\delta}$ doped with GDC for IT-SOFCs, *J. Power Sources*, 162, 279–285 (2006).
9. S. Livermore, J. Cotton, and R. Ormerod, Fuel reforming and electrical performance studies in intermediate temperature ceria–gadolinia–based SOFCs, *J. Power Sources*, 86, 411–416 (2000).
10. U.B. Pal, S. Gopalan, and W. Gong, *FY 2004 Annual Report* (Office of Fossil Energy Fuel Cell Program 2004).
11. N.M. Sammes, Y. Du, and R. Bove, Design and fabrication of a 100 W anode supported micro-tubular SOFC stack, *J. Power Sources*, 145, 428–434 (2005).
12. J.W. Fergus, Electrolytes for solid oxide fuel cells, *J. Power Sources*, 162, 30–40 (2006).
13. M. Mogensen, S. Primdahl, M.J. Jorgensen, and C. Bagger, Composite electrodes in solid oxide fuel cells and similar solid state devices, *J. Electroceram.*, 5, 141–152 (2000).
14. M. Mogensen and S. Skaarup, Kinetic and geometric aspects of solid oxide fuel cell electrodes, *Solid State Ionics*, 86, 1151–1160 (1996).
15. V.V. Kharton, E.V. Tsipis, I.P. Marozau, A.P. Viskup, J.R. Frade, and J.T.S. Irvine, Mixed conductivity and electrochemical behavior of $(La_{0.75}Sr_{0.25})_{0.95}Cr_{0.5}Mn_{0.5}O_{3-\delta}$, *Solid State Ionics*, 178(1–2), 101–113 (2007).
16. J.Y. Yi and G.M. Choi, The effect of mixed conductivity on the cathodic overpotential of $LaGaO_3$-based fuel cell, *Solid State Ionics*, 175, 145–149 (2004).

17. X.J. Chen, Q.L. Liu, K.A. Khor, and S.H. Chan, High-performance (La, Sr)(Cr,Mn)O$_3$/(Gd,Ce)O$_{2-\delta}$ composite anode for direct oxidation of methane, *J. Power Sources*, 165(1), 34–40 (2007).
18. T. Suzuki, T. Yamaguchi, Y. Fujishiro, and M. Awano, Fabrication and characterization of micro tubular SOFCs for operation in the intermediate temperature, *J. Power Sources*, 160, 73–77 (2006).
19. COMSOL 3.2 User Guide (COMSOL Inc., Burlington, MA, 2006).
20. C. Xia and M. Liu, Microstructures, conductivities, and electrochemical properties of Ce$_{0.9}$Gd$_{0.1}$O$_2$ and GDC–Ni anodes for low-temperature SOFCs, *Solid State Ionics*, 152, 423–430 (2002).
21. X. Xue, J. Tang, N.M. Sammes, and Y. Du, Dynamic modeling of single tubular SOFC combining heat/mass transfer and electrochemical reaction effects, *J. Power Sources*, 142, 211–222 (2005).

CURRENT STATE OF R&D ON MICRO TUBULAR SOLID OXIDE FUEL CELLS IN JAPAN

YASUNOBU MIZUTANI
*Fundamental Research Department,
Technical Research Institute
Toho Gas Co., Ltd., Tokai-city, Aichi, Japan*

Abstract: The development of novel type solid oxide fuel cells (SOFC) based on micro-tubular cells has been progressing in Japan. Micro-tubular SOFCs may open the door to wide applications of fuel cells because of their large reactive surface area per unit volume, the possibility of quick start-up, innovative cost reduction through high volumetric power densities, etc. In this paper, various aspects of micro-tubular SOFCs are reviewed, including target applications, materials selection, bundle/stacking designs, manufacturing process, and state-of-the-art achievements.

1. Introduction

It is well known that the downsizing of unit cells is an effective way to enhance the performance of fuel cells. In the case of tubular SOFCs, the tube diameter and current path length (mainly the cathode support tube) dominate their internal resistances, and a long current path causes the power density and efficiency to drop. To improve the performance of a tubular SOFC, therefore, "shape modification," or "downsizing," is effective way, and using flat-plate-shaped tubular cells and/or downsizing the cell diameter have been proposed. For example, Siemens Power Generation developed "HPD cells and Delta cells", Kyocera's 1 kW system achieved an efficiency of 45% (HHV) with flat-plate tubular cells, and TOTO improved their tubular cell by downsizing its diameter. The first micro-tubular cell concept was proposed by Kendall et al.[1] and recently, micro-tubular SOFC diameters of several millimeters have been realized worldwide.[2-6] Also, NEDO (the New Energy Development Organization) have been organized the Advanced Ceramic Reactor Project starting in

---

*E-mail: master@tohogas.co.jp

FY2005. Advanced ceramic reactor cells are based on the following concepts: (1) lower operating temperatures (500–650°C), (2) bundled cell modules of multi-cells with diameters of 0.5–2 mm, (3) an extremely high volumetric power density of 2 W/cm^3, and (4) the application of the advanced (continuous) ceramic processing to the manufacture of modules.[7]

2. Market Applications

Suitable power outputs for micro-tubular SOFCs seem to be from several tens of watts to 10 kW, categories including stationary applications [CHP (Combined Heat and Power)], automotive applications [APU (Auxiliary Power Unit)], and portable generators. In the stationary applications category, 1 kW-class CHP systems are reasonable for residential use, and from 5 to 10 kW CHP systems are applicable for commercial use such as restaurants, convenience stores, nursing homes, etc. In Japan, as for the residential CHP systems driven by IC (Internal Combustion) engine systems have already been commercialized, and PEFC (Polymer Electrolyte Fuel Cell) CHPs are currently undergoing large-scale field tests (Figure 1). The New Energy Foundation Japan conducted these field tests, and they showed an energy saving rate of 17.6% and CO_2 reduction of 83.1 kg/month on average in 2006.[8] SOFCs are expected to be used as highly efficient CHP systems for low heat demand customers in the near future.

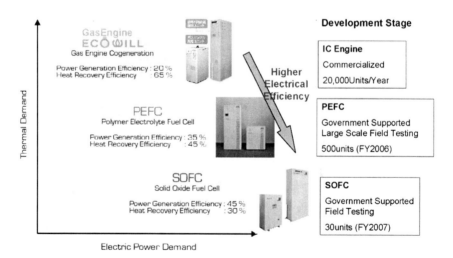

Figure 1. Energy efficiency balance (electricity/heat) of stationary gas co-generation (CHP) systems for residential use and their development stages.

The automotive APU is an attractive market for micro-tubular SOFCs, and the power output range is several kW to 10 kW. Heavy-duty trucks, refrigerator vans, and other vehicles with idle stop requirements will be early customers of SOFC APUs, whose benefits include an independent electricity supply, high efficiency, lower emissions and silent operation (Figure 2).[9–11] On the other hand, requirements for SOFCs are most severe in on-board automotive applications, which put stringent demands on compact size, low cost, robustness and instant start-up capability. In addition, APUs require robustness against vehicle vibration, lower surface temperatures for safety, and clean exhaust gas for environmental concerns. Fuel compatibility in SOFCs is an advantage in view of the fuel issues. The application of bio-methanol or DME (Dimethyl Ether) as the fuel for SOFC APUs seems to be problem-free. The reforming system is a considerable issue, and POx (Partial Oxidation Reforming), ATR (Autothermal Reforming) and SR (Steam Reforming) have been considered. Direct internal reforming is an advantage for SOFCs, but it is difficult to use with higher hydrocarbon fuels in case of reduced-temperature SOFCs.

Portable applications in a wide power range of from 10 to 100 W have interesting market possibilities. Cellular phones and notebook PCs are huge market applications, but the conventional power source (Li-ion battery) and competitive fuel cell technologies (DMFC and PEFC) are preferable for

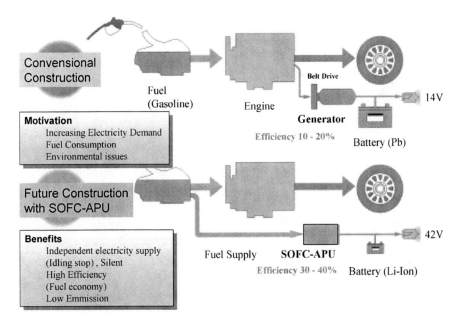

Figure 2. Construction of conventional passenger car and future construction with SOFC-APU system.

these applications. SOFCs offer advantages such as high efficiency (very long runtime), and fuel flexibility (hydrocarbon fuels) and Pt-free construction (lower cost), but they need to maintain low temperatures and low emissions (CO, etc.) to meet the high level of safety required for indoor use. Therefore, the electric wheelchairs, electric power-assisted bicycles, robots, and other devices for outdoor use seem to be preferable applications for SOFCs, which should be used in combination with rechargeable batteries for start-up and load leveling.

3. Design Concept of Micro Tubular SOFC

A large advantage of the micro-tubular SOFC is that it enhances not only single cell performance but also volumetric power densities, as shown in Figures 3 and 4. The reactive electrode surface area per unit volume is significantly increased by downsizing the cell diameter and with increasing the number of cells. However, when the cell diameter is less than 0.1 mm, the differential pressure used to flow the reactant gases also increases drastically (Figure 4). Of course, since these tiny single cells are weak against mechanical loads, a module of bundled cells (referred to as a "cube") with diameters of 0.5–2 mm has been proposed as part of the Advanced Ceramic Reactor Project.[7]

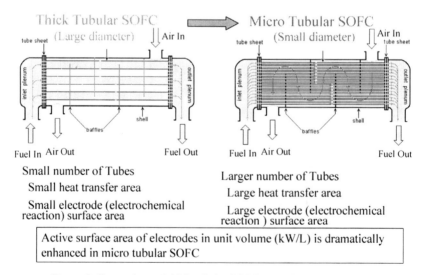

Figure 3. Comparison of thick tubular SOFC and micro-tubular SOFC.

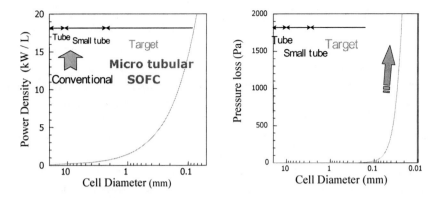

Figure 4. Estimated power densities and pressure loss against cell diameter.

Parallel and/or serial electric connections among multi-cells under minimum IR losses should be considered, and an innovative gas sealing construction is required. Mechanical reliability against thermal stress caused by mismatched thermal expansion coefficients (TEC) is the primary task for the connection issues. X-ray stress measurement, which is a non-destructive measurement method, can be applied to evaluate internal stresses.

4. Cells and Materials

The reduced (intermediate)-temperature SOFC is the main trend for the development of cells and materials, because it has many advantages such as cost effectiveness, short start-up time, and a relatively small amount of heat insulation (Table 1).

TABLE 1. Advantages, disadvantages and approaches for reduced-temperature SOFCs.

Advantages	Lower cost of surrounding materials
	1,000°C (ceramics) → 750°C (stainless steel)
	Short start-up time, less energy required for heat-up
	Less heat leakage and fewer heat insulators
	Long life (sintering avoided by slow mass transfer)
Disadvantages	Temperature mismatch in reforming/gas turbine system
	Prone to carbon deposition (thermodynamic expansion)
	Short life due to use of metallic materials
Approaches	Alternative electrolyte (ScSZ, Doped CeO_2, $LaGaO_3$)
	Thinner electrolyte and process; enhanced electrode areas
	Active anode and cathode

An anode (Ni-base) substrate-supported thin-film electrolyte structure [ASC (Anode-Supported Cell)] has an advantage at lower operating temperatures compared with an electrolyte-supported or cathode-supported SOFC. A temperature range of around 750–800°C has been successfully developed by many developers, and a range of 500–650°C is not yet at the practical development stage (Table 2). Also, electrolytes with higher ionic conductivities and highly reactive electrode materials (especially cathodes) are required in the lower temperature ranges, and many kinds of materials have been applied (Table 3). ScSZ (Sc_2O_3–ZrO_2), LSGM (La(Sr)Ga-(Mg)O_3), and GDC (Gd_2O_3–CeO_2) are candidate materials for the SOFC electrolyte, and doped La (Mn/Fe/Co)O_3 perovskites are often used for the cathode. To enhance the electrochemical performance, ceria/perovs-kite composites or metal/perovskite composites have been tried as highly reactive cathode materials. In addition, doped ceria, which have higher ionic conductivities at lower temperatures, have been evaluated as candidate electrolyte materials that will take the place of YSZ, but doped ceria are difficult to control as SOFC electrolytes because they show electron leakage and efficiency loss when the temperature rises above 600°C.[13] Several developers have proposed the use of multi-layered electrolytes to avoid the electron leakage or interfaces to suppress the electron leakage.

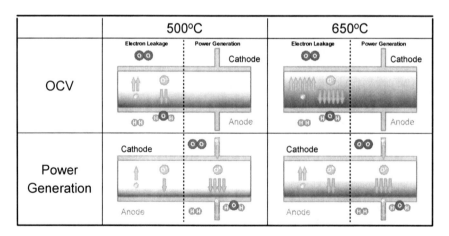

Figure 5. Electron leakage and oxide ion migration phenomena in CeO_2 electrolyte.

TABLE 2. Reduced-temperature SOFCs and their developers.

Temp. (°C)	Shape	Support materials	Developer	Process for electrolyte and electrode
1,000–900	Tubular	Porous cathode	Siemens Power TOTO	EVD or spray Dip/sintering
	Flat tubular Tubular	Porous ceramics	Rolls–Royce, MHI(Mitsubishi Heavy Industry)	Screen printing/sintering
	Planar	Dense electrolyte	MHI, Hexis, InDEC	Tapecasting/sintering
900–800	Tubular	Porous anode	Acumentrics	Dip/sintering
	Planar	Dense electrolyte	CFCL, Staxera	Tapecasting/sintering
800–700	Flat tubular	Porous anode	Kyocera	–
	Flat tubular	Porous ceramics	Tokyo Gas	–
	Planar	Porous anode	Forschungszentrum Julich	Dip/print/sintering
		Porous anode	VersaPower, GE, InDEC, TopsoeFC, HT Ceramics NTT, NGK, NTK	Tapecast/print/sintering
		Porous electrolyte	MMC, Toho Gas	Tapecast/print/sintering
700–600	Small tubular	Porous anode	TOTO	Dip/sintering
650–500	Micro tubular	Porous anode	Adv. Ceramic Reactor PJ	Dip/sintering
600–500	Planar	Porous metal	Ceres Power	–

TABLE 3. Component materials for SOFCs in several temperature range.

Temp. (°C)	Electrolyte	Anode	Cathode	Interconnector
1,000–800	YSZ	Ni-YSZ	LSM LCM	LC, $SrTiO_3$ Cr based metal
800–700	YSZ ScSZ LSGM	Ni-YSZ Ni-SDC Ni-ScSZ	LSCF, LSC LSF, LNF	Ferritic stainlesssteel
700–600	LDC-LSGM	Ni-YSZ Ni-ScSZ	LSCF	
600–500	GDC	Ni-YSZ Ni-GDC	LSCF BSCF	

YSZ: Y_2O_3 doped ZrO_2, ScSZ: Sc_2O_3 doped ZrO_2, LDC: La-doped CeO_2, SDC: Sm-doped CeO_2, GDC: Gd-doped CeO_2, LC: $LaCrO_3$, LSGM : $La(Sr)Ga(Mg)O_3$, LCM: Ca-doped $LaMnO_3$, ,BSCF: BaSrCoFe, LNF: $LaNiFeO_3$

5. Fabrication Process

Continuous ceramics manufacturing processes are important for the realization of low-cost micro-tubular SOFCs. Extrusion molding is applicable to the accurate manufacture of support tubes for cells, not only in centimeter sizes but also in sub-millimeter sizes. Also, dip-coating is a qualified as a candidate for the coating process of the thin-film electrolytes and electrodes (Figure 6).

The drying and firing processes also play key roles in the making of accurate cells. Sequential or simultaneous firing must be considered in view of cell performance and mechanical reliability of the cells. In fact, micro-tubular cells with diameters of 0.8–2 mm and fine honeycomb-shaped monolithic cells with accurate and dense electrolyte films have already been demonstrated in the Advanced Ceramic Reactor Project using extrusion molding, dip-coating and firing processes.[14] Also, future innovative advanced ceramics manufacturing processes of the future are under development, as well as a sol-gel molding process, an automation process for assembling the multi-cells in cubes, and a novel 3D integration process.

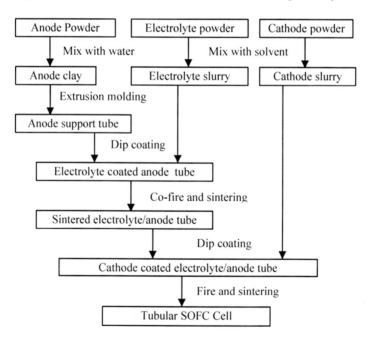

Figure 6. Typical fabrication process for micro tubular SOFC cells by extrusion molding and dip coating process.

6. Electrochemical Performance of the Micro Tubular Cells

As part of the Advanced Ceramic Reactor Project, micro-tubular SOFCs (0.8–1.6 mm in diameter) fabricated from candidate materials (anode: Ni-GDC; electrolyte: GDC; cathode: LSCF) were examined, and an extremely high power density of 1 W/cm^2 was demonstrated even at a reduced temperature of 570°C. These results show the future possibility of micro-tubular SOFCs combined with cell downsizing and materials development aimed at reduced-temperature operations.

Also important, in consideration of actual system operation, are the range of operating temperatures and the electrochemical performance at the higher fuel utilization rates. Figure 7 shows the typical electrochemical performance of a micro-tubular SOFC with a GDC electrolyte. Maximum electric conversion efficiencies of 35–40% were obtained at 550–600°C with hydrogen fuel, and those values are sufficient for actual applications. However, in the case of a GDC electrolyte, the thickness of the electrolyte should be optimized to suppress the electron leakage described above.

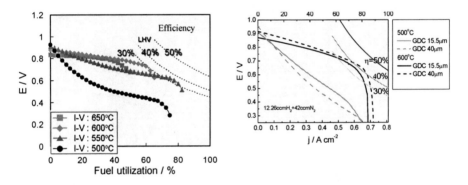

Figure 7. Electrochemical performance under high fuel utilization conditions (left), and the effect on thickness of GDC electrolyte (right).

7. Integrating Single Cells and Stacking the Bundles

Recent topics and technical hurdles for the micro-tubular SOFC are the integration of single cells and stacking. To realize the high volumetric power densities, current collection and gas flow designs have to be optimized. Short current pathways and highly conductive (current collecting) materials have been examined. Porous ceramics (LaCoO$_3$), metallic wires (Ag, Ni), and their composites are candidate materials for current collection.[15] It should be mentioned that high electric conductivity and good gas permeability are often in a trade-off relationship in porous ceramics.

Interface materials for gas sealing, electrical connections and electrical insulation are key technologies for stacking bundles. All of the component materials should be carefully selected from the viewpoint of thermal expansion mismatch, mechanical strength, and chemical stability.

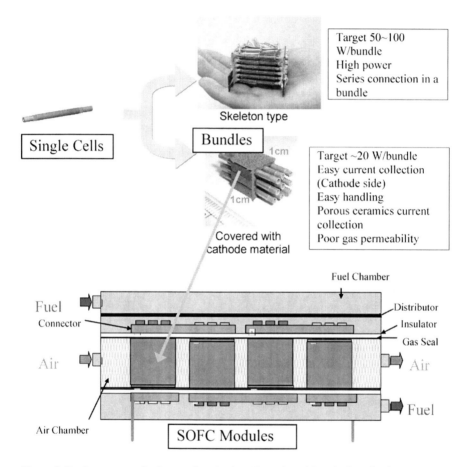

Figure 8. Design concept for integrating single cells and stacking the bundles in "Advanced Ceramic Reactor Project".

8. Future Perspective

Intermediate-temperature SOFCs in the temperature range of 500–650°C are now realistic through the application of alternative materials and advanced fabrication processes. The final operating temperature of the micro-tubular SOFC should be considered in each application because of their fuel selection. An extremely high volumetric power density of 2

W/cm^3 has already been achieved with a 1 cm^3 cube in the Advanced Ceramic Reactor Project, and multi-cube connections for eliminating gas leakage and minimizing IR losses and temperature controls are next-stage subjects. Quick start-up and shutdown procedures and properties of micro-tubular cells must be evaluated, along with the mechanical reliability of cells, cubes, and modules. Recent results and achievements of the micro-tubular SOFC will open the gateway to applications of mini-micro SOFCs in distributed power sources, portable devices, and auxiliary power sources for vehicles.

Acknowledgements

A part of this work was performed as an R&D program of the New Energy and Industrial Technology Development Organization (NEDO). The authors would like to acknowledge and thank the following supporters: Mr. K. Hayashi, Nanotechnology and Materials Technology Development Department (NEDO); Dr. M. Awano, Advanced Industrial Science and Technology (AIST), Project Manager of the Advanced Ceramic Reactor Project; and Professor S. Kakac, University of Miami.

References

1. M. Lockett, M.J.H. Simmons, and K. Kendall, CFD to predict temperature profile for scale up of micro-tubular SOFC stacks, *J Power Sources*, 71, 268 (1998).
2. T. Kawada, in; *Bulletin of Ceramic Society JAPAN*, 36(7) (2001), pp. 489–492.
3. Y. Du, N.M. Sammes, Fabrication and properties of anode-supported tubular solid oxide fuel cells, *J. Power Sources*, 136, 66 (2004).
4. N.M. Sammes, Y. Du, and R. Bove, Design and fabrication of a 100 W anode supported micro-tubular SOFC stack, *J Power Sources*, 145, 428 (2005).
5. A. Crumm, Adaptive Materials and Portable SOFCs: From One Cell to 50 Systems, *Lucern Fuel Cell Forum, Fuel Cell for a Sustainable World*, CD-ROM A051 (2006).
6. J.L. Martin, C. Martin, and R.J. Kee, Portable SOFCs under 300 Watts Operating on Hydrocarbon Fuels, *Lucern Fuel Cell Forum, Fuel Cell for a Sustainable World*, CD-ROM A052 (2006).
7. M. Awano, Y. Fujishiro, T. Suzuki, T. Yamaguchi, and K. Hamamoto, Development of Advanced Ceramic Reactors, *Proceeding of the 7th European SOFC Forum* (2006), p. 44.
8. New Energy Foundation, Japan, http://www.NEF.or.jp
9. J. Zizelman, J. Botti, and J. Tachtler, Solid Oxide Fuel Cell Auxiliary Power Unit – A Paradigm Shift in Electric Supply for Transportation – *SAE Technical Papers*, 2000-01-C07 (2000).
10. J. Tachtler, T. Dietsch, and G. Goetz, Fuel Cell Auxiliary Power Unit – Innovation for the Electric Supply of Passenger Cars?, *SAE Technical Papers*, 2000-01-0374 (2000).

11. N. Lustey, C. Brodrick, D. Sperling, and H.A. Dwyer, Markets for Fuel Cell Auxiliary Power Units in Vehicles: A Preliminary Assessment, *TRB 2003 Annual Meeting CD-ROM* (2003).
12. H. Sumi, K. Ukai, M. Yokoyama, Y. Mizutani, Y. Doi, S. Machiya, Y. Akiniwa, and K. Tanaka, Changes of internal stress in solid-oxide fuel cell during red-ox cycle evaluated by in situ measurement with synchrotron radiation, *J Fuel Cell Sci Technol*, 3, 68 (2006).
13. T. Otake, M. Yokoyama, K. Nagai, K. Ukai, and Y. Mizutani, Electrochemical Evaluation of Micro-Tubular SOFC and Module for Advanced Ceramic Reactor, *Proceedings of the 31st International Conference on Advanced Ceramics and Composites*, in press.
14. T. Suzuki, T. Yamaguchi, Y. Fujishiro, and M. Awano, Improvement of SOFC performance using a microtubular, anode supported SOFC, *J Electrochem Soc*, 153(5), A925–928 (2006).
15. Y. Funahashi, T. Shimamori, T. Suzuki, Y. Fujishiro, and M. Awano, Fabrication and characterization of components for cubic shaped micro tubular SOFC bundle, *J Power Sources*, 163, 731–736 (2007).

INDEX

Activation polarization, 17, 19, 20, 277, 364, 365, 388
Amine boranes, 175
Anode catalysts, 97, 212, 219, 221, 226, 229, 236, 239, 244, 245, 247, 248, 250, 253, 258, 259, 267, 279, 280
Anode methanol molarity, 240
Autonomous test units, 103, 114
Autothermal reforming, 325, 330, 332, 333, 354, 409
Auxiliary power unit, 304, 310, 319, 392, 408
Average
    fracture strength, 400
    micro-hardness, 399
    nano-hardness, 399

Butler-Volmer equation, 17, 19, 62, 109, 211, 249, 365, 370

Capillary flux, 181, 182
Carbon deposition, 320, 324, 326, 328, 329, 331, 333, 337, 339, 340, 342, 355, 411
Carbon-support oxidation, 225, 230
Catalysis and performance, 210, 219
Catalyst, 2, 18–21, 25, 27, 36, 42–44, 53, 57, 60, 62, 63, 67, 95, 97, 104, 106–108, 113, 135, 137, 139, 140, 142, 142, 145–148, 151, 152, 160, 162, 176, 177, 180–182, 209, 211–214, 218, 219, 221, 222, 225–229, 232, 235–237, 239, 240, 244–253, 255, 257–259, 266–268, 270, 271, 275–280, 306, 326–329, 331–335, 339, 340, 342, 343, 355
Catalyst degradation, 225
Cathode catalyst, 60, 62, 180, 209, 213, 219, 221, 222, 225–227, 229, 236, 244, 245, 247–253, 279, 280
Cells and materials, 411
Cellular phones, 272, 273, 409
Chemical hydrides, 173, 174, 176, 186
Components of DMFC, 271

Computational model, 52, 156, 179, 181, 241, 243, 244
Concentration polarization, 20, 21, 23, 276, 277, 362, 364, 388
Conductivity of Nafion®, 205, 278
Consumer electronics, 1, 3, 5, 47, 48, 70, 99, 257, 304
Current density, 17–19, 21, 22, 24, 25, 27, 35, 37, 42, 48, 57, 58, 61–63, 65–70, 109–111, 135, 138–151, 153, 178–180, 182–184, 186, 210–213, 215, 218, 220–222, 226, 227, 230, 238–240, 249–251, 254, 255, 258, 259, 273, 276, 277, 283, 336, 361, 364, 365, 367
Cyclic voltammetry, 228

Deconvolution, 217, 218
Desulfurization, 320, 324, 328, 329, 331, 335, 355
Diffusion, 21, 24, 35, 37, 39–44, 56–58, 60, 63, 67, 69, 71, 104, 106, 107, 109–113, 118, 119, 131, 133, 135, 137, 140, 141, 145, 153, 154, 176–178, 180–182, 185, 186, 189–191, 194, 195, 200–205, 207, 210, 216, 217, 220, 225, 228–232, 236, 241, 244–248, 251, 253, 271, 271, 278–280, 282, 308, 309, 315, 328, 330, 340, 362, 367–369, 387, 388
Diffusion coefficient, 57, 58, 110, 145, 190, 195, 201–205, 207, 216, 217, 245, 362, 367–369, 387, 388
Direct methanol fuel cell (DMFC), 1–3, 5–12, 87, 90–92, 95, 96, 99, 100, 125, 186, 189, 190, 204, 205, 207, 209, 219–222, 225–227, 229, 230, 232, 235, 236, 238–245, 250, 251, 255, 257–267, 269–273, 279, 282–285, 288, 291–293, 295, 297
DMFC
    Charger, 7
    polarization, 238, 365
    rechargeable battery, 2

Droplet dynamics, 160, 161, 180
Durability of cells, 352
Durability of fuel cell system, 75

Easytest cell, 114, 115, 133–140, 146, 147, 149
Effective exchange current density, 18
Electrochemical
    cell, 104, 108, 134, 135, 235, 387
    energy converter, 13, 103, 106, 108, 133–135, 142, 143, 145, 148, 151
    impedance spectroscopy (EIS), 105, 195, 284
    kinetics, 241, 249, 253, 255
    performance, 269, 350, 412, 415
    processor (ECP), 135
    reactions, 13, 17–21, 24, 27, 37, 42, 60, 95, 99, 109, 138, 145, 148, 171, 235, 236, 324, 339, 361–363, 367–369, 410
Electrolyte conductivity, 372, 374, 377, 382
Electro-osmotic drag, 63, 153, 177, 178, 185, 247, 248, 251
Energy density, 3, 5, 8–12, 47–49, 70, 88, 90, 91, 94, 95, 99, 236, 242–244, 257, 266, 270, 293, 306, 308–310, 312, 313, 316
Energy efficiency, 173, 303, 305, 307–309, 313, 314, 316, 319, 330, 392, 408
Engineering durability, 75–77, 85
Evaluations of energy density, 3

Fabrication process of SOFC, 403, 414, 416
Flooding and two-phase transport, 180
Flow field visualization, 59
Flow resistance coefficient, 167
Fluid flow visualization, 123, 243
Flux of protons, 205, 206
Fourier transform infrared, 190, 192
Fuel cartridge, 1, 10, 11, 91, 92, 128, 130, 131, 261, 265
Fuel cell
    basics, 13
    cooling system, 117, 128
    efficiency, 16, 23–25, 49, 130
    modeling, 52, 60, 67, 244
    notebook, 293, 294
    stack, 2, 10, 11, 27–29, 37–39, 44, 49, 53, 105, 106, 118, 125, 171, 172, 176, 186, 218, 262, 269, 271–274, 278, 282, 283, 292, 294–297, 316, 320, 321, 337
    system, 1, 3, 5, 7, 9–12, 47, 51, 75, 76, 85, 89, 90, 117, 143, 171, 173, 176, 226, 232, 257, 267, 269, 296, 297, 311, 321, 330, 350, 359, 391
    technologies, 3, 11, 117, 409
    thermal management, 43, 117, 118, 121, 125, 128, 132, 172, 243
    and batteries, 88, 309
Fuel storage and delivery, 49, 53, 89, 171, 172, 186, 235
Fuel utilization, 24, 25, 89, 91, 98, 258–260, 339, 353, 354, 415
Functional state changes, 80

Gas diffusion electrode, 104, 106, 107, 111, 113, 133, 135, 137, 176
Gas diffusion layer, 35, 37, 39–44, 60, 107, 118, 119, 135, 153, 154, 178, 225, 244, 270, 271, 278, 280
Generic DMFC system, 262
Gibbs free energy, 14–16, 25
Governing equations, 56, 58, 250, 252, 361, 367, 386, 393
Gravimetric sorption, 190–193, 196

Heat management, 30, 37, 50, 54, 236, 321
Heat pipe, 117–123, 125, 126, 128, 130–132
Heat pipes-fuel cells, 117
Heat transfer, 38, 50, 51, 54–57, 60, 64–67, 70, 118–123, 125–128, 131, 176, 323–325, 332, 334, 336, 340, 342, 361, 392, 393, 395, 410
Hybrid
    power system, 313
    concentration distribution, 325, 397, 398
    economy, 303, 305, 307–309, 315, 319
    electrode, 135, 138–140, 145, 148, 150
    storage, 11, 49, 94, 171, 173–176, 186, 264, 307, 309, 320
    test unit, 103, 113, 114
Hydrophilic, 29, 155, 163, 177, 180, 181, 197
Hydrophobic, 153, 155, 157, 166, 168, 177, 180, 181, 197, 236, 248, 280

INDEX 421

Image velocimetry, 154, 156, 159
Infrared spectra, 192, 198, 201
Inlet-outlet boundary conditions, 162, 250
Integrating single cells, 415, 416

Kinetic losses, 211, 212

Leaky membranes, 258, 259
Leverette function, 181
Loop heat pipe, 119, 121, 123, 130
Loop thermosyphons, 120, 121

Market applications, 408, 409
Market niches, 304, 308, 309, 311, 313, 315, 319
Materials of SOFC, 371, 392
Mathematical model, 52, 243, 360–362, 392
Membrane, 13, 20, 22, 24, 27, 29, 32, 39, 40, 42–44, 50, 51, 53, 56, 57, 59–66, 68–71, 83–85, 94, 95, 103–108, 112, 113, 133, 135, 136, 138–141, 153, 171, 176–179, 182–186, 189–195, 197, 199, 201–207, 209–211, 214, 216–218, 220–222, 225, 228, 229, 231, 236, 239–245, 247, 248, 250–252, 254, 257–262, 270–272, 278, 280
Membrane degradation, 231
Membrane-electrode assembly, 37, 40, 50, 52, 53, 60, 61, 64, 66–68, 103–108, 112–115, 133, 135, 136, 138–140, 178, 209, 210, 212–214, 216–218, 220–222, 231, 236, 262–264, 269, 270, 272, 278–284, 343
Methanol
 concentration, 96, 192, 193, 196, 197, 201, 204, 205, 207, 239, 240, 247, 249–251, 253, 254, 259–262, 277, 282, 291
 crossover, 220, 221, 236, 238-241, 243, 247, 251, 255, 258–261, 271
 fuel, 1–3, 5, 6, 8–11, 49, 87, 117, 186, 189, 209, 217, 219, 225, 235, 239, 241, 243, 255, 257, 260, 264, 269, 272, 273, 278, 291, 293, 295, 309, 311, 334
 transport, 236, 238, 239, 247, 251, 253
Micro fuel cell stack, 11, 262, 272–274, 278, 282, 283
Micro fuel cells, 1, 3, 5, 7, 9–12, 43, 49, 123, 125, 128, 243, 257, 261, 262, 264, 265, 269, 272–274, 278, 282, 283, 349
Microchannel fuel processors, 334, 335
Micro-channels, 153, 155, 168, 370
Micro-DMFC, 7, 11, 257, 262
Microfabrication, 51, 52
Micro-mini heat pipes, 117, 118
Microstructured fuel cells, 47, 52, 53, 55, 61, 66, 68–71
Micro-tubular SOFC, 336–338, 342, 391, 392, 397, 401–403, 407, 409, 410, 414–416
Military perspective, 1
Mini thermosyphon, 126
Miniaturization of fuel processing, 324
Miniaturization of SOFC, 308, 313, 315, 321
Model validation, 63, 156, 179, 180, 360
Momentum and mass transport, 56, 361, 367, 368, 393
Multicomponent sorption, 193, 197, 199, 200, 207
Multicomponent transport, 189, 200
Multi-layered structure, 351
Multiphysics equations, 76–78
Multiphysics modeling, 77, 85

Natural convection modeling, 50, 55, 58, 59
Nernst equation, 15, 20, 61, 110, 361, 362, 364, 365, 371
Nyquist diagram, 284

Ohmic losses, 20, 23, 51, 61, 218, 377
Ohmic resistance, 209, 210, 229, 244, 266, 335, 338, 394
Optimization algorithm, 265–267
Oxidation reaction, 17, 62, 108, 138, 140, 210, 236, 241, 245, 251, 270, 331
Oxygen
 electrode, 142, 143, 145–147
 evolution reaction, 108, 142, 145–148, 152
 reduction, 17–19, 23, 51, 105, 108, 142, 148, 150, 184, 210, 226, 238, 239, 241, 279, 338

Perfluorosulfonic acid (PFSA), 51, 176, 211, 231, 236, 259–261
Performance metrics, 82

Performance of the cell, 29, 32, 37, 44, 50, 95, 98, 145, 225, 226, 230, 251, 261, 263, 266, 300, 302, 349, 361, 373, 386, 392, 394, 395, 403, 410, 414
Phase saturations, 250
Physical state changes, 79
Planar solid oxide fuel cells (Planar SOFC), 304, 335, 349–352, 354, 359, 361, 371, 372, 386
Platinum
  dissolution, 226–229
  mass activity, 211–213, 226, 227
  sintering, 225, 226
  catalysts, 18, 211, 235, 270
Polarization curves, 22, 23, 64–66, 68, 69, 77, 110, 111, 138–141, 143–151, 238, 241, 265, 277, 283, 285, 295, 393, 394, 397
Polymer electrolyte fuel cells (PEFC), 168, 171, 172, 176–181, 184, 185, 261, 304, 306–311, 313–315, 349, 350, 354, 408, 409
Polymer electrolyte membrane, 22, 94, 103, 104, 135, 171, 176, 189, 271, 278
Portable DMFC, 100, 236, 239, 241–244, 293
Portable fuel cells, 87–89, 91, 99, 117, 120, 173, 180, 297
Potential cycling, 229
Potential of fuel cell, 13–16, 23
Proton transport losses, 214
Proton-conducting electrolyte, 135
Prototype of DMFC, 6

Quasi-analytical modeling, 60, 70

Reaction of DMFC, 270
Reagent recovery, 111, 112, 133–135, 139, 143, 152
Reduction reaction, 17, 23, 51, 108, 142, 175, 184, 210, 226, 235, 239, 279
Reforming efficiency, 327, 332, 355, 356
Reforming process, 324–327, 331, 333, 342

Scanning electro microscopy (SEM)
  images of electrodes, 236
Shadow-graph visualization, 50

Single-chamber SOFC, 337
Sodium borohydride, 2, 49, 94, 96, 174, 175
SOFC
  cube stacks, 392
  modules, 321, 357
  technology, 303, 304, 307, 315, 319, 320, 343
Solid oxide fuel cell (SOFC), 75, 77–80, 82, 83, 88, 94, 97–100, 303, 304, 307–316, 319–324, 330, 335–343, 349–352, 354–357, 359–363, 365, 369, 371, 377–386, 391, 392, 397, 401–403, 407–417
Solid oxide fuel cell system, 98, 303, 307, 308, 312–316, 319, 321, 323, 324, 330, 342, 349, 353–357, 391
Sorption heat pipes, 122, 123
Sorption of liquid, 192
Species transport, 57, 58, 245
Stack power, 9, 10, 12, 258, 265, 285, 286
Stack voltage, 285, 286
State-of-the art power, 311
System size for portable fuel cells, 89, 90

Technological niche, 304, 305, 308, 309, 315
Temperature distribution, 38, 59, 339, 353, 361, 373, 376, 394–396
Thermal expansion, 29, 55, 324, 339, 341, 352, 353, 411, 416
Thick tubular SOFC, 410
Tubular SOFCs, 336–338, 342, 343, 391, 392, 397, 401–403, 407, 409, 410, 414–417
Two-phase flow, 153, 154, 156, 168, 180, 241, 243–245

Velocity field, 156, 158–160
Voltage losses, 16, 17, 21, 23, 61, 63, 209, 210, 212, 217, 218, 222, 226, 231, 238

Water
  management, 9, 51, 71, 95, 98, 166, 171, 176–178, 180, 184, 186, 236, 257, 258, 261–263
  transport, 63, 153, 156, 177–185, 216, 239–241, 244, 247, 248, 252, 255
  transport coefficient, 178–180, 182–184, 239, 241, 248